NONEQUILIBRIUM EFFECTS IN ION AND ELECTRON TRANSPORT

NONEQUILIBRIUM EFFECTS IN ION AND ELECTRON TRANSPORT

Edited by

Jean W. Gallagher
National Institute of Standards and Technology
Gaithersburg, Maryland

David F. Hudson
Naval Surface Warfare Center
Silver Spring, Maryland

Erich E. Kunhardt
Polytechnic University
Farmingdale, New York

and

Richard J. Van Brunt
National Institute of Standards and Technology
Gaithersburg, Maryland

PLENUM PRESS ● NEW YORK AND LONDON

Library of Congress Cataloging-in-Publication Data

International Swarm Seminar (6th : 1989 : Glen Cove, N.Y.)
    Nonequilibrium effects in ion and electron transport / edited by
Jean W. Gallagher ... [et al.].
       p.   cm.
    "Proceedings of the Sixth International Swarm Seminar, held August
2-5, 1989, in Glen Cove, New York"--T.P. verso.
    Includes bibliographical references and index.
    ISBN 0-306-43713-9
    1. Ion swarms--Congresses.   2. Electron swarms--Congresses.
I. Gallagher, Jean W.   II. Title.
QC702.7.I57I57   1989
537.5'32--dc20   90-7893
                                                                  CIP

Proceedings of the Sixth International Swarm Seminar,
held August 2-5, 1989, in Glen Cove, New York

ISBN 0-306-43713-9

© 1990 Plenum Press, New York
A Division of Plenum Publishing Corporation
233 Spring Street, New York, N.Y. 10013

Printed in the United States of America

PREFACE

This volume presents the contributions of the participants in the
Sixth International Swarm Seminar, held August 2-5, 1989, at the Webb
Institute in Glen Cove, New York. The Swarm Seminars are traditionally
held as relatively small satellite conferences of the International
Conference on the Physics of Electronic and Atomic Collisions (ICPEAC)
which occurs every two years. The 1989 ICPEAC took place in New York
City prior to the Swarm Seminar. The focus of the Swarm Seminars has
been on basic research relevant to understanding the transport of charged
particles, mainly electrons and ions, in weakly ionized gases. This is a
field that tends to bridge the gap between studies of fundamental binary
atomic and molecular collision processes and studies of electrical
breakdown or discharge phenomena in gases. Topics included in the 1989
seminar ranged the gamut from direct determinations of charged-particle
collision cross sections to use of cross sections and swarm parameters to
model the behavior of electrical gas discharges. Although the range of
subjects covered was in many respects similar to that of previous
seminars, there was an emphasis on certain selected themes that tended to
give this seminar a distinctly different flavor. There was, for example,
considerable discussion on the meaning of "equilibrium" and the
conditions under which nonequilibrium effects become important in the
transport of electrons through a gas. It is evident from work presented
here that under certain gas discharge or plasma conditions nonequilibrium
effects can be significant; therefore, application of swarm or transport
parameters determined under equilibrium conditions to the modeling of
such discharges or plasmas must be considered questionable. The
discussions at this seminar, as represented by several of the invited
papers, has helped to remove some of the confusion about the
applicability of equilibrium assumptions and provided guidance for
attempts to deal with nonequilibrium situations. The seminar also
included discussions about the meaning and determination of higher order
"diffusion coefficients" in electron transport and limitations on the
range of validity of "modified effective range theory." Interesting new

developments on both topics were presented. Several of the invited papers were concerned with the peculiarities of ion transport in sulfur hexafluoride, a gas that has become increasingly important because of use in plasma processing of electronic materials and as a gaseous dielectric in electrical power systems. An attempt was made for the first time to include papers on electron transport in dense media, namely high-pressure gases and liquids.

The 1989 Swarm Seminar was sponsored jointly by the Polytechnic University of New York, the National Institute of Standards and Technology, and the Naval Surface Warfare Center. Financial support for the seminar was also provided by the U.S. Air Force Office of Scientific Research.

CONTENTS

Non-Equilibrium Electrons Transport: ,A Brief Overview 1
    L.C. Pitchford, J.P. Boeuf, P. Segur and E. Marode

Beam, Swarm and Theoretical Studies of Low-Energy
        Electron Scattering: Some Exemplars 11
    R.W. Crompton

Coupled Solutions of Boltzmann Equation, Vibrational and
        Electronic Nonequilibrium Kinetics 37
    C. Gorse

Higher-Order Electron Transport in Gases 49
    B.M. Penetrante and J.N. Bardsley

Non-Local Descriptions of Electron Swarms in Space-Time 67
    H.A. Blevin and L.J. Kelly

A Description of the Non-Equilibrum Behavior of Electrons
        in Gases: Macro-Kinetics 83
    E.E. Kunhardt

Nonequilibrium Effects in Electron Transport at High E/n 99
    Y.M. Li

Electron Collision Cross Sections for Processing Plasma Gases from
        Swarm and Discharge Data 121
    L.E. Kline

When Can Swarm Data Be Used to Model Gas Discharges? 143
    M.J. McCaughey and M.J. Kushner

Non-Equilibrium Effects in DC and RF Glow Discharges 157
    D.B. Graves and M. Surenda

Measurement of Attachment Coefficients in the
        Presence of Ionization 177
    D.K. Davies

Ion Transport and Ion-Molecule Reactions of Negative
        Ions in SF_6 197
    Y. Nakamura

A Survey of Recent Research on Ion Transport in SF_6 211
    J. de Urquijo, I. Alvarez, C. Cisneros, and H. Martínez

Collisional Electron-Detachment and Ion-Conversion
    Processes in SF$_6$ 229
    J.K. Olthoff, R.J. Van Brunt, Y. Wang, L.D. Doverspike
    and R.L. Champion

A Close Encounter Between Theory and Experiment
    in Electron-Ion Collisions 245
    M.A. Hayes

Electron-Ion, Ion-Ion, and Ion-Neutral Interactions 261
    R. Johnson

Electron-Ion Recombination in Dense Molecular Media 275
    K. Shinsaka and Y. Hatano

The Mobility of Electrons in Liquid Argon; Some Differences and
    Some Similarities with the Motion of Electrons in
    Crystals and Gases 291
    G. Ascarelli

Ultrafast and Ultrasensitive Dielectric Liquids/Mixtures: Basic
    Measurements and Applications 313
    L.G. Christophorou, H. Faidas, and D.L. McCorkle

Diffusion of Electrons in a Constant Field:
    Steady Stream Analysis 329
    J.H. Ingold

Diffusion of Electrons in a Constant Filed: TOF Analysis 333
    J.H. Ingold

A Monte Carlo Simulation of Electron Drift Limited by Collisions
    in Gas Mixtures Using the Null Collision Method 337
    D. Ramos, E. Patrick, D. Abner, M. Andrews and A. Garscadden

An Analysis of Transient Velocity Distribution of Electrons 339
    N. Ikuta, S. Nakajima and M. Fukutoku

An Exact Theory for Transient Behavior of Electron
    Swarm Parameters 343
    P.J. Drallos and J.M. Wadehra

Electron Transport Property Under Electric and
    Magnetic Fields Calculated by FTI Method 345
    N. Ikuta and Y. Sugai

A Multigroup Approach to Electron Kinetics 349
    S. Clark and E.E. Kunhardt

Integral Expansion Often Reducing to the Density Gradient Expansion,
    Extended to Nonmarkov Stochastic Processes. Consequent
    Stochastic Equation for Quantum Mechanics More Refined
    Than Schrodinger's 351
    G. Cavalleri and G. Mauri

Generalized Diffusion Coefficients and 1/f Power Spectral Noise . . . 353
    G.Cavalleri and G. Mauri

Fokker-Planck Calculation of the Electron Swarm
    Energy Distribution Function 355
    N.J. Carron

Nonlinear Diffusion 357
    E. E. Kunhardt

Sensitive High-Temporal-Resolution TOF Electron
        Drift Tube; Asymmetrical Current Pulse
        Observation and Determination of V_d, D_L and D_3 359
    C.A. Denman and L.A. Schlie

The Characteristic Energy of Electrons in Hydrogen 361
    W. Roznerski, J.Mechlinska-Drewko,
    K. Leja and Z.Lj. Petrovic

Electron Swarm Parameters in Krypton and Its Momentum
        Transfer Cross Sections 363
    Y. Nakamura

The Electron-Mercury Momentum Transfer Cross Section
        at Low Energies 367
    J.P. England and M.T. Elford

Longitudinal Diffusion to Mobility Ratios
        for Electrons in Noble Gases 371
    J.L. Pack, R.E. Voshall, A.V. Phelps and L.E. Kline

Relations Between Electron Kinetics in DC ExB
        and Microwave Discharges 373
    G. Schaefer and P. Hui

Computer Simulation of a Discharge in Crossed Electric
        and Magnetic Fields 375
    G.R. Govinda Raju and M.S. Dincer

Structures of the Velocity Distributions and Transport
        Coefficients of the Electron Swarm in CH_4 in
        a DC Electric Field 377
    N. Shimura and T. Makabe

Negative Ion Kinetics in BCL_3 Discharges 381
    Z.Lj. Petrovic, W.C. Wang, L.C. Lee, J.C. Han and M. Suto

On the Mechanism of Thermal Electron Attachment to SO_2 385
    H. Shimamori and Y. Nakatani

Electron Attachment to NF_3 387
    S.R. Hunter

Electron-Energy Dependence of Electron Attachment to Molecules
        as Studied by a Pulse Radiolysis Microwave
        Cavity Technique 389
    H. Shimamori and Y. Nakatani

Negative Ion Pulses Induced by Laser Irradiation
        of DC Discharge Media 393
    L.C. Lee, J.C. Han and M. Suto

The Attachment of Electrons in Water Vapour
        at Low Values of E/N 395
    J.C. Gibson and M.T. Elford

Isotope Studies and Energy Dependences of Rate Constants for the
        Reaction $O^- + N_2O$ at Several Temperatures 399
    R.A. Morris, A.A. Viggiano and J.F. Paulson

Recent FALP Studies of Dissociative
      Recombination and Electron Attachment 401
  N.G. Adams, D. Smith and C.R. Herd

Recent Sift Studies of Ion-Molecule Reactions:
      A Deuteration of Interstellar Molecules 403
  K. Giles, D. Smith and N.G. Adams

Reactions of Several Hydrocarbon Ions with Atomic Hydrogen
      and Atomic Nitrogen 407
  W. Lindinger, A. Hansel, W. Freysinger and E.E. Ferguson

Quenching of NO^+ (v=1,3) in Low Energy
      Collisions with CH_4 and He 409
  A. Hansel, N. Oberhofer and W. Lindinger

Rotational Energy Effects in Ion-Molecule Reactions 411
  A.A. Viggiano, R.A. Morris, T. Su
  and J.F. Paulson

The Application of a Selected Ion Flow Drift Tube to the
      Determination of Proton Affinity Differences 413
  M. Tichy, G. Javahery, N.D. Twiddy, E.E. Ferguson

Energy Dependences of Rate Constants for the Reaction
      $^{22}Ne^+ + {}^{20}Ne$ at Several Temperatures 415
  R.A. Morris, T. Su, A.A. Viggiano and J.F. Paulson

Deexcitation of $He(2^1P)$ in Collisions with Rare Gas Atoms 417
  M. Ukai, H.Yoshida, Y. Morishima, H. Nakazawa,
  K. Shinsaka and Y. Hatano

Measurements of D_T/K for Sodium Ions Drifing in Argon 419
  M.J. Hogan and P.P. Ong

Transverse Diffusion of Neon Ions in Neon 421
  T. Stefansson

Helium Ion Clusters He_n^+ (n≤16) Formed in a
      Liquid Helium Cooled Drift Tube 423
  T. Kojima, N. Kobayashi and Y. Kaneko

Ion and Fast Neutral Model for Nitrogen at Very High E/N 427
  A.V. Phelps

Study of the Electron Transport in an RF Discharge
      in Ar and CH_4/H_2 by Optical Emission Spectroscopy 429
  F. Tochikubo, T. Kokubo and T. Makabe

H_β Line Shapes in RF Discharges in $CHClF_2$ and H_2 433
  S. Radovanov, S. Vrhovac, Z. Petrovic and
  B. Jelenkovic

Time-Resolved Investigations of H_2 and $H_2:CH_4$ RF Plasmas 435
  S.C. Haydon, W. Hugrass and H. Itoh

Scattering of Electrons of High-Molecular Rydbergs
      in Dense Atomic and Molecular Gases 439
  U. Asaf, K. Rupnik, W.S. Felps and S.P. McGlynn

A Constant Ratio Approximation Theory of the Cylindrical
        Positive Column of a Glow Discharge 441
    T. Dote and M. Shimada

Laser-Induced Opto-Galvanic Studies of Pre-Breakdown
        Swarm Phenomena 445
    A. Ernest, M. Fewell and S.C. Haydon

Influence of Negative Ion and Metastable Transport on the
        Stochastic Behavior of Negative Corona (Trichel) Pulses . . . 449
    S.V. Kulkarni and R.J. Van Brunt

Improvement in the Breakdown Strength of SF_6/N_2
        Mixtures-A Physical Approach 451
    D. Raghavender and M.S. Naidu

Geometry-Dependent Displacement Current in the
        Pulsed-Townsend Drift Tube 453
    E. Patrick, D.Abner, D. Ramos, M. Andrews and A. Garscadden

List of Attendees 455

Index 461

NON-EQUILIBRIUM ELECTRON TRANSPORT: A BRIEF OVERVIEW

L. C. Pitchford[a], J. P. Boeuf[a], P. Ségur[a] and
E. Marode[b]

[a]Centre de Physique Atomique de Toulouse
Toulouse, France
[b]Laboratoire de Physique des Décharges
Gif-sur-Yvette, France

INTRODUCTION

Traditional electron swarm studies have focused on the range of
conditions for which the electron transport and rate coefficients can be
well-parameterized by the local value of the ratio of the electric field
to the neutral density, $E(r,t)/N$. We have rather informally referred to
this condition where the electron velocity distribution function (evdf)
at any point in space or time can be defined by the local reduced field
as "equilibrium with the field" or "local field equilibrium". Over the
years, and driven to a large extent by the need for accurate analyses of
swarm experiments for the determination of cross sections, a rather
complete theory of electron transport in weakly ionized gases has been
developed (Kumar et al., 1980, 1984) subject to the condition of local
field equilibrium. In its usual form, this theory involves an expansion
of the space and time dependent evdf in powers of the gradient of the
electron density, and it provides a computational procedure for obtaining
the space-time evolution of the electron density in terms of electron
transport and rate coefficients which are functions of the local value of
$E(r,t)/N$ (Kumar et al., 1980, 1984). By analogy with theories in other
areas of transport phenomena, this has been referred to as "hydrodynamic"
electron transport, and the terms "hydrodynamic" and "local field
equilibrium" have been used synonymously.

The development of the hydrodynamic theory of electron transport has
allowed the determination of electron drift velocities and diffusion
coefficients to remarkable levels of accuracy from measurements of $N(r,t)$
the space and/or time evolution of a pulse of electrons released from the

cathode and pulled through a neutral gas under the influence of an electric field (Huxley and Crompton, 1974). The extraction of highly accurate electron scattering cross sections from swarm experiments has generally relied on the interpretation of n(r,t) in terms of hydrodynamic transport and rate coefficients. A large body of experimental and computational data exist in this regime (Dutton, 1975; Gallagher et al., 1983), and the data are referred to as "electron swarm" data, in analogy with a drifting, spreading swarm of bees. If the electric field is space or time dependent, but only slowly changing, the transport and rate coefficients can still be parameterized by the local value of the field, E(r,t)/N. Although the hydrodynamic description is restrictive, there is a wide range of electron transport phenomena and many applications which are well described by these concepts. These subjects have been a dominant theme in previous Swarm Seminars.

A description of electron transport based on coefficients which are functions of E/N is of wide, but not universal validity, as is emphasized by M.J. Kushner in these proceedings. In particular, the local value of E(r,t)/N is insufficient to describe, for example, the electron transport near physical boundaries such as electrodes, in the presence of external sources of ionization, and when E/N varies rapidly in space or time. Electron transport phenomena under these conditions can depart significantly from hydrodynamic electron transport where the local value of E(r,t)/N is all that one need specify.

A number of attempts to measure and calculate electron transport or related phenomena in the absence of a local field equilibrium have been made since the 1920's. It is only in the past 10 or so years, however, that detailed calculations based on numerical solutions of the Boltzmann or Monte Carlo simulations have been practical under non-equilibrium conditions. One impetus for the development of these methods has been the need for detailed models of electron transport phenomena in the cathode fall region of gas discharges especially for the application of low-pressure plasma processing of semiconductor devices as is discussed by D.B. Graves in these proceedings. This has led to a renewed and practical emphasis on such phenomena, and it is timely that one of the main themes of this meeting is non-equilibrium electron transport.

It should be pointed out that "non-equilibrium" is a term that is also used in the context of weakly ionized gases to denote either 1) non-Maxwellian electron energy distribution functions or 2) electron transport in background neutral gases with significant internal excitation which cannot necessarily be described by a single temperature. The former has been studied for many years and is well understood. The latter is a more difficult area largely because of the lack of the basic

2

data needed to describe electron interactions with radicals and excited states. C. Gorse and her colleagues at the University of Bari in Italy have devoted considerable attention to this topic over the past ten years and she describes the status of their very interesting work in these proceedings.

In this brief article, we will provide a few comments on our view of the status of non-equilibrium electron transport studies, although more from a computational than from an experimental or theoretical point of view. We will also attempt to provide a perspective for those contributions at this meeting which fall in the general category of non-equilibrium electron transport.

STATUS OF NON-EQUILIBRIUM ELECTRON TRANSPORT STUDIES

Survey of Non-Equilibrium Phenomena

In local field equilibrium, the shape of the evdf depends only on $E(r,t)/N$ and is such that the energy and momentum gained from the field are balanced by collisions in each element in velocity space at each position and time (Allis, 1956); i.e., there is no net electron flux in velocity space at a given r and t. When this delicate balance is perturbed, for example, by emitting or absorbing boundaries, rapidly varying fields or external sources of ionization, there are unbalanced fluxes in velocity space which act to restore the local field equilibrium. The electron velocity distribution function then depends on space and/or time as well as E/N and the gas composition, and the electron transport is said to be non-equilibrium (Kumar et al., 1980, 1984). It should be emphasized that we are discussing the normalized distribution function in the above; a changing electron number density is consistent with equilibrium electron transport provided the rate of change; i.e., ionization or attachment rate coefficient, is independent of time or space.

The space or time dependent evolution of the non-equilibrium evdf (except for the simple and very special case of a constant electron-neutral scattering frequency) is a complex process. The acceleration of the electrons by the field depends locally and instantaneously on $E(r,t)$ while the collisional redistribution of electron velocities is a slower, nonlocal process. And, because the collisional energy losses are dependent on the electron energy, different parts of the distribution will readjust to perturbations with different rates.

The high-energy tail of a steady-state but space dependent evdf is largely populated by the electrons that have "escaped" from the body of the distribution by traveling along the field for several mean free paths without an inelastic collision. Thus, the tail of the evdf at a point x is strongly influenced by both the field and the body of the evdf many volts upstream from x. Deviations of the observable transport and rate coefficients from their local field values are more pronounced in moments of the evdf which weigh more heavily the high-energy tail (for example, the ionization rate coefficient), while moments reflecting the bulk electrons are better approximated by their local field values (the average electron energy) (Moratz et al., 1987).

In contrast, the high-energy tail of a time dependent but spatially homogeneous evdf responds more quickly than does the body of the distribution because the collision frequency or energy exchange frequency is usually higher in that range of energies. The energy regions for which electron energy exchange with the background gas is inefficient are the slowest to respond. The overall effect is a very complex time dependent evdf which passes through a series of transient shapes which are unlike any equilibrium evdf, and for which the high-energy tail is modulated much faster than the body of the distribution (Wilhelm and Winkler, 1979).

Electron transport under high field conditions is complicated by the existence of a runaway electron component, and it is natural to assume that the local field approximation is invalid under these conditions. Building on previous work, recent detailed studies have shown that, in spite of this tendency for individual electrons to runaway, the continuous flux of newly born secondary electrons dominates the evdf and is such that velocity averages over the distribution are functions of the local field. It is not clear, however, that the density gradient expansion and the usual hydrodynamic formalism is valid or useful under these conditions. In these proceedings, Y.M. Li describes recent work on high field electron transport, and A.V. Phelps presents his recent results on the ion and fast neutral behavior under high field conditions.

Observations of diverse non-equilibrium phenomena such as the nature of the oscillations in light intensity near the cathode in rare gases (Holst and Oosterhuis, 1921) and the energy of the electron beam from the cathode fall entering the negative glow (Brewster and Westhaven, 1937) have been made since the 1920's and 30's. In their classic review paper, Druyvesteyn and Penning (1940) comment on these and a number of other cases where E/N is no longer a good parameter to describe the electron transport. In spite of this half century of recognition that there exists a broad range of non-equilibrium electron transport phenomena and their observation, it is hard to draw detailed generalizations. Non-equilibrium phenomena have been difficult to quantify experimentally on a systematic basis because of their dependence on initial and experimental conditions. More importantly, these phenomena are highly dependent on which of the observables is under consideration (Phelps 1983). There are only a few recent examples of experiments or calculations which attempt to systematically quantify well-defined aspects of non-equilibrium behavior (Haydon and Williams, 1976; Hays et al., 1987; Ponomarenko et al., 1985; Pitchford, 1985).

Although a number of empirical generalizations based on the existing data could be cited, there are only three firm conclusions we wish to emphasize. (See E. Marode and J.P. Boeuf, 1983, for a recent review of existing data.) First, in the absence of external sources of electrons, the time and distance variables in the Boltzmann equation scale in products with the neutral density; i.e., Nx and Nt are the scaling parameters for relaxation towards the equilibrium distribution function. Second, there is no general correlation between the non-equilibrium values of the observables. This means, for example, that the average electron energy cannot in general be used to parameterize the ionization rate coefficient when the local field approximation fails. Lastly, the observables can attain values outside their range of equilibrium values. The vibrational excitation rate coefficient in nitrogen, for example, in a particular non-uniform field reached values up to four times the maximum equilibrium value (Moratz et al., 1987).

COMPUTATIONAL METHODS

There are a number of techniques which have been developed over the years and a vast literature describing methods for solving the Boltzmann equation to obtain the electron velocity distribution function under hydrodynamic conditions. Extremely accurate and efficient solution techniques are now available for electron transport in the hydrodynamic regime (Lin et al., 1979; Pitchford et al., 1981; Ségur et al., 1983; Winkler et al., 1984). If the full space and/or time dependence of the evdf must be taken into account, the solution techniques and, in partic- ular, solutions of the Boltzmann equation become considerably more complicated. Advances have been made recently in the development of techniques for time and space dependent situations. Notably, the technique developed by Ségur, et al., (1986) for the solution of the Boltzmann equation in the cathode fall region of discharges is capable of handling physical boundaries and non-uniform fields. P.J. Drallos and J.M. Wadehra describe in these proceedings a promising technique for the solution of the time-dependent Boltzmann equation (Drallos and Wadehra, 1988), and N. Ikuta and his colleagues present results from their flight

time integral method (Ikuta and Murakami, 1987) which is a powerful
method for time dependent (and potentially space dependent) solutions of
the Boltzmann equation. S. Clark and E.E. Kunhardt are developing a
multigroup approach to electron kinetics, and they present a numerical
solution to an initial value problem using this technique.

Monte Carlo simulations of electron transport in gases provide the
same information as do solutions of the Boltzmann equation; i.e., electron transport and rate coefficients and the electron velocity distribution function (Itoh and Musha, 1960; Boeuf and Marode, 1982). In
principle, the simulation results are as accurate as the physical
description of the electron motion and interaction with the background
gas included in the simulation. Monte Carlo simulations obviate the need
for introducing the hydrodynamic approximation which simplifies considerably solutions of the Boltzmann equation. On the other hand, the
difficult in interpreting Monte Carlo simulations is in distinguishing
small physical effects from statistical fluctuations.

The more recent simulation algorithms (Boeuf and Marode, 1982; Li et
al., 1989) make use of several features which enhance the statistical
accuracy and reduce the computational time. These include the null
collision technique (Skullerud, 1968), sampling before collisions
(Friedland, 1977), and artificial attachment or ionization (Li et al.,
1989). Detailed comparisons between the numerical solutions of the
Boltzmann equation and the simulation results have been performed (Ségur
et al., 1986; Braglia et al., 1982), and such comparisons have served to
validate (or invalidate) the various assumptions and approximations which
are necessary in the numerical solution of the Boltzmann equation.

VELOCITY MOMENT APPROACHES

In most cases of technological interest it is the averages over the
distribution function such as the electron drift velocity, diffusion
coefficient, ionization or excitation rate coefficients, that are the
desired results of the calculation. The full evdf from Boltzmann
solutions or simulation results contains far more information than is
used or required, and with present day computing power it is still
difficult to couple Poisson's equation to describe the electric field
with a Boltzmann or Monte Carlo calculation as illustrated by D.B. Graves
in his contribution to this meeting. An alternative to descriptions of
electron behavior based on the solutions of the Boltzmann equation of
Monte Carlo simulations of the evdf is the velocity moment description
which is based on approximate solutions of finite number of velocity

moments of the Boltzmann equation, the first three being the electron continuity equation and the momentum and energy balance equations.

For most applications, we are interested in the solution of the electron continuity equation which in turn depends on the electron flux or momentum balance. In many of the recent models (Boeuf and Ségur, 1987, for example) the average electron velocity from the steady-state momentum balance equation is expressed as the sum of a convective term which depends on the electric field and a diffusion term which depends on the gradient of the electron density. If the coefficients of these terms and the reaction frequency are known functions of E/N, then the continuity equation can be solved independently of the higher order moments.

When the coefficients are unknown functions of E/N or if they can no longer be described by the local field, E(r,t)/N, some approximation to the distribution function is needed to effect a truncation of the moment series. Many different approximations have been used, for example, one (Mason and McDaniel, 1988; Bayle et al., 1986; Yee et al., 1986) or two (Vriens et al., 1978; Morgan and Vriens, 1980) temperature distributions; displaced spherical shells where the displacement is related to the drift velocity and the radius of the shell to the random energy (Ingold, 1978); single-beam distributions where the distribution is monoenergetic and in the field direction (Phelps et al., 1987); multibeam distributions where each secondary electron created in ionization is described by a single-beam distribution (Phelps et al., 1987; Muller, 1962; Friedland, 1974); and combinations such as a local field description of the bulk electrons and a single-beam for the high energy tail (Boeuf and Ségur, 1987; Kushner, these proceedings), etc. In all of these parameterizations, the edf is expressed as a function of the average velocity or average energy or both, thus reducing the problem to the solution of two or three coupled partial differential equations. While there is no general parameterization which is suitable for all non-equilibrium phenomena, the velocity moment approach, judiciously applied, can often provide a good qualitative description of the electron behavior.

J. H. Ingold has long been an advocate of the moment approach, and in these proceedings he makes use of the first three velocity moments to deduce a difference between the transverse and longitudinal diffusion coefficients without recourse to the usual explanation via the density gradient expansion. Also in these proceedings, E. E. Kunhardt describes his recent approach to non-equilibrium electron transport phenomena which is intended to be a general theory when the density gradient expansion is insufficient to describe the space and time dependence of the evdf.

CONCLUDING COMMENTS

We will conclude this very brief outline of our view of the status of the non-equilibrium electron transport studies with a few of the issues which were discussed during the 6th International Swarm Seminar.

· Lacking at present an alternate to the hydrodynamic theory for relating drift tube measurements to transport coefficients, non-equilibrium phenomena must be avoided for the accurate extraction of cross sections from swarm data. Perhaps with suitable theories these non-equilibrium phenomena will one day provide a rich source of data on the microscopic scale.

· Applications such as low-pressure plasma processing are demanding more and more sophisticated models of the cathode fall regions of discharges, regions where equilibrium models fail in even their qualitative predictions. It is imperative that we develop techniques to deal with the highly non-equilibrium nature of electron transport for these and other applications.

· Given the wide and varied range of non-equilibrium phenomena, it is tempting to speculate that new applications may be developed which, for example, require tailoring the distribution of the field in time or space to achieve a desired result, just as gas mixtures are tailored for specific needs.

Contributions elsewhere in this book describe these and other issues in more detail.

REFERENCES

Allis, W. P., 1956, in "Handbuch der Physik," Flugge, S. (Ed.), Springer, Berlin.
Bayle, P., J. Vacquie, and M. Bayle, 1986, Phys. Rev. A 34, 360.
Boeuf, J. P., and E. Marode, 1982, J. Phys. D 15, 2069.
Boeuf, J. P., and P. Ségur, 1987, in "Interactions Plasmas Froids Materiax," C. Lejeune, (Ed.), Les Editions de Physique, Les Ulis, France.
Braglia, G. L., L. Romano, and M. Dilligenti, 1982, Lett. Nuovo Cim. 35, 193.
Brewster, A. Keith, and J. W. Westhaven, 1937, J. Appl. Phys. 8, 779.
Drallos, P. J., and J. M. Wadehra, 1988, J. Appl. Phys. 63, 5601.
Druyvesteyn, M. J., and F. M. Penning, 1940, Rev. Mod. Phys. 12, 87.
Dutton, J., 1975, J. Phys. Chem. Ref. Data 4, 577.
Friedland, L., 1974, J. Phys. D 7, 2246.
Friedland, L., 1977, Phys. Fluids 20, 1461.
Gallagher, J. W., E. C. Beaty, J. Dutton, and L. C. Pitchford, 1983, J. Phys. Chem. Ref. Data 12, 109.
Haydon, S. C., and O. M. Williams, 1976, J. Phys. D 9, 523.
Hays, G. N., L. C. Pitchford, J. B. Gerardo, J. T. Verdeyen, and Y. M. Li, 1987, Phys. Rev. A 36, 2031.
Holst, G., and E. Oosterhuis, 1921, Physica 1, 78.

Huxley, L. G. H., and W. R. Crompton, 1974, "The Diffusion and Drift of
    Electrons in Gases," John Wiley & Sons, Inc., New York.
Ikuta, N., and Y. Murakami, 1987, J. Phys. Soc. Jpn. 56, 115.
Ingold, J. H., 1978, in "Gaseous Electronics, Electrical Discharges,
    M. N. Hirsch, and H. J. Oskam, (Eds.), Academic, New York.
Itoh, T., and T. Musha, 1960, J. Phys. Soc. Jpn. 15, 1675.
Kumar, K., H. R. Skullerud, and R. E. Robson, 1980, Aust. J. Phys. 33,
    343.
Kumar, K., 1984, Phys. Repts. 112, 319.
Kushner, M. J., these proceedings.
Li, Y. M., L. C. Pitchford, and T. A. Moratz, 1989, Appl. Phys. Lett. 54,
    1403.
Lin, S. L., R. E. Robson, and E. A. Mason, 1979, J. Chem. Phys.
    71, 3483.
Marode, E., and J. P. Boeuf, 1983, in "Proceedings of the XVI
    International Conference on the Physics of Ionized Gases, invited
    talks," W. Botticher, H. Wenk, and E. Schulz-Guide, (Eds.), p. 206.
Mason, E. A., and E. W. McDaniel, 1988, "Transport Properties of Ions in
    Gases," John Wiley & Sons, Inc., New York.
Moratz, T. J., L. C. Pitchford, and J. N. Bardsley, 1987, J. Appl. Phys.
    61, 2146.
Morgan, W. L., and L. Vriens, 1980, J. Appl. Phys. 51, 5300.
Muller, K. G., 1962, Z. Phys. 169, 432.
Phelps, A. V., 1983, in "Electrical Breakdown and Discharges in Gases,"
    E. E. Kunhardt, and L. L. Luessen, (Eds.), Plenum, New York.
Phelps, A. V., B. M. Jelenkovic, and L. C. Pitchford, 1987, Phys. Rev. A
    36, 5327.
Pitchford, L. C., S. V. O'Neil, and J. R. Rumble, Jr., 1981, Phys. Rev. A
    23, 294.
Pitchford, L. C., 1985, Technical Report No. AFWAL-TR-85-2016.
Ponomarenko, A. G., V. N. Tishchenko, and V. A. Shveigert, 1985, Sov. J.
    , Plasma Phys. 11, 288.
Segur, P., M. Yousfi, J. P. Boeuf, E. Marode, A. J. Davies, and C. J.
    Evans, 1983, in "Electrical Breakdown and Discharges in Gases,"
    , E. E. Kunhardt, and L. L. Luessen, (Eds.), Plenum, New York.
Segur, P., M. Yousfi, M. H. Kadri, and M. C. Bordage, 1986, Transport
    Theory and Stat. Phys. 15, 705.
Skullerud, H. R., 1968, J. Phys. D 1, 1567.
Vriens, L., R. A. J. Keijser, and F. A. S. Ligthart, 1978, J. Appl. Phys.
    49, 3807.
Wilhelm, J., and R. Winkler, 1979, J. de Phys. C7, 251.
Winkler, R., G. L. Braglia, A. Hess, and J. Wilhelm, 1984, Beitr. Plasma
    Phys. 24, 657.
Yee, J. H., R. A. Alvarez, D. J. Mayhall, D. P. Byrne, and J. DeGroot,
    1986, Phys. Fluids 29, 1238.

BEAM, SWARM AND THEORETICAL STUDIES OF LOW-ENERGY ELECTRON SCATTERING:
SOME EXEMPLARS

R. W. Crompton

Research School of Physical Sciences
Australian National University
Canberra

INTRODUCTION

This series of satellite meetings of ICPEAC began ten years ago in
Tokyo to provide a forum for the results of electron and ion swarm
research in the areas covered by ICPEAC itself. The Swarm Seminars now
cover a field wider than was perhaps originally foreseen, but nevertheless the original theme remains an important element. It has received
particular emphasis at some meetings, notably at the inaugural Tokyo
meeting and at the Lake Tahoe meeting in 1985 at which there was a joint
session with the Inelastic Electron-Molecule Collisions Symposium.

Increasingly there has been overlap between the work of those
engaged in single-collision and swarm research, while advances in the
theory of low-energy electron-atom and electron-molecule collisions have
both stimulated, and been stimulated by, new experimental work. This
paper in the session on cross section determinations from swarm data
describes some examples of recent work that is aimed at obtaining a
consistent body of information on low-energy electron scattering from the
two complementary experimental techniques and from theory. The paper is
intended to be illustrative rather than comprehensive; other papers in
this session will provide further examples. The first part of the paper
deals with elastic scattering from atoms, and the second part with an
update on the two simplest test-cases in electron molecule scattering:
$e-H_2$ and $e-N_2$.

LOW-ENERGY ELASTIC CROSS SECTIONS FOR ATOMIC GASES

In recent years there has been a remarkable convergence in the
integral cross sections obtained from beam and swarm experiments, and of
experimental and theoretical results. Nevertheless, when attempts have

Nonequilibrium Effects in Ion and Electron Transport
Edited by J. W. Gallagher *et al.*, Plenum Press, New York, 1990

been made to compare total cross sections (σ_T) from beam experiments with momentum transfer cross sections (σ_m) derived from data taken in swarm experiments, there have sometimes appeared differences that seem too large to be ignored. But a loophole has always made it difficult to make definitive statements. This loophole arises because such comparisons can be made only by invoking Modified Effective Range Theory (MERT) (O'Malley, 1963), and the unanswered question has been whether or not this theory was being applied outside the range of energies where the MERT expansions are valid.

The agreement between σ_m and σ_T is reasonable but by no means always as good as might be expected given the accuracy claimed for the two types of experiment. In the case of the swarm-derived momentum transfer cross sections, it is not straightforward to assign error limits for a cross section that shows an energy dependence as strong as those in the heavier monatomic gases. Nevertheless, the compatibility of the data from the two types of experiment can readily be checked by comparing the transport coefficients measured in swarm experiments with values calculated with a σ_m derived via MERT from a σ_T measured in beam experiments. For example, when such comparisons are made for argon using the data from recent beam experiments (Ferch et al., 1985; Buckman and Lohmann, 1985) the differences between measured and calculated values of the transport coefficients are often 3 times and sometimes 5 times the experimental uncertainty.

Are the primary experimental data incompatible, or is the method of comparison flawed? Buckman and Mitroy set out to answer this question for neon, argon and krypton in a recently published paper (Buckman and Mitroy, 1989). Their conclusions for Ar are the most definitive and therefore the most interesting. The next section summarizes their approach to this problem and their conclusions.

The second topic I wish to discuss is the determination of σ_m for krypton. The emphasis is different here in that the discussion is about the results of applying two different swarm methods to derive the cross section rather than a comparison between the results of beam and swarm experiments.

Thirdly, the cross sections obtained from beam and swarm experiments in Ne are compared with some very recent theoretical results. In this case, direct definitive comparisons of beam and swarm results cannot be made (Buckman and Mitroy, 1989).

Finally, I shall briefly mention a new swarm study of elastic scattering in mercury; this study is the subject of another paper in this Seminar.

An Analysis of the Range of Validity of MERT; Implications for Comparisons of Low-Energy Integrated Cross Sections for Ar

The MERT expansions (in atomic units) for the phase shifts that have usually been used to compare σ_m and σ_T comprise the following expressions without underlined terms (see Buckman and Mitroy, 1989 and references therein):

$$\tan \eta_0 = -Ak\left\{1+(4\alpha_d/3)k^2\ln(k)\right\} - \left(\pi\alpha_d/3\right)k^2 + Dk^3 + Fk^4 \tag{1}$$

$$\tan \eta_1 = a_1\alpha_d k^2 - A_1 k^3 + \left(b_1\alpha_d^2 + c_1\alpha_q\right)k^4 + Hk^5 \tag{2}$$

$$\tan \eta_1 = a_1\alpha_d k^2 + \left(b_1\alpha_d^2 + c_1\alpha_q\right)k^4 \qquad 1 \geq 2 \tag{3}$$

where α_d is the dipole polarizability and

$$a_1 = \frac{\pi}{(2l+3)(2l+1)(2l-1)}.$$

For convenience Buckman and Mitroy named this 4-parameter expansion MERT4. However, in order to obtain satisfactory fits to the theoretical phase shifts of McEachran and Stauffer (1983) which were used to test the range of validity of the MERT expansions, Buckman and Mitroy found it necessary to include the underlined terms if data up to 1 eV were to be included. In these additional terms

$$b_1 = \frac{\pi\left[15(2l+1)^4 - 140(2l+1)^2 + 128\right]}{\left[(2l+3)(2l+1)(2l-1)\right]^3(2l+5)(2l-3)}$$

$$c_1 = \frac{3a_1}{(2l+5)(2l-3)},$$

and α_q is the "effective" quadrupole polarizability. This 5-parameter expansion they named MERT5.

In order to investigate the applicability of MERT for comparing the results of beam and swarm experiments, Buckman and Mitroy adopted the following strategy.

1. They first used the theoretical calculations of McEachran and Stauffer (1983) to test the range of validity of MERT by:

(a) determining a set of MERT coefficients by a least squares fit to the phase shifts, then comparing the phase shifts calculated using these coefficients with the original data;

(b) determining the MERT coefficients through a dual fit to the σ_T and σ_m calculated from the theoretical phase shifts, then comparing the σ_T and σ_m calculated using these coefficients with the original data.

13

2. Having established the range of validity for a particular system, Buckman and Mitroy sought the limits of uncertainty for a momentum transfer cross section derived by application of MERT to a total cross section that had been measured at relatively few points with typical experimental uncertainty. To do this they generated "experimental" data from McEachran and Stauffer's σ_T at energies corresponding to published experimental data by multiplying each value by a random number chosen from a normal distribution representing the experimental uncertainty at that energy. They fitted each of about 100 such "experimental" data sets using MERT5 and then generated the corresponding σ_m. They also calculated the mean and standard deviation of σ_m at each energy.

3. The beam-derived σ_m, with its uncertainty determined as in (2) above, was then compared with the swarm-derived σ_m.

Their conclusions for argon can be summarized as follows:

(i) MERT4, which has usually been used for such intercomparisons, is not adequate to represent the p-wave phase shift accurately above 0.5 eV. Since the p-wave dominates scattering at the Ramsauer-Townsend (R-T) minimum at about 0.25 eV, its accurate representation at low energies is essential.

(ii) From the results in Fig. 1, when the MERT parameters are derived from a dual fit to the σ_T and σ_m data up to 1 eV, MERT4 leads to errors in σ_m greater than 5% in the range 0.2 to 0.5 eV and greater than 10% near the R-T minimum, although σ_T is represented more accurately. On the other hand, both σ_m and σ_T calculated with MERT5 agree everywhere to within 3% with values calculated directly from the phase shifts.

From (i) and (ii) Buckman and Mitroy concluded that the MERT expansions including all terms in equations (1), (2), and (3) (i.e., MERT5) can provide an adequate representation of the phase shifts and cross sections for e-Ar scattering in the range 0 to 1 eV.

(iii) A spread of between $\pm$ 5 and $\pm$ 10% in the MERT-derived σ_m results from typical uncertainties in an experimental σ_T. This result is illustrated in Fig. 2. The shaded area represents the uncertainty in the σ_m derived from McEachran and Stauffer's theoretical data using the procedure described in (2) above. Their "true" σ_m lies, as expected, within the range of uncertainty of the derived cross section.

(iv) The momentum transfer cross section derived from Buckman and Lohmann's 1985 total cross section lies well outside the cross section derived from swarm data by Haddad and O'Malley (1982). As shown in Fig. 3., the differences between the two σ_m's are larger than 20% in the range 0.5 to 0.8 eV, and there appears to be a significant displacement

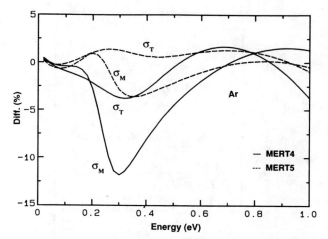

Fig. 1. Differences between total and momentum transfer cross
sections calculated directly from the phase shifts of
Stauffer et al. and those calculated with MERT4 and MERT5
from the parameters obtained from a dual fit to these cross
sections.

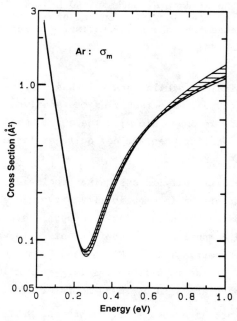

Fig. 2. A comparison of Stauffer et al.'s theoretical momentum
transfer cross section (full curve) with the MERT5 derived
σ_m. The MERT parameters were obtained by fitting to
"experimental" σ_T data generated by radomizing the theoretical σ_T by typical experimental uncertainties. The
hatched area covers the derived σ_ms that are within 1 SD of
the mean (see text).

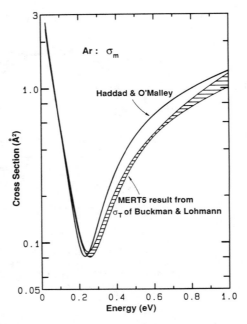

Fig. 3.   The momentum transfer cross section derived from a MERT5
          fit to the Ar data of Buckman and Lohmann compared with the
          swarm cross section (full curve) of Haddad and O'Malley.
          The hatched area shows the estimated uncertainty in the
          MERT5 σ_m.

between the two minima. A similar analysis has also been applied to the
σ_T of Ferch et al. (1985). The resulting beam-derived momentum transfer
cross section is also incompatible with the swarm momentum transfer cross
section although the differences are not quite as large as they are for
Buckman and Lohmann's cross section.

     Because of the difficulty in assigning precise error limits to the
swarm σ_m, the extent of the incompatibility between beam and swarm data
can best be illustrated by calculating transport coefficients using the
beam-derived σ_m and comparing these coefficients with values obtained in
swarm experiments (Robertson, 1977; Milloy and Crompton, 1977). Both
beam experiments give rise to differences larger than 10%; at many values
of E/N the differences are 3 to 5 times the uncertainty in the swarm data.

     Finally, attempts have been made (within the MERT5 formalism) to
derive a set of phase shifts that can fit both the total cross section
and the raw transport data. It is not possible to generate a set of
phase shifts that fits the beam data and the transport data within the
quoted experimental uncertainties.

     Despite apparently incompatible results for the integral cross
sections obtained from beam and swarm experiments at and above the R-T

16

minimum, scattering lengths derived from the two types of experiment agree to within 3%, as can be seen from Table 1.

O'Malley (1989) has suggested that the effect of multiple scattering on transport coefficients measured in the high pressure experiments of Robertson (1977) and Milloy and Crompton (1977) is responsible for the somewhat larger value of the scattering length derived from these data (Haddad and O'Malley, 1982). He noted that even though this cross section, fitted with the aid of MERT in the energy range 0 to 1 eV, optimized the fit to the transport data over the full range of E/N for which data were available, the differences between measured and calculated transport coefficients increased sharply and systematically at the lowest values of E/N where the highest pressures were used (Haddad and O'Malley, 1982). However, such effects are unlikely to have affected the Ar-H_2 mixture experiments of Petrovic and Crompton (1987) which were carried out using gas number densities 5 to 10 times smaller than those used for the earlier experiments in pure Ar.

Krypton

At the Fifth International Swarm Seminar in Birmingham, Hunter et al. (1987) gave a preliminary account of their work in Kr and Xe. Using pressures up to 600 kPa at room temperature (that is, almost 6 atm), they measured drift velocities in the E/N range between 0.002 and 3 Td and then used their data to derive σ_m. They found no evidence of pressure dependence, indicating that their pressures were sufficiently high to eliminate significant diffusion effects but insufficiently high to show significant multiple scattering effects. In Kr at the lowest values of E/N used by these authors, the electron energy distribution was essentially thermal (kT~ 0.025 eV), so they could derive from their drift data a cross section for energies down to 10 meV.

In the preceding year Koizumi et al. (1986) had published their determination of the e-Kr cross section from analysis of measured data

Table 1

source	type of experiment	scattering length (A/a_o)
Haddad & O'Malley, 1982	swarm	$- 1.49 \pm 0.015$
Ferch et al., 1985	beam	$- 1.45$
Petrovic & Crompton, 1987	swarm	$- 1.48 \pm 0.015$
Buckman & Mitroy, 1989	beam	$- 1.44 \pm 0.02$

17

for D_T/μ, such data being more sensitive to the cross section in the vicinity of a R-T minimum than data for v_{dr} (Milloy et al., 1977). Koizumi et al.'s cross section gives calculated values of v_{dr} that differ by up to 20% from those of Pack et al. (1962) although they agreed to within 8% with the earlier drift velocity data of Bowe (1960). One motivation for Hunter et al.'s work was to resolve this inconsistency, perhaps by showing the earlier drift velocity data to be inaccurate.

Unfortunately, as pointed out by Hunter et al. at the Birmingham meeting, their new measurements reduced but did not eliminate the inconsistency between the v_{dr} and D_T/μ data. When the new drift data were compared with values calculated using Koizumi et al.'s cross section, differences of up to 10% remained. Hunter et al. therefore derived a new cross section based on their drift data; this σ_m is shown along with that of Koizumi et al. in Fig. 4. The large difference at the R-T minimum (almost a factor of 2) may reflect the insensitivity of the analysis based on the v_{dr} measurements; nevertheless, the differences between the measured drift velocities and those calculated from Koizumi et al.'s cross section were well outside the claimed experimental uncertainty of between 1 and 2%.

In a further attempt to discriminate between the conflicting data sets, England and Elford, (1988) measured and analyzed v_{dr} data in mixtures of H_2 and Kr.

The most obvious advantage of mixing a molecular gas with a heavy monatomic gas is that it enables one to investigate scattering at low energies without having to use either very low electric field strengths or excessively high pressures. As pointed out for Ar, one possible

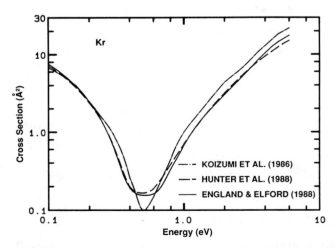

Fig. 4.   Momentum transfer cross sections in Kr from recent swarm experiments.

disadvantage of using high gas pressures is the introduction of multiple scattering, with consequent modifications of the energy distribution function and transport coefficients (O'Malley, 1989). For krypton the Oak Ridge group, and others, found that such effects were not significant at the pressures used by them (Hunter et al., 1988). Nevertheless, using large volumes of high pressure gas incurs another penalty: the high cost of the gas.

A less obvious advantage of mixture measurements, however, is that the measurement and analysis of drift velocity data in mixtures improves the uniqueness of the derived cross section in the region of the R-T minimum (England and Elford, 1988). The reason for this is the following.

In the two-term approximation (Huxley and Crompton, 1974), the expressions for the drift velocity and energy distribution function $f(\varepsilon)$ are:

$$v_{dr} = -\frac{eE}{3N}\left(\frac{e}{m}\right)^{1/2}\int_0^\infty \frac{\varepsilon}{\sigma_m(\varepsilon)}\frac{df}{d\varepsilon}\,d\varepsilon \tag{4}$$

and

$$f(\varepsilon) = A\,\exp\left\{-\int_0^\infty\left[\frac{ME^2e^2}{6m\ N^2\ \varepsilon'\ \sigma_m^2(\varepsilon')} + \kappa T\right]^{-1}d\varepsilon'\right\}. \tag{5}$$

From Eq. (5), the derivation of $f(\varepsilon)$ is

$$\frac{df(\varepsilon)}{d\varepsilon} = f(\varepsilon)\left[\frac{Me^2}{6m}\left|\frac{E}{N}\right|^2\frac{1}{\varepsilon\ \sigma_m^2(\varepsilon)} + \kappa T\right]^{-1}. \tag{6}$$

Now for krypton, if one chooses values of E/N such that the most probable energies occur near the R-T minimum at about 0.5 eV, then near the minimum the following inequality holds:

$$\left[(E/N)^2(Me^2/6m)\left\{\varepsilon\sigma_m^2(\varepsilon)\right\}^{-1}\right] \gg \kappa T.$$

It follows that, at a given E/N, $df(\varepsilon)/d\varepsilon$ is proportional to $\varepsilon\,f(\varepsilon)$ $\sigma_m^2(\varepsilon)$. Thus the integrand in Eq. (4) is proportional to $\varepsilon^2\,f(\varepsilon)\,\sigma_m(\varepsilon)$. Since $f(\varepsilon)$ is a monotonically decreasing function of ε, one might expect the integrand to reflect the R-T minimum in σ_m. Figure 5 shows that this is indeed the case.

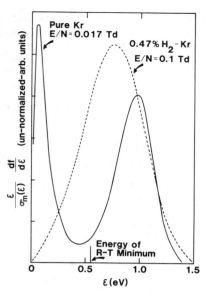

Fig. 5. The integrand in the formula for v_{dr} (Eq. 4) plotted as a function of energy in pure Kr for E/N = 0.017 Td, and in an 0.47 H_2 - Kr mixture for E/N = 0.1 Td. In each case the value of E/N was chosen so that the most probable energy was close to the energy of the R-T minimum.

A quite different situation obtains in a gas mixture containing H_2. Because of the rapid energy exchange in inelastic ro-vibrational collisions, the presence of H_2 causes the distribution function to much more nearly approximate a normal Maxwellian distribution at low values of E/N. In such a distribution, ϵ df/dϵ changes by only 30% when ϵ changes from $\kappa T/2$ to 2 κT, passing through its maximum value at $\epsilon = \kappa T$. Hence, if the electron "temperature" corresponds to the energy of the R-T minimum, one would expect the maximum of the integrand to occur at about this energy, provided the contribution to the aggregate σ_m from the σ_m for H_2 is not too large. The curve in Fig. 5 for the integrand for a mixture of about 0.5% H_2 in Kr confirms this expectation.

In practice, for such a mixture the momentum transfer cross section for H_2 contributes about 15% of the total σ_m at the minimum so the sensitivity is somewhat reduced. Nevertheless, for the range of E/N over which the cross section near the R-T minimum has most influence on the drift velocities, those in the mixtures are more sensitive to the cross section than those in the pure gas. This sensitivity is shown in Fig. 6, where differences between calculated drift velocities for the two cross sections shown in the inset are plotted for pure Kr and for this mixture.

A third advantage pointed out by England and Elford is the greatly reduced sensitivity of drift velocities in mixtures to trace impurities.

20

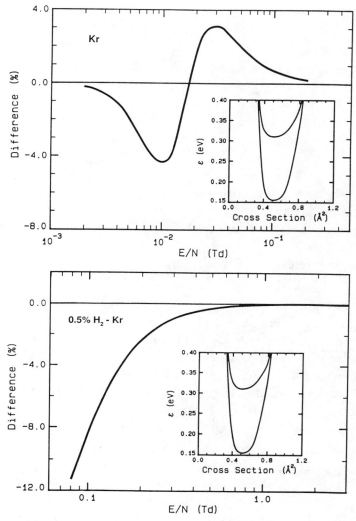

Fig. 6. Differences [{v_{dr} (trial) $-$ v_{dr} (ref)}/v_{dr} (ref)] between calculated drift velocities in pure Kr and in the Kr-H$_2$ mixture (see text) for cross sections with different R-T minima (see inset). Values calculated with the upper curve are used as the reference set.

In the heavier monatomic gases, molecular impurities at the ppm level introduce significant errors in the measured drift velocities (Robertson, 1977). As pointed out by Hunter et al. (1987, 1988), who paid particular attention to this problem in their work, impurities may have been responsible for some of the differences between their measured drift velocities and those obtained in earlier work. In the mixture measurements, a molecular "impurity" is introduced at the 1% level, so the influence of other likely molecular impurities at the ppm level will be negligible.

England and Elford made measurements in mixtures containing about 0.5 and 1.5% of H_2 and analyzed their data using the recently published low-energy cross sections for H_2 of England et al. (1988). The new cross section is shown in Fig. 4 in comparison with those of Hunter et al. and Koizumi. Several recent studies (Buckman and Mitroy, 1989) have shown that it is questionable to use MERT for Kr above about 0.4 eV, that is, at sufficiently high energies to include the R-T minimum. Hunter et al. used MERT to constrain the shape of the cross section only up to 0.35 eV. Their cross section at the minimum is therefore less accurate in this region than a cross section for Ar derived from v_{dr} data of similar accuracy where such a restraint can be applied. However, the arguments presented above and the data shown in Fig. 6 suggest that the cross section derived from the mixture data should be more accurate than that derived from data in pure Kr, although this may not, in fact, be the case for England and Elford's cross section.

Figure 7 shows the differences between the measured v_{dr} data in the mixtures and those calculated with the three cross sections in Fig. 4. The cross section of Koizumi et al., which is consistently larger than the other two cross sections above the minimum, is reflected in the steady increase in the difference between calculated and measured values with increasing E/N. Since this difference increases to 20 times the claimed experimental uncertainty in the mixture data, the cross section of Koizumi et al. is clearly incompatible with these data, confirming Hunter et al.'s conclusion that it is too large above about 0.8 eV.

Hunter et al.'s cross section predicts values of v_{dr} in the mixtures to within 3% of the measured values. This difference is outside the experimental uncertainty of the measurements, and occurs where the data are sensitive to the R-T minimum. Because of the reduced sensitivity of the v_{dr} data in pure Kr to this part of the cross section, it may be possible to modify Hunter et al.'s cross section to fit the mixture data without significantly affecting the fit to the data in the pure gas. However, England and Elford's cross section is likely to need modification also. Examination of Fig. 6 suggests that, because of experimental limitations, England and Elford's data did not extend to sufficiently low values of E/N (<0.08 Td) to define the cross section in the minimum with the accuracy inherently possible by their mixture technique. Their cross section might lead to unacceptably large discrepancies between calculated and measured values of v_{dr} at lower values of E/N, requiring a revision of the cross section not only at energies below the minimum, but also at the minimum itself. The differences (up to 6%) between Hunter et al.'s v_{dr} data and values calculated with England and Elford's cross section at values of E/N where

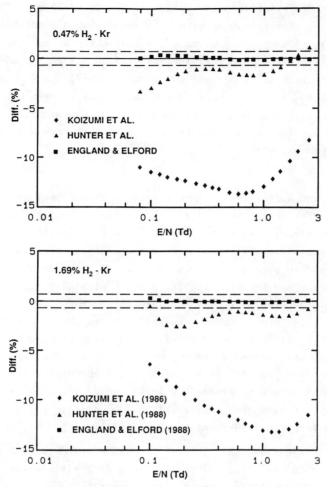

Fig. 7.   Differences $[\{v_{dr} \text{ (calc)} - v_{dr} \text{ (exp)}\}/v_{dr} \text{ (exp)}]$ between measured drift velocities in the mixtures and those calculated with the cross sections of Koizumi et al., Hunter et al., and England and Elford.

v_{dr} is sensitive primarily to the cross section below the minimum [see Fig. 7(a) of Hunter et al. (1988)] adds weight to this suggestion. Thus it has been suggested (England, 1989) that a more definitive result could be obtained by making a simultaneous fit to the results from both experiments, thereby taking advantage of the greater sensitivity of one or the other data set to particular features of the cross section.

Although MERT cannot be used to compare integral elastic cross sections for krypton from beam and swarm experiments except over a very limited range at low energies, it can be used to determine the scattering length. The results of such determinations from a number of beam and swarm experiments lie within the range – 3.28^* to –3.54 a_o. The two most recent swarm results of Hunter et al. (1988) (–3.36 a_o) and England and Elford (1988) (–3.43 a_o) lie close to, but slightly above, the much earlier result of Frost and Phelps (1964) (–3.32 a_o) and straddle the average of the results of Buckman and Lohmann (1987) and Weyhreter et al. (1988) (–3.38 a_o^{**}). Thus there is good agreement between the results for the scattering length obtained from beam and swarm experiments via MERT despite its limited usefulness in comparing the cross sections at the R–T minimum and above.

Neon

Buckman and Mitroy (1989) showed that, without supplementary information, MERT cannot be used to make meaningful comparisons between total and momentum transfer cross sections at low energies for Ne. They showed, for example, that the typical uncertainty in the experimental σ_T of 3 or 4% lead to uncertainties in the derived σ_m above 0.2 eV as large as 40%. Buckman and Lohmann (1985) reached a similar conclusion for He. In each case, the reason is that the integrated cross sections have insufficient structure to enable the MERT-derived phase shifts to be determined with sufficient accuracy from experimental data to provide a useful transformation from σ_T to σ_m, or vice versa. This situation contrasts with that for Ar where the structure associated with the R–T minimum enables such a determination to be made. The problem for Ne was recognized by O'Malley and Crompton (1980) who constructed their p-wave phase shifts from the experimental data of Williams (1979).

A test of the compatibility of beam- and swarm-derived integrated cross sections for Ne, like He, is therefore most satisfactorily performed by comparing both with theory. Until recently the best theoretical data for such a comparison were the cross sections of McEachran and Stauffer (1985) which are based on their application of the dipole adiabatic-exchange approximation. In the case of Ne, however, the cross section at low energies is extremely sensitive to the model potential. In their first calculation, both σ_T and σ_m below 15 eV lay well above the available experimental data; the scattering length was, in fact, over 40%

---

* See Table 4 of (England and Elford, 1988). Buckman and Lohmann's value of –3.19 a_o was subsequently revised by Buckman and Mitroy (1989) to –3.28 $\pm$ 0.08 a_o.

** Weyhreter et al.'s value of –3.478 a_o, obtained from a MERT fit up to 0.5 eV, was used here. In the light of Buckman and Mitroy's analysis, the value of –3.536 a_o from a fit only up to 0.3 eV may be more accurate.

higher than that of O'Malley and Crompton. McEachran and Stauffer (1985) subsequently attributed this to the fact that for Ne, but to a much lesser degree for the other gases they studied, their approximation did not represent polarization sufficiently accurately; the clue was an 11% error in the calculated polarizability. Scaling the polarization potential by a factor [α (experimental)/ α (calculated)] led to cross sections that agree with the swarm-derived σ_m of O'Malley and Crompton (1980) and Robertson (1972) to within 1% from 0.5 to 5 eV; below 0.5 eV the calculated σ_m lie increasingly below the experimental value, with a difference of 7% at 0.025 eV. McEachran and Stauffer claimed the agreement with Ferch's unpublished results for σ_T to be excellent (McEachran and Stauffer, 1985).

The very large difference between McEachran and Stauffer's original value of the scattering length and the value they obtained after scaling the polarization potential shows the sensitivity of low-energy cross sections to polarization effects. In a very recent paper, Saha (1989) has published phase shifts and cross sections for Ne based on a multi-configuration Hartree-Fock method. In this ab initio calculation, polarization is, according to the author, treated more accurately than previously.

The comparison between the theoretical cross sections of Saha and experimental results is shown in Fig. 8. For the few tabulated theoretical results between 0.136 eV (the lowest energy at which these data are available) and 5 eV, the agreement between the calculated σ_m and the swarm-derived cross section is to within 3%, the claimed uncertainty of the experimental results. Koizumi et al. (1984) derived σ_m from their measured data for D_T/μ, which also agrees with the theoretical σ_m to within their claimed uncertainty of 4%. Ferch's σ_T data appear to be generally within a few percent of the theoretical cross section, although the theoretical σ_T are tabulated only at the extremes of the energy range covered by the experimental data. The unpublished σ_T of Buckman (1987) are also within a few percent of the theoretical cross section above 2 eV, but appear to be 6 to 10% smaller below 0.6 eV. However, in each case accurate comparisons are difficult because of the relatively widely spaced energies at which the theoretical cross sections are tabulated.

In summary, the agreement between the results for the integrated cross section for Ne from beam and swarm experiments and theory is approaching the situation for He, where theory and experiment generally agree to within the claimed experimental uncertainty: for Ne, 3% for σ_m from swarm experiments, and 3 to 5% for σ_T from time-of-flight beam experiments.

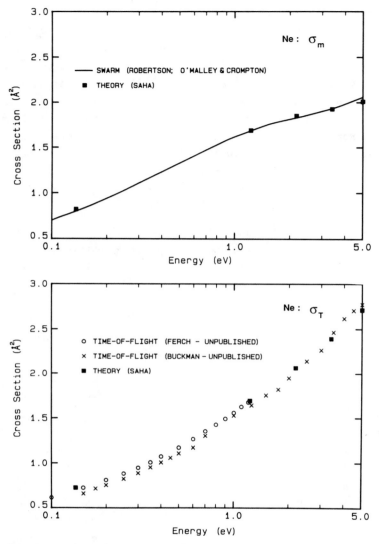

Fig. 8. Comparison between total and momentum transfer cross
sections for Ne calculated by Saha and the experimental
results of Ferch and of Buckman (for σ_T), and of Robertson
and O'Malley and Crompton (for σ_m).

Mercury

The measurement by swarm techniques of low-energy electron scattering cross sections for metallic vapors presents special difficulties.
This is because the low vapor pressures of these elements at temperatures
that are easily accessible prevent the establishment of low swarm energies when moderate electric field strengths are used in the experiments –
an essential feature of such techniques. Nakamura and Lucas's (1978)

solution to this problem was to use a heat-pipe technique by which they achieved gas number densities (N) in Hg of 8×10^{18} cm^{-3}. Elford (1980), using a more conventional high temperature static technique, achieved densities approaching 2×10^{18} cm^{-3}. But in both cases, despite the relatively high number densities that were achieved, the unfavorable electron-neutral mass ratio resulted in relatively high swarm energies at the lowest values of E/N, preventing an accurate determination of the cross section at low energies. In addition, an unexpected complication arose in these experiments: a relatively large pressure dependence in the measured drift velocities attributed to energy losses through inelastic collisions with mercury dimers. For example, Elford recorded an overall systematic increase of 20% in v_{dr} in the gas density range from 1.02 to 1.83×10^{18} cm^{-3}; he estimated the dimer concentration at his highest pressure to be approximately 300 ppm.

Nakamura and Lucas's and Elford's experiments in pure Hg vapor provided valuable information about the elastic cross section at the resonance at approximately 0.5 eV and to several eV above the resonance, although the agreement between the cross sections above 1 eV is not good. Below the resonance, however, the two cross sections disagree, at some energies differing by an order of magnitude. Moreover, neither cross section looks plausible when compared with Walker's theoretical cross section as reported in Elford's paper (Elford, 1980a).

In order to examine this problem further, England and Elford (1989) have recently applied the mixture technique to this problem also, using helium or nitrogen as the additional component; their results are being presented in detail in another paper at this conference.

The addition of either gas as a thermalizing agent has two consequences: (1) it enables v_{dr} measurements to be made for electron swarms with mean energies considerably lower than those in pure Hg, providing data more sensitive to the cross section below the resonance; and (2) it reduces the influence of dimers, because the energy loss in electron-dimer collisions are now a much smaller fraction of the total energy loss. England and Elford's mixture measurements have resulted in a derived σ_m in much better agreement with theory (England and Elford, 1989).

ELASTIC AND INELASTIC CROSS SECTIONS FOR MOLECULAR GASES

As the simplest molecule, hydrogen has served as the molecular analogue of helium for testing theoretical and experimental techniques for calculating or measuring low-energy cross sections for electron-molecule collisions. At the Fourth International Swarm Seminar, Crompton and Morrison (1987) reported the surprising outcome of a joint theoretical and experimental program to determine benchmark rotational and vibrational excitation cross sections for H_2. Since that conference, considerable further theoretical and experimental research has been carried out on this system, as will be reported in the first section.

Morrison et al. (1987) subsequently extended key aspects of their theoretical formulation for H_2 to nitrogen, resulting in rotational excitation cross sections for N_2 that are expected to be at least as accurate as those for H_2. The surprising outcome of a test of compatibility of these cross sections with swarm data was presented by Haddad and Phelps (1987) to the 1987 Electron Molecule and Photoionization Satellite Conference of XV ICPEAC. The second section discusses the implications of the results of that test.

Hydrogen

At the Lake Tahoe meeting, Crompton and Morrison (1987) compared cross sections for elastic scattering and rotational and vibrational excitation derived from beam or swarm experiments with theoretical results based on approximate treatments of exchange (for inelastic collisions only) and polarization. Left unresolved at that time was a striking difference between near-threshold theoretical and swarm-derived cross sections for vibrational excitation from the ground state.

Since that time further theoretical and experimental work have been undertaken aimed at resolving this problem. On the theoretical side, Morrison and Saha, (1986) have implemented an exact treatment of exchange (based on a separable representation of the kernel) and have further investigated the adequacy of their model polarization potential. On the experimental side, England et al. (1988) have measured drift velocities in H_2-Ne mixtures, thereby providing additional data to test the theoretical and experimental cross sections reported in Crompton and Morrison (1987). Because the momentum transfer cross for neon is strongly energy dependent at energies near the threshold for vibrational excitation in hydrogen (0.5 eV), drift velocities in Ne-H_2 mixtures are more sensitive to the threshold behavior of the vibrational cross section than those in the He-H_2 mixtures used previously (Petrovic and Crompton, 1987a). An important consequence of England et al.'s work, which included a detailed analysis by England of all the transport data in H_2 and in mixtures with He, Ne, and Ar (Haddad and Crompton, 1980) was the revision of the set of rotational and vibrational cross sectionss to optimize the fit to all the data. In doing so he took account of the fact that the relative error in the measured drift velocities (~0.25%) is very much less than the quoted overall uncertainty (~1% or less).

Somewhat surprisingly, this fact has tightened the constraints that can
be placed on the rotational cross sections above the vibrational
threshold derived from these data.

In summary, the results of the new theoretical and experimental
research have improved agreement between the theoretical and swarm-
derived σ_m in the range 0 to 1 eV and for $j_o=0$ to $j=2$ and $j_o=1$ to $j=3$
rotational excitation from threshold to about 0.4 eV (where the swarm
cross sections begin to lose uniqueness); but this work has increased
slightly the difference between the vibrational cross sections!

Morrison and his colleagues (Morris and Saha, 1986) have compared
vibrational cross sections calculated with the BTAD polarization
potential (Gibson and Morrison, 1984) to those based on the
correlation/polarization approach of O'Connell and Lane, 1983 (see also
Padial and Norcross, 1984; Padial, 1985). The comparison in Fig. 9
illustrates the sensitivity of vibrational excitation cross sections to
the theoretical treatment of polarization - a sensitivity not seen in
elastic or rotational excitation cross sections (Morrison and Saha,
1986). Clearly the correlation/polarization model potential does not
represent the effect of polarization as well as the BTAD potential, the
cross section based on the former being larger than either the beam or
swarm cross section. Nevertheless, despite the tests that have already
been carried out on the BTAD potential, it is possible that even it may
not adequately represent polarization effects for vibrational excitation
despite its success for rotational excitation.

Despite this considerable body of new work, the incompatibility
persists between theory and the results of swarm experiments in the
region where vibrational excitation dominates. This remains an important
problem to resolve because of possible implications for theory or for the
application of swarm experiments to normalize inelastic cross sections
from beam experiments (Crompton, 1983). As a further check on the
analysis of the experimental swarm data, a Monte Carlo simulation to
determine the transport coefficients directly from the cross sections is
being considered (Brennan, 1989). Previous simulations based on the Reid
"ramp" model (Reid, 1979) have essentially verified the results for the
transport coefficients calculated with a variety of Boltzmann codes for a
model gas at 0 K with a single inelastic cross section with characteristics similar to the H_2 vibrational cross section, and other tests
have been made with more realistic models with a multiplicity of
inelastic cross sections (Segur et al., 1984). Nevertheless, a check has
not been made of the Boltzmann results for "real" hydrogen at the
temperatures used in the experiments.

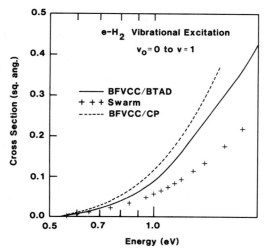

Fig. 9. Vibrational cross sections calculated with different treatments of polarization compared with the swarm-derived cross
section. The theoretical curves are calculated by the
body-frame vibrational close coupling method with (a) the
"better-than-adiabatic" polarization potential of Morrison
et al. and (b) the correlation model potential of O'Connell
and Lane.

Nitrogen

The comparison of theoretical and swarm-derived rotational
excitation cross sections for nitrogen has some advantages and some
disadvantages as compared with similar comparisons for hydrogen. On the
positive side, the separation between the thresholds of rotational and
vibrational excitation in N_2 is much larger than in H_2. This feature
extends somewhat the range of E/N for swarm experiments in N_2 that are
sensitive to one inelastic process but not to the other. Second, the
closely spaced rotational levels in the nitrogen molecule make it
possible to apply adiabatic-nuclei theory to calculate accurate
rotational (but not vibrational) excitation cross sections to within a
few meV of threshold (Morrison et al., 1984). The disadvantages are that
it is not possible to measure near-threshold rotational cross sections
for N2 in beam experiments, nor is it possible to single out one or two
rotational cross sections for individual study in swarm experiments, as
can be done for H_2.

Nevertheless, one can test the compatibility of the transport
coefficients measured in swarm experiments with a theoretically-derived
family of cross sections for rotational excitation from those levels that
are significantly populated (approximately 25 at room temperature and 10
at 77 K). Frost and Phelps (1962) first made such a test and came to the
now surprising conclusion that drift and diffusion data from a number of

30

sources was best fitted with cross sections generated by the simple
quadrupole-Born formula (Gerjuoy and Stein, 1955) un-modified by a dipole
polarization correction (Dalgarno and Moffett, 1963). Since that time,
Morrison et al. (1984) have demonstrated (using e-H_2 as a prototype) that
the quadrupole-Born formula is applicable only within millivolts of
threshold. Fortunately, in the case of N_2, unlike H_2, the adiabatic-
nuclear-rotation approximation can be validly applied (Morrison et al.,
1984; Morrison, 1988). Thus Morrison and co-workers (Morrison et al.,
1987) combined this approach with an exact representation of static and
exchange effects and their BTAD polarization potential in calculations of
e-N_2 rotational excitation cross sections at a high level of numerical
precision.

Both the energy dependence and magnitude of the resulting cross
sections differ significantly from the quadrupole-Born cross sections.
Fig. 10 illustrates this difference for the $j_0=0$ to $j=2$ cross section;
also shown is the recent result of Onda (1985) in which polarization
effects were represented in the polarized orbital approximation.

Haddad and Phelps (1987) subsequently reanalyzed the transport data
on which the original Frost and Phelps results were based, substituting
the cross sections of Morrison et al. (1987) or those of Onda for the
original set of quadrupole-Born cross sections. Surprisingly their
results, shown in Fig. 11, confirmed Frost and Phelps' original conclusions. This analyis has eliminated two possible explanations for the

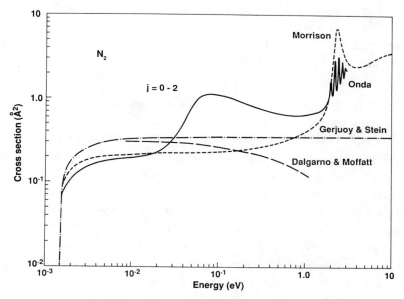

Fig. 10. The theoretical rotational excitation cross sections for
$j_0=0$ to $j=2$ excitation for N_2 of Gerjuoy and Stein,
Dalgarno and Moffatt, Morrison et al. and Onda (see text).

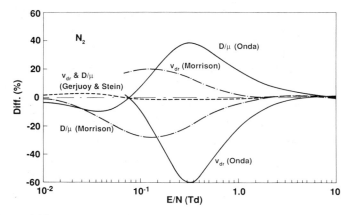

Fig. 11.   Differences between transport coefficient measured in N$_2$
at 77 K and values calculated with the rotational
excitation cross sections of the authors shown in Fig. 11.

incompatibility between theory and experiment: lack of sensitivity of the
transport coefficients to the energy dependence of the rotational cross
sections in the range of E/N for which data were available, or lack of
separation between the influence of the momentum transfer and rotational
cross sections in determining those coefficients.

Figure 11 suggests the energy range over which the cross sections
principally affect the transport coefficients at a particular value of
E/N. In the range of E/N about 0.08 Td, the quadrupole–Born cross sections give negligible differences between calculated and measured transport coefficients, while the differences for both v_{dr} and D_T/μ change
sign when Onda's cross sections are used. Figure 10 shows that these two
cross sections cross at about 30 meV; thus at E/N ~ 0.08 Td the transport
coefficients are most sensitive to the cross sections around 30 meV.

Unfortunately, Lowke's 77 K v_{dr} data (Lowke, 1963), which Haddad and
Phelps used to compare calculated and experimental values, do not extend
below 0.06 Td. Lowke published results of measurements down to E/N ~
0.003 Td, but gave no "best–estimate" values because of an unexplained
pressure dependence. On the other hand, D_T/μ data are available down to
E/N = 0.01 Td (Huxley and Crompton, 1974). In the absence of v_{dr} data,
these data can be used to examine the compatibility of theoretical
rotational cross sections with experiment provided the momentum transfer
cross section is known. The results shown in Fig. 13 were calculated
using the swarm–derived σ_m; these results should be close to those
calculated using the theoretical cross section (Morrison et al., 1987),
since below 1 eV the two cross sections agree to within 5%. The comparisons based on the D_T/μ data below 0.08 Td clearly reflect that, below 30

meV, the rotational excitation cross sections of Onda and those of Morrison et al. are much smaller than the quadrupole-Born values.

The incompatibility between experiment and the theoretical cross sections of Morrison et al. is, in fact, more serious than in shown in Fig. 11. Haddad and Phelps had access to rotational cross sections only at energies above 50 meV, so had to interpolate from this energy down to threshold. Subsequent theoretical calculations (Morrison et al., 1987) produced cross sections between 10 and 50 meV that are smaller than the interpolated values used by Haddad and Phelps; the use of these cross sections will increase the differences between calculated and measured transport coefficients.

There is a high level of confidence in the theoretical rotational cross section for N_2, and the disagreement between theory and experiment must be regarded at least as seriously as that for vibrational excitation of H_2. The disagreement is particularly puzzling in view of the excellent agreement for the e-H_2 rotational cross sections.

One major difference between the determination of these cross sections from swarm data in the two gases is the influence of superelastic collisions on the energy distribution functions and transport coefficients at low values of E/N, since it is from the low E/N data that the rotational cross sections are determined. In H_2 the $j_o=2$ and $j_o=3$ states are only slightly populated (< 0.5%) at 77 K, whereas in N_2 the maximum in the Boltzmann population at this temperature occurs at $j_o=4$, and states below and above this one are well populated. Consequently significant energy is supplied to the lowest energy swarms through rotational de-excitation in N_2, but rather little such energy is supplied in H_2. The possibility that the explanation for the large difference between theoretical and swarm derived rotational cross sections in nitrogen arises from inadequacy of the currently used Boltzmann transport codes is presently being investigated (Mitroy, 1989).

CONCLUDING REMARKS

In this paper I have chosen a few examples to illustrate the present position regarding the benchmarking of low-energy electron-atom and electron-molecule cross sections. For there to be confidence in the results from different experimental approaches the results should be consistent within the claimed experimental uncertainty, but direct comparisons are not always possible. Techniques for applying such consistency tests for the results of beam and swarm experiments have been discussed in the first section which dealt with elastic electron-atom scattering.

In the case of data for electron-molecule systems, tests for consistency are more difficult, and perhaps less convincing, since beam experiments to measure inelastic cross sections at low energies are either not possible or are very much more difficult at the level of accuracy required. One must therefore look for consistency either between the results of different types of swarm experiment, or with theory in those cases where considerable effort has been made to calculate cross sections with an accuracy matching the claimed accuracy of the experiments. The results of such comparisons with theory, which were described in the second section, have been both interesting and, in some cases, controversial.

Hayashi (1987) and his colleagues have taken another approach to the problem of deriving reliable cross sections for electron scattering over a wide range of energies (0 to 1 keV). Their procedure consists of assembling all available data from beam and swarm experiments and making judgements on the weight to be placed on each set of results where the results do not form a consistent whole. They have published results for over 30 atoms and molecules by analyzing the data in this way. A description of their approach and examples of their results are to be found in Hayashi (1987).

ACKNOWLEDGEMENT

I wish to record my grateful thanks to the following for their assistance with the preparation of some of the material for this paper and for many valuable discussions: M. J. Brennan, M. T. Elford, J. P. England, J. Mitroy and M. A. Morrison. I am particularly indebted to Michael Morrison for his critical reading of the manuscript.

REFERENCES

Bowe, J. C., 1960, Phys. Rev. 117, 1411.
Brennan, M. J., 1989, personal communication.
Buckman, S. J., 1987, in Electron Physics Group Quarterly Report 109, Australian National University.
Buckman, S. J., and B. Lohmann, 1987, J. Phys. B 20, 5807.
Buckman, S. J., and B. Lohmann, 1985, J. Phys. B 19, 2547.
Buckman, S. J., and J. Mitroy, 1989, J. Phys. B 22, 1365.
Crompton, R. W., 1983, Proceedings of the 16th International Conference on the Physics of Ionized Gases, (Invited papers), W. Botticher, et al, (Eds.), 58.
Crompton, R. W., and M. A. Morrison, 1987, in "Swarm Studies and Inelastic Electron-Molecule Collisions," L. C. Pitchford, et al. (Eds.), Springer-Verlag, New York, 143.
Dalgarno, A., and R. J. Moffett, 1963, Proc. Natl, Acad. Sci. India A 33, 511.
Elford, M. T., 1980, Aust. J. Phys. 33, 231.
Elford, M. T., 1980a, Aust. J. Phys. 33, 251.
England, J. P., 1989, personal communication.
England, J. P., and M. T. Elford, 1988, Aust. J. Phys. 41, 701.

34

England, J. P., and M. T. Elford, 1990, in "Non-equilibrium Processes in
    Gas Discharges," J. W. Gallagher, et al. in (Eds.), Plenum, New
    York, 367.
England, J. P., M. T. Elford, and R. W. Crompton, 1988, Aust. J. Phys.
    41, 573.
Ferch, J., [see (McEachran and Stauffer, 1985) and personal
    communication].
Ferch, J., B. Granitza, C. Masche, and W. Raith, 1985, J. Phys. B 18,
    967.
Frost, L. S., and A. V. Phelps, 1964, Phys. Rev. 136, A1538.
Frost, L. S., and A. V. Phelps, 1962, Phys. Rev. 127, 1621.
Gerjuoy, E., and S. Stein, 1955, Phys. Rev. 97, 1671.
Gibson, T. L., and M. A. Morrison, 1984, Phys. Rev. A 29, 2497.
Haddad, G. N., and R. W. Crompton, 1980, Aust. J. Phys. 33, 975.
Haddad, G. N., and T. F. O'Malley, 1982, Aust. J. Phys. 35, 35.
Haddad, G. N., and A. V. Phelps, 1987, personal communication.
Hayashi, M., 1987, in "Swarm Studies and Inelastic Electron-Molecule
    Collisions," L. C. Pitchford, et al., (Eds.), Springer-Verlag,
    New York, 167.
Hunter, S. R., J. G. Carter, and L. G. Christophorou, 1987, Proceedings
    of the 5th International Swarm Seminar, University of Birmingham, 5.
Hunter, S. R., J. G. Carter, and L. G. Christophorou, 1988, Phys. Rev. A
    38, 5539.
Huxley, L. G. H., and R. W. Crompton, 1974, "The Diffusion and Drift of
    Electrons in Gases," Wiley: New York.
Koizumi, T., H. Murakoshi, S. Yamamoto, and I. Ogawa, 1984, J. Phys. B
    17, 4387.
Koizumi, T., E. Shirakawa, and I. Ogawa, 1986, J. Phys. B 19, 2331.
Lowke, J. J., 1963, Aust. J. Phys. 16, 115.
McEachran, R. P., and A. D. Stauffer, 1985, Phys. Lett. 107A, 397.
McEachran, R. P., and A. D. Stauffer, 1983, J. Phys. B 16, 4023,
    Analytical expressions for the phase shifts were given in A. D.
    Stauffer, T. H. V. T. Dias, and C. A. N. Conde, 1986, Nucl. Instrum.
    Methods A 242, 327.
Milloy, H. B., and R. W. Crompton, 1977, Aust. J. Phys. 30, 51.
Milloy, H. B., R. W. Crompton, J. A. Rees, and A. G. Robertson, 1977,
    Aust. J. Phys. 30, 61.
Mitroy, J., 1989, personal communication.
Morrison, M. A., 1988, in "Advan. Atom. Molec. Phys.," B. Bederson, and
    D. Bates, (Eds.), 24, 51.
Morrison, M. A., A. N. Feldt, and D. Austin, 1984, Phys. Rev. A 29, 2518.
Morrison, M. A., R. W. Crompton, B. C. Saha, and Z. Lj. Petrovic, 1987,
    Aust. J. Phys. 40, 239.
Morrison, M. A., and B. C. Saha, 1986, Phys. Rev. A 34, 2786.
Morrison, M. A., B. C. Saha, and T. L. Gibson, 1987, Phys. Rev. A 36,
    3682.
Nakamura, G., and J. Lucas, 1978, J. Phys. D 11, 325.
O'Connell, J. K., and N. F. Lane, 1983, Phys. Rev. A 27, 1893.
O'Malley, T. F., 1963, Phys. Rev. 130, 1020.
O'Malley, T. F., 1989, personal communication; see also O'Malley, T. F.,
    1980, J. Phys. B 13, 1491; O'Malley, T. F., 1983, Phys. Lett. 95A,
    32, and reference therein.
O'Malley, T. F., and R. W. Crompton, 1980, J. Phys. B 13, 3451.
Onda, K., 1985, J. Phys. Soc. Jpn. 54, 4544.
Pack, J. L., R. E. Voshall, and A. V. Phelps, 1962, Phys. Rev. 127, 2084.
Padial, N. T., and D. W. Norcross, 1984, Phys. Rev. A 29, 1742, ibid,
    1590; see also Padial, N. T., 1985, Phys. Rev. A 32, 1379.
Petrovic, Z. Lj., and R. W. Crompton, 1987, XV International Conference
    on the Physics of Electronic and Atomic Collisions, J. Geddes, et
    al., (Eds.), Contributed Papers, 128.
Petrovic, Z. Lj., and R. W. Crompton, 1987a, Aust. J. Phys. 40, 347.
Reid, I. D., 1979, Aust. J. Phys. 32, 231.
Robertson, A. G., 1972, J. Phys. B 5, 648.
Robertson, A. G., 1977, Aust. J. Phys. 30, 39.

Saha, H. P., 1989, Phys. Rev. A 39, 5048.
Segur, P., M. Yousfi, and M. C. Bordage, 1984, J. Phys. D 17, 2199.
Weyhreter, M., B. Barzick, A. Mann, and F. Linder, 1988, Z. Phys. D 7, 333.
Williams, J. F., 1979, J. Phys. B 12, 265.

COUPLED SOLUTIONS OF BOLTZMANN EQUATION, VIBRATIONAL AND ELECTRONIC NONEQUILIBRIUM KINETICS

C. Gorse

Dipartimento di Chimica dell'Università di Bari
Centro di Studio per la chimica dei Plasmi del CNR
Traversa Re David 200 n°4 – 70126 BARI (Italy)

INTRODUCTION

To understand many properties of molecular plasmas widely used these days in advanced technologies (lasers, plasma etching and deposition, negative ion production) the nonequilibrium plasma kinetics is of paramount importance. In the active medium, the different species are affected by processes involving electrons and heavy particles (atoms, molecules and ions in ground and excited states). In these conditions plasma modeling requires a self consistent solution of the Boltzmann equation (electron kinetics) together with the vibrational and electronic master equations (heavy component kinetics). Although the coupling between Boltzmann equation (BE) and vibrational kinetics has been in common use (Capitelli, 1986; Boeuf and Kunhardt, 1986; Loureiro and Ferreira, 1986; Gorse et al., 1986) only recently has self consistent coupling been extended to account for the electronic kinetics too (Bretagne et al., 1987; Gorse et al., 1988; Gorse and Capitelli, 1987; Gorse and Capitelli, 1988).

THE MODEL

We solve the time-dependent Boltzmann equation (Rockwood, 1973) in the two-term approximation

$$\frac{\partial n(\varepsilon, t)}{\partial t} = -\left(\frac{\partial J_f}{\partial \varepsilon}\right) - \left(\frac{\partial J_{el}}{\partial \varepsilon}\right) + In + Sup_{(v)} + Sup_{(e)} + Rot + Ion. \qquad (1)$$

This equation describes the temporal evolution of electrons between ε and $\varepsilon + d\varepsilon$ under the action of the electric field ($\partial J_f/\partial \varepsilon$), elastic ($\partial J_{el}/\partial \varepsilon$), inelastic (In), superelastic vibrational (Sup$_v$), superelastic electronic (Sup$_e$), ionization (Ion) and rotational (Rot) collisions. It is worth noting that the Boltzmann equation (BE), via superelastic and inelastic collisions, depends on the heavy particle kinetics and at the same time affects this kinetics, via electronic excitation processes. The vibrational and electronic master equations

$$\frac{dN_v}{dt} = \left(\frac{dN_v}{dt}\right)_{e-V} + \left(\frac{dN_v}{dt}\right)_{V-V} + \left(\frac{dN_v}{dt}\right)_{V-T} + \left(\frac{dN_v}{dt}\right)_{e-D} + \left(\frac{dN_v}{dt}\right)_{e-I}$$

$$+ \left(\frac{dN_v}{dt}\right)_{e-E} + \sum_{X^*} \left(\frac{dN_v}{dt}\right)_{X^*} \tag{2}$$

$$\frac{dN_X^*}{dt} = \left(\frac{dN_X^*}{dt}\right)_{e-E} + \left(\frac{dN_X^*}{dt}\right)_{e-I} + \left(\frac{dN_X^*}{dt}\right)_X + \sum_{X' \neq X^*} \left(\frac{dN_X^*}{dt}\right)_{X'} \tag{3}$$

describe the temporal evolution of the vth vibrational level of the ground X state and of the electronic excited X^* state under the influence of e-V (electron – vibration) processes, V-V (vibration – vibration) and V-T (vibration – translation) energy transfers, e-D /I /E (electron – dissociation /ionization /electronic excitation) processes and exchanges with other electronic states X'. In the self consistent model we simultaneously solve the BE and the system of master equations starting (at t=0) with the cold gas approximation, i.e., only the fundamental vibrational level of the ground electronic state is populated.

Such a model can be applied to describe different discharge devices: DC discharges and post-discharges, RF discharges and multipolar magnetic discharges.

RESULTS

a)   DC N_2 flowing discharge

The first case we discuss is the solution of the Boltzmann equation in the presence of both vibrational and electronic superelastic collisions in a nitrogen DC discharge under fixed values of the reduced electric field (E/N = 60Td), electron number density ($n_e = 10^{10}$ cm^{-3}) and gas temperature (T_g = 500K). In this calculation the V-T (vibration-translation) energy exchange rates between atoms and molecules have been considered equal to the corresponding V-T rates between two N_2 molecules. Work is now in progress (Ciccarelli and Laganà, 1988) to better estimate all of those energy exchange rates in nitrogen. The e-V (electron – vibration) rate coefficients are derived averaging the cross sections of Chandra and Temkin, 1976, and Newmann and Detemple, 1976, with the electron energy distribution function (EEDF) solution of the BE. Due to the lack of data, the Gryzinski method (Cacciatore et al., 1982) is used in the calculation of the electronic excitation, dissociation and ionization

cross sections of ground molecules N_2 (v=0) and vibrationally excited
molecules N_2 (v≠0). The kinetics for the states A (A=A$^3\Sigma_u^+$), B (B=B$^3\Pi_g$)
and C (C=C$^3\Pi_u$) discussed in Gorse and Capitelli, 1987, includes the most
important collisional and radiative processes acting on these states.

Figure 1 reports the temporal evolution of the vibrational distribution N_v. We note that this vibrational distribution, after some residence time in the discharge (t>15ms), is highly non Boltzmann presenting
a long plateau due to the redistribution of the vibrational quanta by V-V
exchange processes. The increasing efficiency in vibrational excitation
ends in the growth of both pseudo-vibrational temperature (θ_1) and
population densities of nitrogen atoms (Fig. 2). The concentrations of A
and B states are strictly connected to the vibrational content of the
ground state N_2(X). Until about 15ms A and B states are more and more
populated as the electronic excitation processes (e-E) from vibrationally
excited ground molecules are promoted. After this residence time the
vibrational content of N_2(X,v) is high enough to excite the N_2(A) molecules towards the B state, and nitrogen atoms will contribute to the

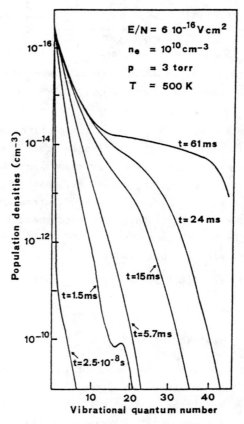

Fig. 1.   Temporal evolution of the ground nitrogen vibrational
          distribution in a DC discharge.

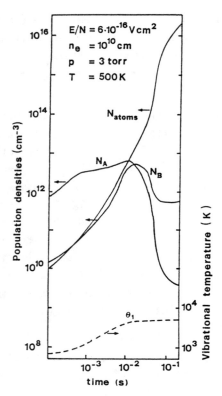

Fig. 2.   Temporal evolution of vibrational temperature θ_1, nitrogen
atoms, A and B electronically excited states densities.

deexcitation of the A state.   The B state is deactivated by the vibrationally excited molecules and by radiative losses.   The temporal evolution of the EEDF (Fig. 3) reflects the corresponding temporal evolution
in the concentrations of excited N_2 molecules, respectively the bulk
($2 < \varepsilon < 14eV$) and the tail ($\varepsilon \geq 14eV$) of the EEDF increases (or decreases)
as the concentrations of vibrationally and electronically excited mole-
cules increase (or decrease) (compare the EEDF at 15 and 64ms).   Electrons heated in superelastic collisions

$$e(\varepsilon) + N_2(X,w) \longrightarrow e(\varepsilon + \varepsilon^*) + N_2(X,v) \qquad w > v \qquad (4)$$

$$e(\varepsilon) + N_2(A) \longrightarrow e(\varepsilon + \varepsilon^*) + N_2(X,v) \qquad \varepsilon^* = 6.2eV \qquad (5)$$

$$e(\varepsilon) + N_2(B) \longrightarrow e(\varepsilon + \varepsilon^*) + N_2(X,v) \qquad \varepsilon^* = 7.3eV \qquad (6)$$

are carried from the low energy part of EEDF to the high one.   The
enhancement of the bulk and the tail of the EEDF will promote those
processes with high energy thresholds (ε^*) like electronic excitation,
dissociation and ionization.   It is worth noting that the effect of
superelastic electronic collisions will disappear both at higher values

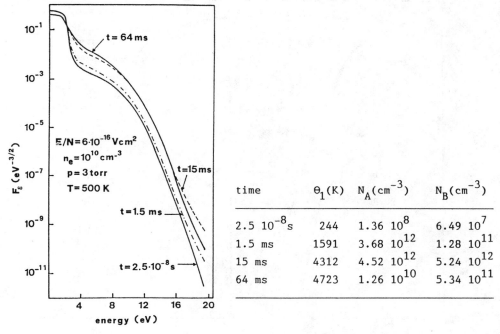

Fig. 3. Electron energy distribution functions at different residence times in a nitrogen discharge.

time	θ_1(K)	N_A(cm^{-3})	N_B(cm^{-3})
$2.5\ 10^{-8}$s	244	$1.36\ 10^{8}$	$6.49\ 10^{7}$
1.5 ms	1591	$3.68\ 10^{12}$	$1.28\ 10^{11}$
15 ms	4312	$4.52\ 10^{12}$	$5.24\ 10^{12}$
64 ms	4723	$1.26\ 10^{10}$	$5.34\ 10^{11}$

of the pseudo vibrational θ_1 (promoting selectively superelastic vibrational effects) and at high values of the reduced electrical field (E/N).

b)   Radio frequency discharges

We are interested in the behavior of electrons submitted to the
influence of periodically oscillating electric field (E) with maximum
amplitude (E_o) and frequency ω. We, therefore, have to couple the BE and
the master equations to the new equation

$$E = E_o \cos \omega t \tag{7}$$

giving the instantaneous value of the electrical field. We use the self
consistent model to describe the microscopic effects on a laser mixture
Ne-Xe-HCl (99.5:0.44:0.06) at a 3 atm pressure, submitted to the appli-
cation of an electric field such as E_o = 5840 Vcm^{-1} and $\omega = \pi\ 10^9 s^{-1}$. At
this field frequency ω, the effective field approximation is no longer
valid so that we must solve the time dependent Boltzmann equation. All
the reactions included in the kinetics are listed elsewhere (Gorse et
al., 1988).

Figures 4a-b report the temporal evolution of the EEDF on half-cycle
at early (4a) and long (4b) residence times in the discharge. We note in
the energy range $\varepsilon \gtrsim 10$eV, where the inelastic processes involving Ne and

41

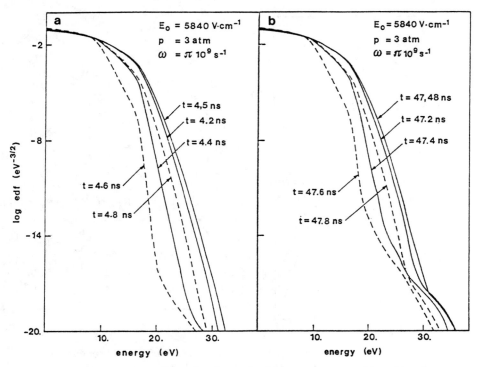

Fig. 4a-b. Temporal evolution of electron energy distribution functions on half-cycle at short (a) and long (b) residence time in the RF discharge.

Xe are acting, a large modulation of the EEDF as the electric field is oscillating. At lower values of ε the inelastic losses are too small to permit an instantaneous rearrangement of the electrons to the new values of E/N. Once again we note the importance of the superelastic electronic collisions responsible for the enhancement of the tail of the EEDF ($\varepsilon \gtrsim$ 17eV), the effect being more and more evident at longer residence times when the concentrations of the excited states Xe^* and Ne^* increase and at low values of the instantaneous electric field. The oscillating frequency of the rate coefficients which are depending on the electrons is twice the field frequency. Moreover, the rate coefficients for electronic processes, slightly evolve with time if their threshold energies are very low as for dissociative attachment (Fig. 5) or ionization from excited states (Fig. 6), or show a pronounced oscillating behavior when their threshold energies are higher as for ionization coefficients (Fig. 6). As the residence time increases in the discharge, new created species and electrons ($n_e = 10^9 cm^{-3}$ at starting time t=0) are more and more populated (Fig. 7).

All the results reported here and in earlier works (Capitelli et al., 1987) well evidence the fact that in this range of field frequencies

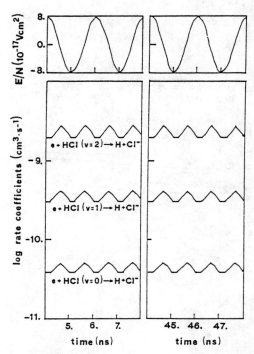

Fig. 5. Dissociative attachment rate coefficients for HCl as a
function of time.

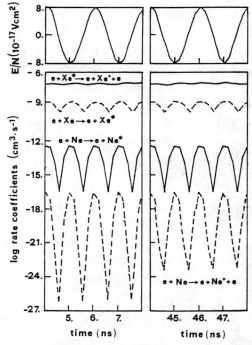

Fig. 6. Electronic excitation and ionization rate coefficients of
Ne and Xe as a function of time.

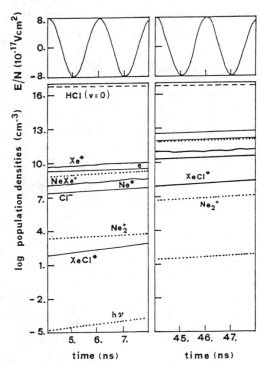

Fig. 7. Temporal evolution of the densities of some representative species for early and late residence times in the RF discharge.

only a self consistent treatment of the kinetics and of the BE can give realistic information on the bulk plasma. However, this model has to be improved to account for some important processes occurring in the sheath region where fast electrons created by secondary emission and accelerated by the potential fall (Belenguer and Boeuf, 1989) play an important role in determining plasma characteristics.

c) H_2 magnetic multicusp discharges

It has been shown (Bacal et al., 1984) that in this kind of discharge, the concentration of negative ions is quite high. Beams of negative ions can be accelerated to high energies and converted into beams of neutral atoms used, for instance, for heating in magnetic fusion devices. Until now experimental and theoretical work on those plasmas have been mainly devoted to the production of H^- negative ions (Gorse et al., 1985; 1987).

The self consistent model utilized for DC and RF discharges is used here to describe a H_2 magnetic multicusp discharge. In this device electrons emitted by tungsten filaments and then accelerated, impinge on

hydrogen molecules. Dissociative attachment processes involving vibra
tionally excited molecules

$$e + H_2(v) \longrightarrow H_2^- \longrightarrow H + H^- \tag{8}$$

produce the high densities of negative ions H^-. Then the vibrational
distribution of H_2 molecules must be accurately calculated accounting for
the most important microscopic processes working in such a plasma. In
addition to the processes reported for the general model (e-V, V-V, V-T,
e-D, e-E, e-I) we will introduce here dissociative attachment (e-da)
(Bardsley and Wadehra, 1979), the excitation of high vibrational levels
of the ground state through the decay of some excited electronic states
(E-V) (Hiskes, 1980)

$$e + H_2(v=0) \longrightarrow e + H_2^*(B^1\Sigma_u^+, C^1\Pi_u) \longrightarrow e + H_2(v') + h\nu \quad v'>0 \tag{9}$$

and the wall effects.. In fact, vibrationally excited molecules can be
produced on the wall both through atomic recombination (Hall et al.,
1988) and electron attachment to molecular ions (H_2^+) (Hiskes et al.,
1985) or can be deexcited.

In the theoretical modeling we disregard the coupling between the
electronic kinetics and the BE. Figure 8 shows the temporal evolution of
the EEDF for a selected plasma condition (volume: 8.81, loss surface:
831 cm^2, pressure: 2 m torr, filament current: 5A, plasma potential: 1.96
V, voltage: 50 V. We assume that the translational temperatures of
hydrogen molecules and atoms are both equal to 1000 K). The source term
is well apparent at high energies ($45 \leq \varepsilon \leq 50eV$). Inelastic processes
degradate the electrons which form a plateau in the range 10-40eV. Most
of the electrons are concentrated in the low energy range ($0<\varepsilon<10eV$)
where the distribution is not far from Maxwellian with an associated
electronic temperature T_e. Figure 9 reports the temporal evolution of
the vibrational distribution. Here again the distribution is far from a
Boltzmann one, but in this case the plateau is due to E-V processes which
populate the high vibrational levels. Table 1 shows a satisfactory
agreement between theory and experiment (Bacal) when we compare the
electronic temperatures T_e, the electron number densities n_e, and the
ratio of negative ions to electron densities.

The differences in the input data compared with our earlier work can
be found in the use of new sets of electronic cross sections. Once again
use has been made of the semi-classical Gryzinski theory to obtain electronic dissociation and ionization cross sections involving all the
vibrational levels. A comparison between those cross sections and the

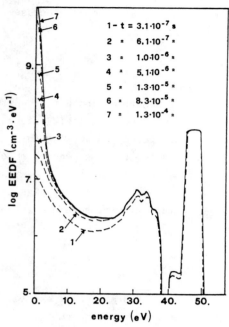

Fig. 8. Electron energy distribution function at different residence times in a magnetic multicusp H_2 discharge (pressure:
2 m torr, accelerating voltage: 50 V, filament current: 5A,
total volume: 8.81, $T_H = T_{H_2} = 1000K$).

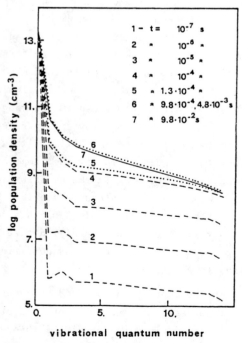

Fig. 9. Vibrational distribution of H_2 at different residence times
in the multipolar discharge (same conditions as in Fig. 8).

Table 1. Comparison between theoretical and experimental data

	T_e (eV)	n_e (cm^{-3})	N_{H^-}/n_e
experiment	2.25 10^{10}	0.6	0.14
theory	9.8 10^9	0.47	0.04

corresponding ones derived from more accurate methods when available (Rescigno and Schneider) shows differences not exceeding a factor of 2.

CONCLUSIONS

The data we report here show unambiguously that, in all the cases studied, both the electrons and the heavy components are far from equilibrium conditions. The self-consistent coupling between BE and all the different kinetics acting in a plasma is a promising model able to describe qualitatively the microscopic processes present in different kinds of plasmas. The next step toward a better description requires an improvement of the input data relative to the cross sections for energy exchanges (V-V, V-T, e-E and e-I). Some new accurate or more extended data are now available for N_2 (Ciccarelli and Laganá, 1988; Huo, 1989; Launay and LeDourneuf, 1987), H_2 (Rescigno and Schneider; Cacciatore et al., 1989) and HCl (Bardsley and Wadehra, 1983). Another important point to clarify is the problem of the wall effect. In the N_2 DC discharge we totally disregard this problem due to the lack of data. In RF discharges, it is worth noting that our model is applied only to describe the behavior of the bulk plasma far from the boundaries. For the deexcitation of vibrationally excited H_2 molecules on the wall container, the data reported in the specialized literature for iron and copper surfaces do not agree well.

All these remarks should emphasize the qualitative nature of our results which, however, represent an important guideline for future improvements.

ACKNOWLEDGMENT

The author enjoyed many helpful discussions with M. Capitelli.

REFERENCES

Bacal, M., private communication.
Bacal, M., A. M. Bruneteau, and M. Nachman, 1984, J. Appl. Phys. 55, 15.
Bardsley, J. N., and J. M. Wadehra, 1979, Phys. Rev. A 20, 1398.
Bardsley, J. N., and J. M. Wadehra, 1983, J. Chem. Phys. 78, 7227.
Belenguer, P. and J. P. Boeuf, 1990, Proc. of "Non Equilibrium Processes in
    Partially Ionized Gases," NATO ASI Maratea, Capitelli, M. and J. N.
    Bardsley, (Eds.), Plenum, New York; Dilecce, G., M. Capitelli, S.
    DeBenedictis, and C. Gorse, 1989, Proc. of ISPC 9 Pugnochiuso, Italy.
Boeuf, J. P., and E. E. Kunhardt, 1986, J. Appl. Phys. 60, 915.
Bretagne, J., M. Capitelli, C. Gorse, and V. Puech, 1987, Europhys. Lett. 3,
    1179.
Cacciatore, M., M. Capitelli, and G. D. Billing, 1989, Chem. Phys. Lett.
    157, 305.
Cacciatore, M., M. Capitelli, and G. D. Billing, 1989, Surface Science, 217,
    391.
Cacciatore, M., M. Capitelli, and C. Gorse, 1982, Chem. Phys. 66, 141.
Capitelli, M., R. Celiberto, C. Gorse, R. Winkler, and J. Wilhelm, 1987, J.
    Appl. Phys. 62, 4398; 1988, J. Phys. D: Appl. Phys. 21, 691; 1988,
    Plasma Chem. Plasma Process. 8, 175; 1988, Plasma Chem. Plasma Process.
    8, 399.
Chandra, N., and A. Temkin, 1976, Phys. Rev. A 13, 188, 14, 507, NASA TN
    D-8347.
Ciccarelli, L., and A. Lagana, 1988, J. Phys. Chem 92, 932.
Gorse, C., M. Cacciatore, M. Capitelli, S. DeBenedictis, and G. Dilecce,
    1988, Chem. Phys. 119, 63.
Gorse, C. and M. Capitelli, 1987, J. Appl. Phys. 62, 4072.
Gorse, C. and M. Capitelli, 1988, 41st Annual Gaseous Electronics
    Conference, Minneapolis, 84.
Gorse, C., M. Capitelli, J. Bretagne, and M. Bacal, 1985, Chem. Phys. 93, 1;
    1987, 117, 177.
Gorse, C., R. Caporusso, and M. Capitelli, 1988, New Laser Technologies and
    Applications, 1st GR-I International Conference, Carabelas, A. A., and
    T. Letardi, (Eds.), Olympia 439; Gorse, C., 1990, Proc. of "Non
    Equilibrium Processes in Partially Ionized Gases," NATO ASI Maratea,
    Capitelli, M. and J. N. Bardsley, (Eds.), Plenum, New York.
Gorse, C., F. Paniccia, J. Bretagne, and M. Capitelli, 1986, J. Appl. Phys.
    59, 4004; 1986, 59, 4731.
Hall, R. I., I. Cadez, M. Landau, F. Pichou, and C. Schermann, 1988, Phys.
    Rev. Lett. 60, 337.
Hiskes, J. R., 1980, J. Appl. Phys. 51, 4592; Celiberto, R. et al., to be
    published.
Hiskes, J. R., A. M. Karo, and P. A. Willmann, 1985, J. Vac. Sci. Technol. A
    3, 1229; 1985, J. Appl. Phys. 58, 1759.
Huo, W., 1990, Proc. of "Non Equilibrium Processes in Partially Ionized
    Gases," NATO ASI Maratea, Capitelli, M. and J. N. Bardsley, (Eds.),
    Plenum, New York.
Launay, J. M., and M. LeDourneuf, 1987, XV ICPEAC, Brighton 288.
Loureiro, J., and C. M. Ferreira, 1986, J. Phys. D: Appl. Phys. 19, 17.
Newmann, L. A., and T. A. Detemple, 1976, J. Appl. Phys. 47, 1912.
"Non-Equilibrium vibrational kinetics," Top. Curr. Phys. 39; 1986,
    Capitelli, M., (Ed.), Springer-Verlag, New York.
Rescigno, T. N., and B. I. Schneider, UCRL-99210 (to be published in J.
    Phys. B).
Rockwood, S. D., 1973, Phys. Rev. A 8, 2348.

HIGHER-ORDER ELECTRON TRANSPORT IN GASES

B. M. Penetrante and J. N. Bardsley

Lawrence Livermore National Laboratory
Livermore, CA 94550

ABSTRACT

We present calculations of the electron transport coefficients in He, Ne and Ar, based on the density gradient expansion theory. Of particular interest is the skewness coefficient D_3. The arrival time spectra (ATS) are computed for typical drift tube conditions in order to assess the measurability of the higher-order transport coefficients. We show that D_3 is measurable and that present experiments have the capability of resolving it from the ATS. The reported measured data on D_3, however, show large non-hydrodynamic behavior. An analysis of the TOF density profile based on a nonlinear continuity equation is also presented. The significance of including higher-order transport effects, either through the use of higher-order coefficients or the use of nonlinear continuity equation, is discussed.

INTRODUCTION

Before 1967 it was assumed that the diffusion of electrons in gases is isotropic even in the presence of an external electric field. Wagner et al. (1967) showed that the components of the diffusion perpendicular and parallel to the electric field are different. Theories were then developed (Parker and Lowke, 1969; Skullerud, 1969; Huxley, 1972; Huxley and Crompton, 1974; Skullerud, 1974) to take account of this anisotropy at the level of the continuity equation for the electron density

$$\partial_t n(\mathbf{r},t) = -\mathbf{v}_d \cdot \nabla n(\mathbf{r},t) + D : \nabla\nabla n(\mathbf{r},t) \tag{1}$$

where $\mathbf{v}_d$ is the drift velocity and D is the diffusion tensor. The component of D which is perpendicular to the electric field is called the transverse diffusion coefficient D_T, while the one parallel to the field is called the longitudinal diffusion coefficient D_L.

The state of steady or stationary transport in which the electron number density $n(r,t)$ satisfies Eq. (1) with constant coefficients is referred to as the hydrodynamic regime. The basic assumption in transport calculations is that after some suitable time the electron phase-space distribution function is given by

$$f(r,v,t) = n(r,t) \, f^{(0)}(v) + \nabla n(r,t) \cdot f^{(1)}(v) \tag{2}$$

This is the assumption that makes it possible to derive transport coefficients which are independent of time. Eq. (2) is the foundation of the density gradient expansion theory, which is the currently accepted theory in electron and ion swarm transport. All other methods for deriving the transport coefficients, particularly the longitudinal diffusion coefficient, are equivalent to the density gradient expansion in the hydrodynamic regime. A brief critical review of these theories and their equivalence is given in Penetrante and Bardsley (1984). More comprehensive discussions of the current status of swarm theory are given in Kumar et al. (1980).

A natural consequence of the density gradient expansion theory is the formulation of a more general continuity equation

$$\partial_t n(r,t) = \sum_{k=0}^{\infty} \omega^{(k)} \cdot (-\nabla)^k n(r,t) \tag{3}$$

The coefficients of $\omega^{(k)}$ for $k > 2$ are referred to as the higher-order transport coefficients. The calculation of these coefficients is based on the expansion

$$f(r,v,t) = \sum_{k=0}^{\infty} f^{(k)}(v) \cdot (-\nabla)^k n(r,t) \tag{4}$$

The coefficient $\omega^{(3)}$ is called the skewness coefficient (written as D_3 from here on), while the $\omega^{(4)}$ is the kurtosis coefficient (written as D_4 from here on). Until recently there has been no need for Eq. (3) beyond that of the description given by Eq. (1). This is because the higher-order transport coefficients were not measurable with the kind of resolution available in drift tube experiments. Nevertheless, the drift velocity and the diffusion coefficients by themselves provide a very practical means by which collision cross sections could be inferred.

Recently there have been reports (Denman and Schlie, 1989a; Denman and Schlie, 1989b) of high-resolution time-of-flight (TOF) measurements of D_3 for the rare gases He, Ne, Ar, Kr and Xe. Should these measurements prove to be accurate, they could provide an additional transport coefficient that will give a more unique inference of the collision cross sections. It is therefore important to carefully assess these measurements. The skewness coefficient D_3 is particularly sensitive to the rapid variations with energy

50

of the momentum transfer cross section. Thus they would be very useful in determining fine structures such as the position, depth and width of the Ramsauer minimum. It would be very desirable to measure this coefficient as accurately as possible, especially for gases such as Xe and Kr, for which there still exist substantial disagreement with regard to the position and depth of the Ramsauer minimum.

The aims of this work are (1) to provide Boltzmann and Monte Carlo calculations of the skewness coefficient for various rare gases, (2) to compute the arrival time spectra of the electron swarm for typical drift tube conditions, and (3) to assess the recently reported TOF measurements of D_3.

CALCULATIONS OF TRANSPORT COEFFICIENTS AND ARRIVAL TIME SPECTRA

In this section we will present calculations of the electron transport coefficients in the rare gases He, Ne and Ar. We intentionally deal with E/N's for which the mean electron energies are well below the first in-elastic threshold. In this case the transport is well described by the quasi-Lorentz gas model. In this model one assumes that m/M is very small, that only elastic collisions occur and that the stationary distribution function is nearly isotropic. This model has been treated in detail in the literature. Formulae for both the higher-order distribution functions and the higher-order transport coefficients are explicitly given in Huxley and Crompton (1974). All the transport coefficients can be expressed in terms of the isotropic components $f_0^{(0)}$, $f_0^{(1)}$ and $f_0^{(2)}$, etc. of the density gradient expanded distribution functions.

Figure 1 shows the momentum transfer cross sections we used in the calculations. The He cross section is from Crompton et al. (1970). The Ne cross section is from Robertson (1972). The Ar cross section is from Nakamura and Kurachi (1988). The range of E/N's and corresponding mean energies we deal with are shown in Fig. 2. The usual transport coef-ficients v_d and D_L, as well as D_L/D_T, are shown in Figs. 3-5. For the case of a constant cross section, the quasi-Lorentz gas model is known to give 0.5 for D_L/D_T. One can see from Figs. 1 and 5 that He and Ne are closely approximated by this "hard-sphere" model for the E/N's we are dealing with. For the hard-sphere model, the transport coefficients can be expressed simply by the following formulae:

$$v_d = 3.16 \times 10^7 \left(\frac{2m}{M}\right)^{1/4} \left(\frac{E/N}{\sigma_m}\right)^{1/2} \tag{5a}$$

$$N D_L = 7.15 \times 10^6 \left(\frac{M}{2m}\right)^{1/4} \frac{(E/N)^{1/2}}{\sigma_m^{3/2}} \tag{5b}$$

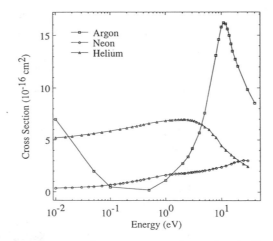

Fig. 1.   Momentum transfer cross sections for helium, neon and argon.

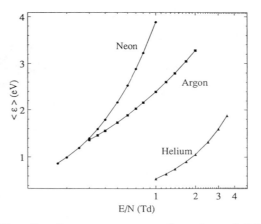

Fig. 2.   Mean energy as a function of E/N.

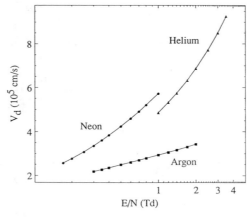

Fig. 3.   Drift velocity as a function of E/N.

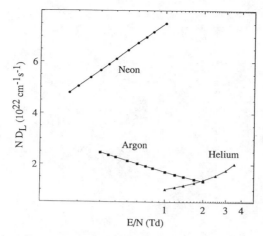

Fig. 4. Product of gas density and longitudinal diffusion
coefficient as a function of E/N.

$$N^2 D_3 = 2.37 \times 10^6 \left(\frac{M}{2m}\right)^{3/4} \frac{(E/N)^{1/2}}{\sigma_m^{5/2}} \qquad (5c)$$

where E/N is in units of V-cm^2 and σ_m is in units of cm^2.

Figures 6 and 7 show the skewness coefficient D_3 for He and Ne,
respectively, with values obtained using Boltzmann calculations, Monte
Carlo calculations and hard-sphere approximation. Figure 8 shows the
calculated D_3 for all three gases; He, Ne and Ar. The calculated energy
distribution functions for Ar are shown in Figs. 9-11.

Having obtained the transport coefficients we can compute the
arrival time spectra for typical drift tube conditions in order to assess
the measurability of the deviations from gaussian. For simplicity we

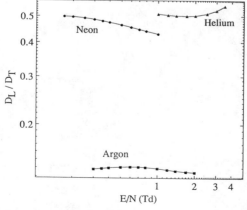

Fig. 5. Ratio of the longitudinal to the transverse diffusion
coefficient as a function of E/N.

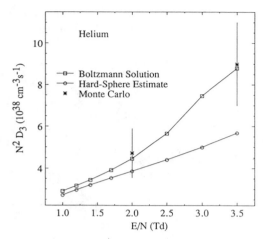

Fig. 6. Product of gas density squared and skewness coefficient as a function of E/N for helium, using the density gradient expansion solution, Monte Carlo simulation and hard-sphere approximation.

consider only the spatial dimension parallel to the field. In the usual first-order approximation the continuity equation, density profile and current profile are given by

$$\partial_t n_1(z,t) = -v_d \partial_z n_1(z,t) + D_L \partial_z^2 n_1(z,t) \tag{6a}$$

$$n_1(z,t) = \frac{n_0}{(4\pi D_L t)^{1/2}} \exp\left[-\frac{(z-v_d t)^2}{4D_L t}\right] \tag{6b}$$

$$I_1(z,t) \propto v_d n_1(z,t) - D_L \partial_z n_1(z,t) \tag{6c}$$

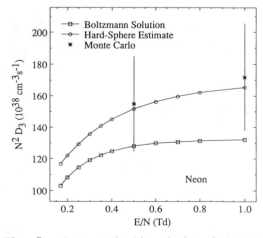

Fig. 7. Same as in Fig. 6, but for neon.

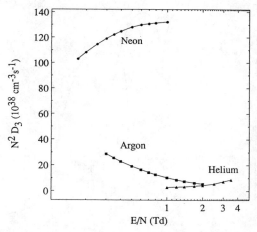

Fig. 8. Boltzmann solutions of the skewness coefficient for helium, neon and argon.

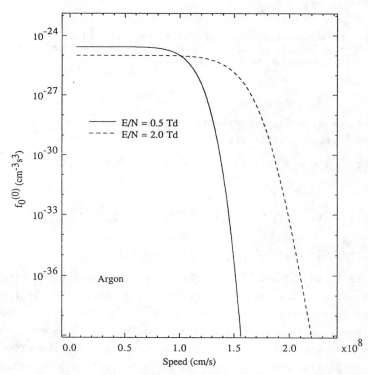

Fig. 9. Isotropic component of the first term in the density gradient expansion of the electron velocity distribution function in argon.

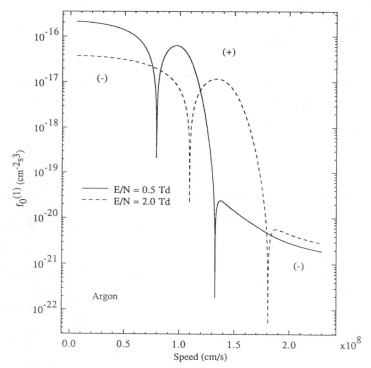

Fig. 10. Isotropic component of the second term (diffusion term) in the density gradient expansion of the electron velocity distribution function in argon.

In the second-order approximation the continuity equation, density profile and current profile are

$$\partial_t n_2(z,t) = - v_d \partial_z n_2(z,t) + D_L \partial_z^2 n_2(z,t) - D_3 \partial_z^3 n_2(z,t) \tag{7a}$$

$$n_2(z,t) = n_1(z,t) \left\{ 1 - \frac{D_3}{4D_L^2 t} \left[3 \ (z - v_d t) - \frac{(z - v_d t)^3}{2D_L t} \right] \right\} \tag{7b}$$

$$I_2(z,t) \ \alpha \ v_d n_2(z,t) - D_L \partial_z n_2(z,t) + D_3 \partial_z^2 n_2(z,t) \tag{7c}$$

Figures 12 and 13 show the computed arrival time spectra for Ne at E/N = 1 Td and z = 15.7 cm. At a pressure of 30 Torr, the spectra obtained from the various orders of approximation to the continuity equation can be distinguished with experimental resolution of about 1%. However, at a pressure of 300 Torr, the current profile cannot discern the effects of D_3 and D_4. Calculations such as these would be helpful in predicting the conditions at which the higher-order transport coefficients could be measured.

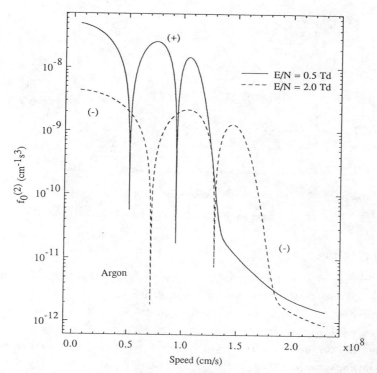

Fig. 11. Isotropic component of the third term (skewness term) in
the density gradient expansion of the electron velocity
distribution function in argon.

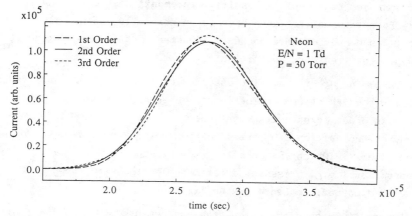

Fig. 12. Computed arrival time spectra for neon at E/N = 1 Td and
P = 30 Torr, using the various order of approximation in
the hydrodynamic continuity equation.

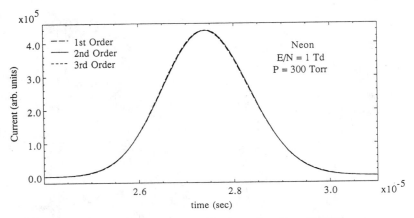

Fig. 13. Same as Fig. 12, but for P = 300 Torr.

ASSESSMENT OF TOF DATA

According to the hydrodynamic theory the quantities v_d, ND_L and N^2D_3 should be independent of pressure. Dependence on pressure would imply non-hydrodynamic effects such as gradients due to metallic boundaries and/or that drift equilibrium simply has not been reached at a sufficient time. In assessing the TOF data it is very important to establish whether or not the data are hydrodynamic. Non-hydrodynamic effects are generally defined as all those that cannot be described by the distribution function (2) or the continuity equation (3) with constant coefficients. These effects occur on short-time scales and near boundaries. They are difficult to measure systematically because the causes are difficult to control. Experiments should therefore be designed to minimize these effects in order to provide significant and reproducible results. The manifestations of hydrodynamic higher-order transport effects have to be carefully disentangled from those due to (1) non-hydrodynamic boundary phenomena and (2) short-time-scale time-dependent transport coefficients.

From the point of view of cross section inference, we want transport coefficients which depend on the least number of parameters. That parameter is E/N. In order to have transport coefficients which are independent of time and depend only on E/N, these coefficients have to be measured under the conditions in which the density gradient expansion is valid. The drift tube conditions, e.g. pressure, drift tube length, etc., need to be set such that non-hydrodynamic effects are negligible.

Figures 14-16 show the calculated and measured (Denman and Schlie, 1989a; 1989b) values of the drift velocity, longitudinal diffusion coefficient and skewness coefficient, respectively, for He. At this point it would be useless to compare the calculations with the measurements, especially for D_3, since the measurements show very large non-hydrodynamic

58

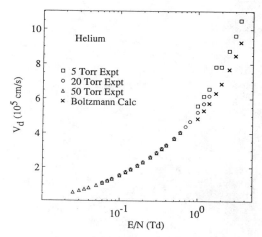

Fig. 14. Calculated and measured drift velocities for helium, using
the apparatus of Denman and Schlie (1989a; 1989b).

behavior. Similar non-hydrodynamic pressure dependence is observed for the
measurements in the other rare gases. It has been suggested by R. W.
Crompton (personal communication) that due to the long drift distance
(15.7 cm) used in the experiments of Denman and Schlie, diffusion cooling
might be modifying the equilibrium transport properties of the swarm; i.e.,
the transverse diffusion of electrons to the walls is distorting the
electron energy distribution at the distance at which the arrival time
spectra is taken.

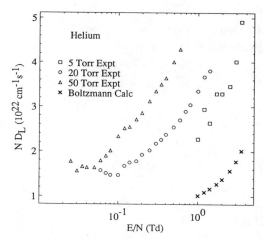

Fig. 15. Calculated and measured longitudinal diffusion coefficients
for helium, using the apparatus of Denman
and Schlie (1989a; 1989b).

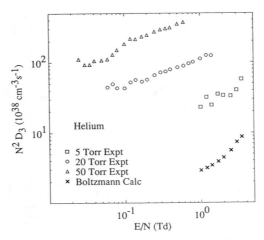

Fig. 16. Calculated and measured skewness coefficients for helium, using the apparatus of Denman and Schlie (1989a; 1989b).

NONLINEAR DRIFT AND DIFFUSION

In this section we will present a procedure for calculating the electron density profile without invoking the assumption of the density gradient expansion. The procedure is based on the TOF analysis of Ingold (1989). This alternative approach to TOF analysis is significant for two reasons: (1) it shows how the longitudinal diffusion coefficient can be calculated without using the density gradient expansion procedure, and (2) the calculated density profile effectively includes the effects of all the higher-order transport coefficients. A similar but more rigorous procedure using the characteristic-time-ordering of Kunhardt et al. (1988) is outlined by Kunhardt (1989).

In the following we will assume the conditions for the quasi-Lorentz gas model. The Boltzmann equation is written in terms of the density $n(z,t)$ and the isotropic component of the speed distribution function $f_0(v,z,t)$ as

$$\frac{\partial}{\partial t}[n(z,t)\,f_0(v,z,t)] = \frac{v}{3N\sigma(v)}\frac{\partial^2}{\partial z^2}[n(z,t)\,f_0(v,z,t)] \tag{8}$$

$$+ \frac{a}{3N\sigma(v)}\frac{\partial}{\partial z}\left[n(z,t)\frac{\partial}{\partial v}f_0(v,z,t)\right] + \frac{a}{3N}\frac{\partial}{\partial z}\left\{n(z,t)\frac{1}{v^2}\frac{\partial}{\partial v}\left[\frac{v^2}{\sigma(v)}f_0(v,z,t)\right]\right\}$$

$$+ \frac{a^2}{3N}n(z,t)\frac{1}{v^2}\frac{\partial}{\partial v}\left[\frac{v}{\sigma(v)}\frac{\partial}{\partial v}f_0(v,z,t)\right] + \frac{m}{M}N\,n(z,t)\frac{1}{v^2}\frac{\partial}{\partial v}[v^4\sigma(v)f_0(v,z,t)]$$

Multiplying Eq. (8) by $4\pi v^2$ and integrating over v, we get a nonlinear continuity equation

$$\frac{\partial}{\partial t}n(z,t) + \frac{\partial}{\partial z}J(z,t) = 0 \tag{9a}$$

where

$$J(z,t) = -\frac{\partial}{\partial z} [D(z,t)\ n(z,t)] + W(z,t)\ n(z,t) \tag{9b}$$

$$D(z,t) = \frac{4\pi}{3}\frac{1}{N} \int_0^\infty \frac{v^3}{\sigma(v)}\ f_0(v,z,t)\ dv \tag{9c}$$

$$W(z,t) = \frac{4\pi}{3}\frac{a}{N} \int_0^\infty \frac{\partial}{\partial v}\left[\frac{v^2}{\sigma(v)}\right]\ f_0(v,z,t)\ dv \tag{9d}$$

$$4\pi \int_0^\infty f_0(v,z,t)\ v^2\ dv = 1 \tag{9e}$$

By multiplying Eq. (8) by $4\pi v^4$ and integrating over v, we get a nonlinear conservation of energy equation

$$\frac{\partial}{\partial t}[n(z,t)\ \varepsilon(z,t)] = -\frac{\partial}{\partial z}[n(z,t)\ B(z,t)] + \frac{\partial^2}{\partial z^2}[n(z,t)\ G(z,t)]$$

$$+ 2\ a\ n(z,t)\ W(z,t) - 2\ a\ \frac{\partial}{\partial z}[n(z,t)\ D(z,t)] - \frac{2m}{M}\ N\ n(z,t)\ \dot{\varepsilon}(z,t) \tag{10a}$$

where

$$\varepsilon(z,t) = 4\pi \int_0^\infty f_0(v,z,t)\ v^4\ dv \tag{10b}$$

$$B(z,t) = \frac{4\pi}{3}\frac{a}{N} \int_0^\infty \frac{\partial}{\partial v}\left[\frac{v^4}{\sigma(v)}\right]\ f_0(v,z,t)\ dv \tag{10c}$$

$$G(z,t) = \frac{4\pi}{3}\frac{1}{N} \int_0^\infty \frac{v^5}{\sigma(v)}\ f_0(v,z,t)\ dv \tag{10d}$$

$$\dot{\varepsilon}(z,t) = 4\pi \int_0^\infty v^5 \sigma(v) f_0(v,z,t)\ dv \tag{10e}$$

Ideally one would solve Eq. (8) in order to evolve the electron density n(z,t) in space and time. The explicit solution of the space–time–dependent Boltzmann equation is in general, difficult to obtain. Kunhardt et al. (1988) has presented a procedure in which the Boltzmann equation is rewritten in equivalent, but more tractable, forms which are valid for different time scales. The application of such a procedure for solving the nonlinear continuity equation is presented by Kunhardt (1989). Since our purpose here is simply to illustrate the virtue of using a nonlinear continuity equation, we will use the simpler procedure of Ingold (1989).

The main assumption in this procedure is that the electron velocity distribution can be described locally in each point in space and time by the equilibrium solution. Thus $f_0(z,v,t)$ is expressed in terms of a single space-time varying parameter $\alpha(z,t)$. Consider the case in which the cross section $\sigma(v)$ is proportional to a power of v, that is,

$$\sigma(v) = \sigma_0 \, v^r$$

where σ_0 and r are constants. Then

$$f_0(v,z,t) = \frac{s}{4\pi[\alpha(z,t)]^3 \, \Gamma(3/s)} \exp\left[-\left(\frac{v}{\alpha(z,t)}\right)^s\right] \tag{11}$$

where $s = 2r + 4$.

Substituting Eq. (11) into Eqs. (9) and (10), one obtains two coupled nonlinear partial differential equations for the density n and the energy $n\alpha^2$. Letting $p = n\alpha^2$, Eqs. (9) and (10) for the case of a constant cross section are transformed into

$$\frac{\partial n}{\partial t} = -W_0 \frac{\partial}{\partial z}\left[n \, (p/n)^{-1/2}\right] + D_0 \frac{\partial^2}{\partial z^2}\left[n \, (p/n)^{1/2}\right] \tag{12}$$

$$\frac{\partial p}{\partial t} = -(B_0 + 2aD_0) \frac{\partial}{\partial z}\left[n \, (p/n)^{1/2}\right] + G_0 \frac{\partial^2}{\partial z^2}\left[n \, (p/n)^{3/2}\right]$$

$$+2aW_0 \left[n \, (p/n)^{-1/2}\right] - \frac{2m}{M} \, N \, \dot{\varepsilon}_0 \left[n \, (p/n)^{3/2}\right] \tag{13}$$

where W_0, D_0, B_0 G_0 and $\dot{\varepsilon}_0$ are constants.

Equations (12) and (13) are solved straightforwardly using the code PDEONE of Sincovec and Madsen (1975). PDEONE provides centered differencing in the spatial variable, giving a semidiscrete system of nonlinear ordinary differential equations which are then solved using the ODE integrator code LSODE of Hindmarsh (1980). The solution of Eqs. (12) and (13) are shown in Figs. 17 and 18 for the case of $\sigma_0 = 6 \times 10^{-16}$ cm^2, $M = 4$ amu and $E/N = 1.4$ Td. Figure 19 shows a comparison of the nonlinear solution with the solution of the linear continuity equation with drift velocity and diffusion coefficient given by the homogeneous, steady-state values

$$v_d = 0.964 \frac{a}{N \, \sigma_0 \, \alpha} \qquad \text{and} \qquad D = 0.272 \frac{\alpha}{N \, \sigma_0}$$

where

$$\alpha = \left[\frac{s}{\sigma_0^2} \frac{M}{3m} \left(\frac{e}{m}\right)^2 \left(\frac{E}{N}\right)^2\right]^{1/s}$$

Figure 20 shows the nonlinear solution in comparison with the fits to a gaussian (first-order continuity equation), a skewed gaussian (second-order) and a skewed-kurtosed gaussian (third-order) profile. The solution can be seen to be well approximated by a skewed gaussian. This is as expected since the situation most likely to prevail, with moderate initial density

62

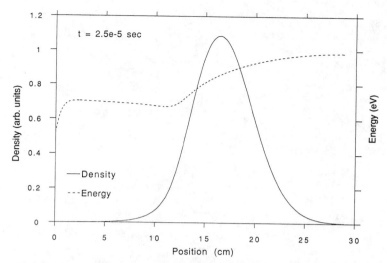

Fig. 17. Solution of the coupled nonlinear partial differential
          equations for the continuity and energy conservation for
          the case of a constant cross section.

distributions, is a small finite field deviation from the first-order diffusion equation and negligible zero-field deviations. By doing a nonlinear
least squares fit to the density profile we could derive effective values
for the transport coefficients v_d, D and D_3. For the diffusion coefficient
we obtain

$$\frac{D \text{ (nonlinear solution)}}{D \text{ (homogeneous, steady-state solution)}} = 0.6$$

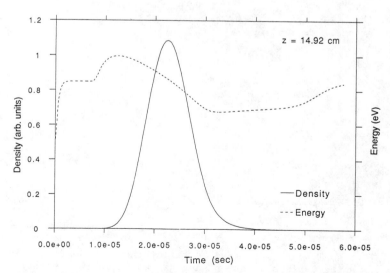

Fig. 18. Arrival time spectra corresponding to the solution shown
          in Fig. 17.

63

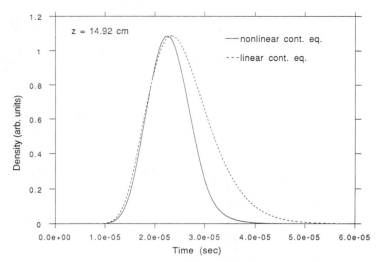

Fig. 19. The arrival time spectra from the nonlinear continuity
equation compared with that from the linear continuity
equation with homogeneous, steady-state coefficients. The
diffusion coefficient used in the linear solution corresponds to the transverse diffusion coefficient. The
effective diffusion coefficient obtained in the nonlinear
solution corresponds to the longitudinal diffusion
coefficient.

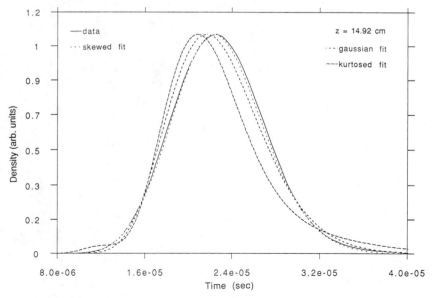

Fig. 20. Nonlinear least squares fit of the nonlinear solution to
the various orders of approximation in the density
gradient expansion.

One would have expected this ratio to be 0.5 as predicted by the density gradient expansion theory. Note however that the diffusion coefficient obtained by fitting to the nonlinear solution includes the effect of the skewness in the density profile.

DISCUSSION

The calculations presented here indicate that the steady-state values of D_3 are measurable. The E/N and pressures at which they are measurable can be predicted. Such measurements could be useful in providing a sensitive transport coefficient for the inference of collision cross sections. Unfortunately, the present TOF data on D_3 show very large non-hydrodynamic behavior. At present these data cannot be used for cross section inference.

We have shown that D_3, and possibly D_4, are measurable and that present experiments have the capability of resolving them for the arrival time spectra. However, there are still some questions about the significance of these higher-order transport coefficients. For instance, can the uncertainties in the measurements of D_3 be made small enough to provide significant additional information on the collision cross sections? Present experience with D_L shows that even though this quantity is more sensitive than v_d to the cross section, the corresponding larger uncertainty in its measured values offsets the accuracy to which the cross sections can be deduced.

We have also indicated that it is possible to calculate the arrival time spectra without resorting to the approximations involved in the density gradient expansion. But we need to ask: do we need to accurately describe the deviations from gaussian of the current profile in order to derive the correct steady-state values for the drift velocity and the diffusion coefficients? Phelps et al. (1960; 1961) had long ago used techniques in which different drift distances were employed to get difference measurements of the drift velocity. It has been suggested by Skullerud (1974) that difference measurements of the mean square width can give experimental values for the diffusion coefficients even under conditions when the pulse form is far from gaussian. Nakamura (1987), for example, has applied such difference measurements to get both the drift velocity and the longitudinal diffusion coefficient. Such difference measurements indicate that it is not necessary to have a gaussian, nor is it necessary to accurately describe deviations from gaussian, in order to get the correct steady-state values for the drift velocity and diffusion coefficients.

ACKNOWLEDGMENT

This work was performed under the auspices of the U.S. Department of Energy by Lawrence Livermore National Laboratory under Contract No. W-7405-Eng-48.

REFERENCE

Crompton, R. W., personal communication.

Crompton, R. W., M. T. Elford, and A. G. Robertson, 1970, Aust. J. Phys. 23, 667.

Denman, C. A., and L. A. Schlie, 1989a, submitted to Rev. Sci. Instr.

Denman, C. A., and L. A. Schlie, 1989b, "Sensitive High-Temporal-Resolution TOF Electron Drift Tube: Assymetrical Current Pulse Observation and Determination of V_d, D_L and D_3," Sixth International Swarm Seminar, Glen Cove, N.Y.

Hindmarsh, A., 1980, ACM Signum Newsletter.

Huxley, L. G. H., 1972, Aust. J. Phys. 25, 43.

Huxley, L. G. H., and R. W. Crompton, 1974, "The Diffusion and Drift of Electrons in Gases," (Wiley, New York).

Ingold, J. H., 1989, "Diffusion Theory of Electrons in a Uniform Electric Field: Time-of-Flight Analysis," Sixth International Swarm Seminar, Glen Cove, N.Y.

Kumar, K., H. R. Skullerud, and R. E. Robson, 1980, Aust. J. Phys. 33, 343.

Kunhardt, E. E., J. Wu, and B. M. Penetrante, 1988, Phys. Rev. A 37, 1654.

Kunhardt, E. E., 1989, "Nonlinear Diffusion," Sixth International Swarm Seminar, Glen Cove, N.Y.

Nakamura, Y., 1987, J. Phys. D 20, 933.

Nakamura, Y., and M. Kurachi, 1988, J. Phys. D 21, 718.

Pack, J. L., and A. V. Phelps, 1961, Phys. Rev. 121, 798.

Parker, J. H., and J. J. Lowke, 1969, Phys. Rev. 181, 290.

Penetrante, B. M., and J. N. Bardsley, 1984, J. Phys. D 17, 1971.

Phelps, A. V., J. L. Pack, and L. S. Frost, 1960, Phys. Rev. 117, 470.

Robertson, A. G. 1972, J. Phys. B 5, 648.

Sincovec, R. F., and N. K. Madsen, 1975, ACM Trans. Math. Software 1, 261.

Skullerud, H. R., 1969, J. Phys. B 2, 696.

Skullerud, H. R., 1974, Aust. J. Phys. 27, 195.

Wagner, E. B., F. J. Davis, and G. S. Hurst, 1967, J. Chem. Phys. 47, 3138.

NON-LOCAL DESCRIPTIONS OF ELECTRON SWARMS IN SPACE-TIME

H. A. Blevin and L. J. Kelly

The Flinders University of South Australia
Bedford Park, South Australia 5042
Australia

INTRODUCTION

The behavior of a group of electrons moving in a neutral gas under the influence of a space-time dependent force can be described by the velocity distribution function $f(\underline{v}, \underline{r}, t)$. In the following discussion it will be assumed that space-charge fields are negligible and only an external electric field $\underline{E}(\underline{r}, t)$ is considered. The distribution function can be obtained from the Boltzmann equation

$$\frac{\partial f}{\partial t} + \underline{v} \cdot \nabla_r f + \frac{q}{m} \underline{E} \cdot \nabla_v f = \left(\frac{\partial f}{\partial t}\right)_{coll.} \tag{1}$$

The solution of this equation is difficult to obtain for space-time varying fields and a realistic collision term $(\partial f / \partial t)_{coll}$. Even when $\underline{E}$ is constant in space and time, only the spatially integrated energy distribution function $f(\varepsilon)$ can be expressed in closed form (for model elastic collision cross-sections and after initial transients have decayed). This distribution function depends only on E/N, where N is the neutral gas concentration and the gas temperature. It is usually referred to as the "equilibrium" distribution, although it may be far from Maxwellian (Huxley and Crompton, 1974).

In spatially uniform electric fields that are also constant in time, "non-equilibrium" distributions can refer to the transient behavior following the production of an electron swarm in the gas (Holst and Oosterhuis, 1921; Fletcher, 1985). It may also refer to boundary regions near electrodes (Kelly et al., 1989; Blevin et al., 1987) or electron runaway at very high E/N values (Phelps et al., 1987). In spatially or time-varying electric fields, non-equilibrium occurs when the field changes appreciably in an energy exchange distance (or time) so that the distribution function $f(\underline{v},\underline{r},t)$ cannot be described in terms of the local field $\underline{E}(\underline{r},t)$. A more detailed description of non-equilibrium conditions has been given by Marode et al. (1983).

Many approaches have been made towards the evaluation of $f(\underline{v},\underline{r},t)$ including Monte Carlo simulations (Braglia and Lowke, 1979), numerical solution of the Boltzmann equation (Segur et al., 1983), and approximate solutions involving expansions of the distribution function in terms of spatial (or temporal) gradients in fields or electron concentration (Aleksandrov et al., 1980; Kumar et al., 1980). It is characteristic of all non-equilibrium situations that the local energy distribution $f(\underline{v},\underline{r},t)$ does not depend on the local quantities $\underline{E}(\underline{r},t)/N$ and the electron concentration $n(\underline{r},t)$ alone. The past history (typically over a few energy exchange times) of all the electrons in an elementary volume must be accounted for including variations of $\underline{E}(\underline{r}',t'$ <t) along their paths. Even for the special case when $\underline{E}$ is constant in space-time, the presence of the field makes space anisotropic so that those electrons moving in the field direction must be treated differently than those travelling against the field. Consequently, spatial variations in electron concentration will modify the local velocity distribution. This example of non-local behavior (although not classified as "non-equilibrium" by Marode et al., [1983]) has been extensively studied (Kumar et al., 1980) by expansion of the local distribution function in terms of gradients in $n(\underline{r},t)$;

$$f(\underline{v},\underline{r},t) = n(\underline{r},t)\ g^o(\underline{v}) - \frac{\partial n}{\partial(\lambda_L z)}\ g^1(\underline{v}) + \frac{\partial^2 n}{\partial(\lambda_L z)^2} \cdot g^2(\underline{v}) - \cdots, \qquad (2)$$

where $\underline{E}$ is in the negative z-direction, and the parameter $\lambda_L (= W/2D_L) \sim$ (energy exchange distance)$^{-1}$, has been introduced as a scaling factor to make the $g^k(\underline{v})$ comparable in magnitude (radial gradients have not been included as they generally have a smaller influence on $f(\underline{v},\underline{r},t)$ than the terms given here). This theory has been successful in explaining the anisotropic diffusion observed in swarm experiments (Parker and Lowke, 1969; Wagner et al., 1967).

The zero-order moment of Eq. (1) in velocity space gives the continuity equation

$$\frac{\partial n}{\partial t} + \nabla \cdot \underline{J} = n(\nu_i - \nu_a) \qquad (3)$$

where the electron flux $\underline{J}(\underline{r},t) = \int \underline{v} f(\underline{v},\underline{r},t)dv$, and ν_i, ν_a are the ionization and attachment frequencies. In the following discussion only ionization will be considered in the source term,

$$n(\underline{r},t)\nu_i(\underline{r},t) = \int (\frac{\partial f}{\partial t})_{coll.}\,d\underline{v}.$$

68

Using the gradient expansion Eq. (2) to determine $\underline{J}$ and ν_i, the continuity equation can be written (Blevin and Fletcher, 1984; Thomas, 1969)

$$\frac{\partial n}{\partial t} + \left(W^o \frac{\partial n}{\partial z} - D^o \frac{\partial^2 n}{\partial x^2} - D^o \frac{\partial^2 n}{\partial y^2} - D'_L \frac{\partial^2 n}{\partial z^2} + S' \frac{\partial^3 n}{\partial z^3} - \cdots \right)$$

$$= n\nu_i^o - \frac{\nu_i^1}{\lambda_L} \frac{\partial n}{\partial z} + \frac{\nu_i^2}{\lambda_L^2} \frac{\partial^2 n}{\partial z^2} - \cdots$$

or

$$\frac{\partial n}{\partial t} + \left(W \frac{\partial n}{\partial z} - D_T \frac{\partial^2 n}{\partial x^2} - D_T \frac{\partial^2 n}{\partial y^2} - D_L \frac{\partial^2 n}{\partial z^2} + S \frac{\partial^3 n}{\partial z^3} - \cdots \right) = n\nu_i^o , \qquad (4)$$

where $W = W^o + \nu_i^1/\lambda_L$, $D_L = D'_L + \nu_i^2/\lambda_L^2$. It will be shown later that the ionization contribution to D_L is significant at high E/N and arises from the second derivative term in the gradient expansion. Note that the bracketed term in equation (4) no longer represents $\nabla \cdot \underline{J}$ although this is often assumed when calculating electron currents at the boundaries (Kelly and Blevin, 1989).

Most experiments have been analyzed using a truncated form of Eq. (4),

$$\frac{\partial n}{\partial t} + \left(W \frac{\partial n}{\partial z} - D_T \frac{\partial^2 n}{\partial x^2} - D_T \frac{\partial^2 n}{\partial y^2} - D_L \frac{\partial^2 n}{\partial z^2} \right) = n\nu_i^o . \qquad (5)$$

There appears to have been little consideration given to the circumstances under which this truncation is appropriate or, equivalently, to the convergence of the gradient expansion of the distribution function. In the following section, experimental results are presented to show that it is possible to observe the influence of second and higher order derivatives in the gradient expansion. In the next section the results of a Monte Carlo simulation are described and compared with the experimental observations.

EXPERIMENTAL PROCEDURES

The observation of photons emitted from two or more excited states has been used to indicate spatial variations in the energy distribution function for steady-state Townsend discharges (Wedding and Kelly, 1989) and in an isolated swarm (Kelly, 1985). These experiments were compared with predictions from the gradient expansion although only terms up to the first derivatives were considered. In both cases, clear evidence of spatial variations in the energy distribution function were obtained, but some features of the experimental results were not fully understood, and it seemed likely that higher order derivatives should be included in the expansion.

69

In order to compare the experimental observations of light output with the theoretical model of electron behavior [Eq. (4) or (5)], it is necessary to determine the relation between photon production and electron concentration. For a simple state with one decay channel and no cascade, the photon production rate is given by:

$$N_{ph}(\underline{r},t) = \frac{1}{\tau} \int_0^t n(\underline{r},t') v_{ex}(\underline{r},t') e^{-(t-t')/\tau} dt' \qquad (6)$$

where τ is the lifetime of the excited state and v_{ex} is the electron excitation frequency for that state. More complicated expressions are required when cascading occurs (Wedding et al., 1985).

The excitation rate can be determined from the gradient expansion Eq. (2), with a change of variable to energy ε;

$$n(\underline{r},t) v_{ex}(\underline{r},t) = \left\{ n(\underline{r},t) \int g^o(\varepsilon) v_{ex}(\varepsilon) d\varepsilon - \frac{\partial n}{\partial(\lambda_L z)} \int g^1(\varepsilon) v_{ex}(\varepsilon) d\varepsilon + \cdots \right\}$$

$$= n(\underline{r},t) v_{ex}^o - \frac{\partial n}{\partial(\lambda_L z)} v_{ex}^1 + \frac{\partial^2 n}{\partial(\lambda_L z)^2} v_{ex}^2 - \cdots$$

i.e.,

$$n(\underline{r},t) v_{ex}(\underline{r},t) = n v_{ex}^o \left\{ 1 - \frac{1}{n} \frac{\partial n}{\partial(\lambda_L z)} \frac{v_{ex}^1}{v_{ex}^o} + \frac{1}{n} \frac{\partial^2 n}{\partial(\lambda_L z)^2} \frac{v_{ex}^2}{v_{ex}^o} - \cdots \right\} \qquad (7)$$

or, to first order in the derivatives,

$$n(\underline{r},t) v_{ex}(\underline{r},t) = v_{ex}^o \left\{ n - \frac{1}{\lambda_L} \left(\frac{v_{ex}^1}{v_{ex}^o} \right) \frac{\partial n}{\partial z} \right\} . \qquad (8)$$

If it is assumed that spatial variation in $v_{ex}(\underline{r},t)$ is small and can be adequately described by the first two terms in Eq. (8), then for one-dimensional case

$$n(z,t) v_{ex}(z,t) \sim v_{ex}^o n(z - \Delta, t) \qquad (9)$$

where $\Delta = \dfrac{1}{\lambda_L} \dfrac{v_{ex}^1}{v_{ex}^o}$.

If Eq. (5) is used to solve for $n(\underline{r},t)$, then the solution for a δ-function dipole source at the cathode is

$$n(x,y,z,t) = \frac{n_o z}{Wt} e^{v_i^o t} \frac{e^{-(x^2 + y^2)/4D_T t}}{4\pi D_T t} \cdot \frac{e^{-(z-Wt)^2/4D_L t}}{\sqrt{4\pi D_L t}} , \qquad (10)$$

70

and

$$n(z,t) = \frac{n_o z}{Wt} e^{v_i^o t} \frac{e^{-(z-Wt)^2/4D_L t}}{\sqrt{4\pi D_L t}} .$$ (11)

Pulsed Time-of-Flight Experiment

Details of the experimental apparatus have been given elsewhere
(Wedding et al., 1985), together with modifications to Eq. (6) and Eq.
(11) to allow for cascading and non-equilibrium conditions near the
cathode boundary. For the pulsed experiment described here, the light
output from an axial interval was observed using a narrow slit collimator,
thus integrating over all transverse directions and giving results
corresponding to a one-dimensional swarm. The light output $N_{ph}(z,t)$
calculated from Eqs. (6), (9), and (11) could be compared with
experimental results for several values of z and parameters such as v_i^o,
W, D_L, Δ, τ obtained to give the best fit with experiment.

Figure 1 shows a comparison of theory and experiment for electrons
in N_2 gas at E/N = 443 Td and a pressure of 0.28 Torr. The light output
was measured for the 337.1 nm band of the second positive system, and the
391.4 nm band of the first negative system at z = 2.2 cm, 3.2 cm, and 4.2
cm. The values of v_i^o, W, D_L are common to both calculated curves, but Δ
and τ are dependent on the excited state under consideration. (The data
shown in Fig. 1 have been normalized to a common maximum for display
purposes.) The parameters obtained in this way were used to calculate
$n\nu_{ex}$ for each state using (9) and (11) and the (normalized) results are
shown in Fig. 2. As expected, the excitation to the $B^2\Sigma_u^+$ state (in the
future referred to as the B^+ state) is relatively greater than the $C^3\Pi_u$
(or C state) at the earlier times or front of the swarm, where the mean
energy of electrons is higher. The peak of the electron density (not
shown in Fig. 2) occurs at later times than the peak of both the C and B^+
excitation rates, and both Δ_C and Δ_{B+} are positive. However, the
magnitude of these quantities (particularly Δ_{B+}) is large enough to
invalidate the assumption that the spatial variations in $\nu_{ex}(\underline{r},t)$ are
small, and it is not possible to use Eq. (9) for the computation of
excitation rates. The experiments were therefore reanalysed using Eqs.
(6), (8), and (11), and the agreement with theory and experiment is shown

in Fig. 3. In this case $\frac{1}{\lambda_L}(\frac{v_i^1}{v^o})_C$ and $\frac{1}{\lambda_L}(\frac{v_i^1}{v^o})_B$ are additional fitting

parameters in Eq. (8), replacing the parameters Δ_C and Δ_{B+} used in the
previous analysis. Again, $n\nu_{ex}$ was calculated for each state and the
results are shown in Fig. 4. Again, the peak of the B^+ excitation occurs

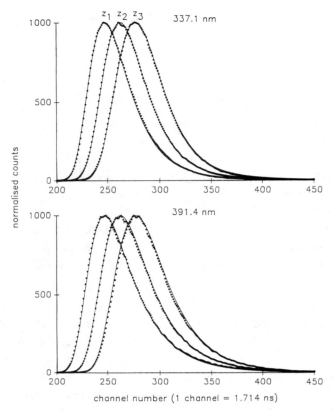

Fig. 1. The temporal photon emission from a swarm (z_1 = 2.2 cm, z_2 = 3.2 cm, z_3 = 4.2 cm) for the 337.1 nm and 391.4 nm bands of molecular nitrogen, E/N = 443 Td, p = 0.28 Torr. The solid line is the fit to the experimental data (dots) using the approximate theory described in the text. The data have been normalized to give equal peak heights.

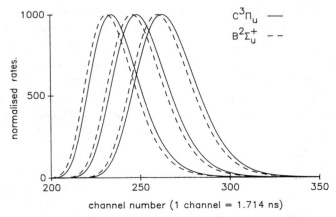

Fig. 2. The time dependent excitation rate $n\nu_C$ and $n\nu_B^+$ determined from the theoretical results shown in Fig. 1.

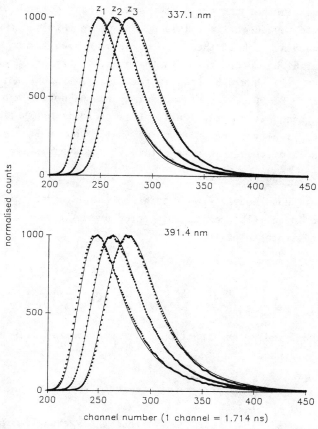

Fig. 3. The temporal photon emission from a swarm for the same
conditions as Fig. 1. The theoretical results are obtained
from the gradient expansion terminated at the first
derivative.

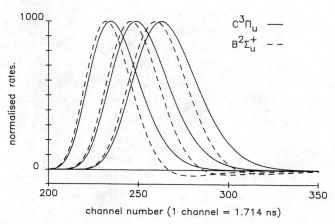

Fig. 4. The time dependent excitation rate $n\nu_C$ and $n\nu_{B^+}$ determined
from the theoretical results shown in Fig. 3.

earlier in time (further forward in the swarm) than the peak of the C
state excitation, but negative values of $n\nu_{B^+}$ were obtained towards the
back of the swarm at each of the three observation positions. This is an
indication that higher order terms in the gradient expansion are required, and that Eq. (7) should be used in the analysis rather than the
truncated form in Eq. (8). This immediately raises doubts about using
Eq. (11) as an adequate description of $n(z,t)$, and it seems likely that a
higher order continuity equation (such as that given in Eq. (4)) should
be used. This is discussed in more detail in the Monte-Carlo section.

Steady-State Townsend Experiment

For a continuous source from a point on the cathode, the pulse
solution given in Eq. (10) can be integrated over all source times to
give the steady-state solution,

$$n(r,z) = \lambda_L z \cdot \exp(\lambda_L z) \cdot K_{3/2}(\phi)/\phi^{3/2}, \qquad (12)$$

where

$$\phi = \lambda_L z \left[1 - 2\nu_i^0/\lambda_L W\right]^{1/2} \left[1 + \frac{D_L}{D_T}(r/z)^2\right]^{1/2} ,$$

$K_{3/2}(\phi)$ is a modified Bessel function, and r is the radial position in
cylindrical coordinates.

Since $\dfrac{1}{n}\dfrac{\partial n}{\partial(\lambda_L z)}$, $\dfrac{1}{n}\dfrac{\partial^2 n}{\partial(\lambda_L z)^2}$, $\cdots$ are functions of r at a fixed

axial position, it follows from Eq. (7) that the excitation rate ν_{ex} is
also a function of r, for example

$$\nu_C(r) = \nu_C^0 \left\{1 - \left(\frac{\nu^1}{\nu^0}\right)_C \frac{1}{n}\frac{\partial n}{\partial(\lambda_L z)} + \left(\frac{\nu^2}{\nu^0}\right)_C \frac{1}{n}\frac{\partial^2 n}{\partial(\lambda_L z)^2} - \cdots\right\}, \qquad (13)$$

and a similar expression for $\nu_{B^+}(r)$.

Experiments have been carried out to measure $\nu_C(r)$, $\nu_{B^+}(r)$ by
scanning laterally across a steady-state discharge with a tubular
collimator and performing an Abel inversion of the observed light output
(Wedding et al., 1986). After normalizing the radially integrated output
from each state, the ratio $\nu_C(r)/\nu_{B^+}(r)$ shows a departure from unity
indicating the presence of radial variations in the energy distribution
function.

The results shown in Fig. 5 for a gas mixture of CO_2:N_2:He:CO
(6:34:54:6) at E/N = 300 Td, p = 0.373 Torr, and z = 0.4 cm show the
normalized ratio varying by more than a factor of four as a function of
radial position. This is an extreme example for small $\lambda_L z(\sim 1)$, and it is
possible that the cathode boundary conditions and initial conditions are

74

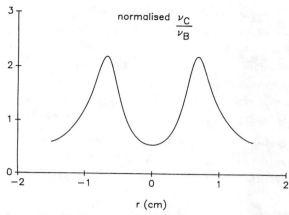

Fig. 5. The observed radial varation of ν_C/ν_B+ (normalized) for a
steady-state Townsend discharge (see text for the discharge
conditions). The large departures from unity indicate a
strong radial dependance of the mean energy for the
small $\lambda_L z$ value (~ 1.0) in this example.

still important. Other experimental results at large $\lambda_L z$ have been
reported (Wedding and Kelly, 1989) where spatial variations of about 20%
are still observed. Attempts to explain the radial dependence of ν_C/ν_B+
based on the first order gradient expansion were unsatisfactory and, in
particular, did not give the off-axis maximum observed in the experiments. A possible explanation of this feature was suggested by including
secondary electron production at the cathode (Wedding and Kelly, 1989).
However, the same maximum in $\nu_C(r)/\nu_B+(r)$ was observed in other gases and
$\lambda_L z$ values, and the results of the Monte-Carlo simulation given in the
next section indicate that the experimental results can be explained by
inclusion of higher order terms in the density gradient expansion. It
would be possible to determine the contribution from secondary electron
production by varying the electrode separation, while observing at a
fixed distance from the cathode, and clearly further experimentation will
be required before quantitative comparisons with theory can be made.

MONTE-CARLO SIMULATION

Both the pulsed and steady-state experiments described in the last
section indicate that it is possible to observe the influence of second
and possibly higher order terms in the density gradient expansion of the
distribution function. Quantitative comparisons of experiment and theory
have not been made as yet, since it appears that it will be necessary to
include many terms in the gradient expansion and the generalized continuity equation. To clarify these matters, a Monte Carlo simulation for

electron swarms in N_2 has been carried out for E/N = 500 Td, (p = 1 Torr) using a cross-section set derived from data given by Tagashira et al., (1980). A more detailed description of the simulation procedure has been given elsewhere (Brennan et al., 1989). The functions $g^0(\varepsilon)$, $g^1(\varepsilon)$ and $g^2(\varepsilon)$, determined by taking spatial moments of the Monte Carlo distribution $g(\underline{r}, \varepsilon, t)$, are shown in Fig. 6. Transport parameters and reaction rates of immediate interest to this work are listed in Table 1.

It is of interest to note that the ionization contribution to D_L (see discussion following Eq. (4)), which is derived from the second order gradient term in the distribution function, accounts for 26% of its magnitude.

From the parameters the electron concentration can be approximated at any time using Eq. (11); we call this $n_1(z,t)$. If third order spatial derivatives are retained in the continuity equation (Eq. (4)), then the electron concentration becomes

$$n(z,t) = n_1(z,t) - St \frac{\partial^3 n_1}{\partial z^3} + \frac{(St)^2}{2!} \frac{\partial^6 n_1}{\partial z^6} - \cdots \tag{14}$$

Table 1. Monte-Carlo Results for Transport and Rate Coefficients
E/N = 500 Td   p = 1 Torr

$W = 46.7 \quad cm \quad \mu s^{-1}$

$D_L = 1.51 \quad cm^2 \quad \mu s^{-1} \quad (D_i = v_i^2/\lambda_L^2 = 0.4 \ cm^2 \ \mu s^{-1})$

$S = 0.034 \ cm^3 \ \mu s^{-1}$

$\alpha_T = 1.62 \quad cm^{-1}$

$v_C^0 = 4.60 \times 10^7 \ s^{-1}$

$v_C^1 = 2.44 \times 10^7 \ s^{-1} \qquad v_C^1/v_C^0 = 0.53$

$v_C^2 = 0.22 \times 10^7 \ s^{-1} \qquad v_C^2/v_C^0 = 0.047$

$v_{B^+}^0 = 0.61 \times 10^7 \ s^{-1}$

$v_{B^+}^1 = 1.05 \times 10^7 \ s^{-1} \qquad v_{B^+}^1/v_{B^+}^0 = 1.72$

$v_{B^+}^2 = 0.98 \times 10^7 \ s^{-1} \qquad v_{B^+}^2/v_{B^+}^0 = 1.61$

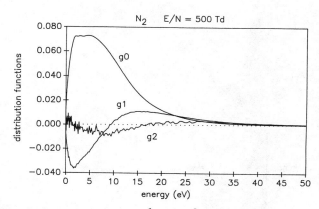

Fig. 6. The functions $g^o(\varepsilon)$, $g^1(\varepsilon)$, $g^2(\varepsilon)$ obtained from the Monte Carlo simulation of electrons in N_2 at E/N = 500 Td.

For $\lambda_L z \gg 1$ the first two terms give a good approximation (except in the "wings" of the distribution) and $n_1(z,t)$, $n(z,t)$ are shown in Fig. 7 for $\lambda_L z = 10.0$. This value of $\lambda_L z$ is approximately equal to the smallest distance used for the experimental results shown in Fig. 1, and it is clear that the skewness coefficient should have been included in the analysis of this experiment. From Eq. (4) (first two terms),

values of $\dfrac{1}{\lambda_L n}\dfrac{\partial n}{\partial z}$, $\dfrac{1}{\lambda_L^2 n}\dfrac{\partial^2 n}{\partial z^2}$ are calculated and are shown in Fig. 8.

These derivatives, together with the excitation rate coefficients given in Table 1, are used to calculate the spatial excitation rates $n\nu_C$, $n\nu_{B^+}$ using either first order $(n\nu_C)_1$, $(n\nu_{B^+})_1$ or second order $(n\nu_C)_2$, $(n\nu_{B^+})_2$ expansions. These are shown in Fig. 9 although $(n\nu_C)_1$ is omitted since it differs only slightly from $(n\nu_C)_2$. The results obtained for $(n\nu_{B^+})_1$ and $(n\nu_{B^+})_2$ are similar to the experimentally derived values shown in Fig. 4. This supports the earlier interpretations of the experimental results, but suggests that even higher order derivatives in the gradient expansion (and the continuity equation) will be required since negative excitation rates are still found at larger times. At larger $\lambda_L z$ the second order expansion may be adequate and permit a determination of the skewness coefficient. It is clear that further experiments over a wider range of $\lambda_L z$ values would be useful in determining the limits of the hydrodynamic regime and the applicability of the density gradient expansion within that regime.

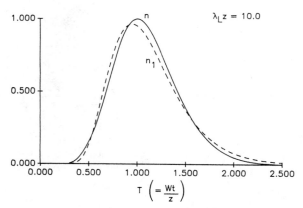

Fig. 7. The electron density at $\lambda_L z = 10.0$ as a function of time for a δ-function source. n_1 – solution for the continuity equation truncated at second order spatial derivatives. n – solution for truncation at third order derivatives.

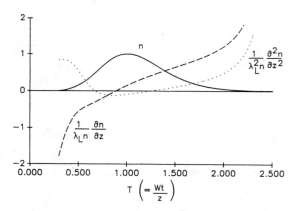

Fig. 8. Values of n, $\dfrac{1}{\lambda_L n}\dfrac{\partial n}{\partial z}$, $\dfrac{1}{\lambda_L^2 n}\dfrac{\partial^2 n}{\partial z^2}$ calculated for the same conditions given in Fig. 7.

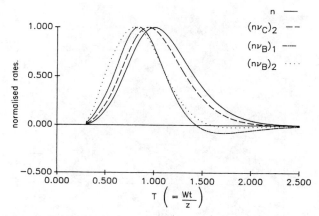

Fig. 9. Values of n, $(n\nu_C)_2$, $(n\nu_B{}^+)_1$, $(n\nu_B{}^+)_2$ calculated for the conditions given in Fig. 7.

The steady-state Townsend discharge can be modelled from the Monte Carlo results. Integrating Eq. (14) (only the first two terms are included in this analysis) over all source times, the electron concentration becomes:

$$n(r,z) = n_1(r,z) - \left(\frac{SW}{4D_L}\right)^2 \frac{1}{[1-2\nu_i^0/\lambda_L W]} \frac{\partial^3(\lambda_L z e^{\lambda_L z} K_{1/2}(\phi)/\phi^{1/2})}{\partial(\lambda_L z)^3}$$

where $n_1(r,z)$ is the solution of the second order continuity equation (cf. Eq. (12)) for a steady source.

The radial dependence of n, $\dfrac{1}{\lambda_L n}\dfrac{\partial n}{\partial z}$, $\dfrac{1}{\lambda_L^2 n}\dfrac{\partial^2 n}{\partial z^2}$ are shown in Fig. 10 for $\lambda_L z = 4.0$. Using these values with the excitation rates shown in Table 1 enable $\nu_C(r)/\nu_B{}^+(r)$ to be calculated from Eq. (13), either up to first or second order derivatives in the expansion. The results for $\nu_C/\nu_B{}^+$ are shown in Fig. 11, and the inclusion of the second order derivatives produces a maximum in the radial dependence, as found experimentally. Even though the experiments were for a different gas than that modelled in the Monte Carlo simulation, the theory suggests that the conclusions are generally applicable; that is, higher order terms are required in the gradient expansion to describe the outer regions of the discharge.

CONCLUSIONS

The observation of the space-time dependence of radiation from swarms and steady-state Townsend discharges has shown that spatial

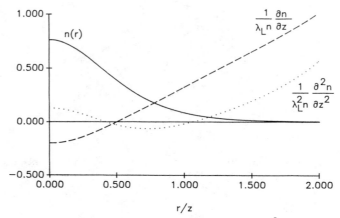

Fig. 10. Calculated values of n, $\frac{1}{\lambda_L n}\frac{\partial n}{\partial z}$, $\frac{1}{\lambda_L^2 n}\frac{\partial^2 n}{\partial z^2}$ for a steady-state discharge, $\lambda_L z = 4.0$.

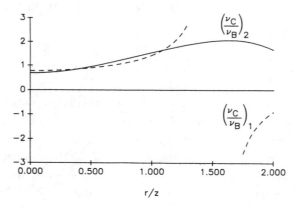

Fig. 11. Calculated values of ν_C/ν_B+ for first order and second order terms in the gradient expansion. Only the expansion including second-order terms gives the off-axis maximum in the excitation ratio, as observed experimentally.

variations in the energy distribution function occur and are in qualitative agreement with the gradient expansion. However, in the outer regions of the swarm (or steady-state discharge), it appears that higher order derivatives (≥ 2) are required to give quantitative agreement with experiment. The swarm experiments described here should be extended to a wider range of $\lambda_L z$ values to determine the skewness coefficient S and demonstrate the range of applicability of the gradient expansion within the hydrodynamic regime.

Although space-time varying fields have not been considered in this work, most experimental studies of swarm parameters under these conditions will necessarily involve concentration gradients. Theories dealing with space-time varying fields should include these gradients otherwise many of the important and interesting phenomena described here will be overlooked. There is very little experimental work directed towards an understanding of the internal structure of swarms for time varying fields, and the methods described here could be applied to these conditions.

REFERENCES

Aleksandrov, N. L., A. P. Napertovich, and A. N. Starostin, 1980, Sov. J. Plasma Phys. 6, 618.
Blevin, H. A., M. J. Brennan, and L. J. Kelly, 1987, ICPIG XVIII contributed papers 1, 92.
Blevin, H. A., and J. Fletcher, 1984, Aust. J. Phys. 37, 593.
Braglia, G. L., and J. J. Lowke, 1979, J. Phys. D 12, 1831.
Brennan, M. J., A. M. Garvie, and L. J. Kelly, 1989, Aust. J. Phys., submitted.
Fletcher, J., 1985, J. Phys. D. 18, 221.
Holst, G., and E. Oosterhuis, 1921, Physica 1, 78.
Huxley L. G. H., and R. W. Crompton, 1974, "The Diffusion and Drift of Electrons in Gases" (Wiley, New York).
Kelly, L. J., 1985, Project Report for Honors Degree, Flinders University.
Kelly, L. J., and H. A. Blevin, 1989, ICPIG XIX, contributed papers, 4, 892.
Kelly, L. J., M. J. Brennan, and A. B. Wedding, 1989, Aust. J. Phys. 42, 365.
Kumar, K., H. R. Skullerud, and R. E. Robson, 1980, Aust. J. Phys. 33, 343.
Marode, E., and J. P. Boeuf, 1983, ICPIG XVI, invited papers, 206.
Parker, J. H., and J. J. Lowke, 1969, Phys. Rev. 181, 290.
Phelps, A. V., B. M. Jelenkovic, and L. C. Pitchford, 1987, Phys. Rev. A 36, 5327.
Segur, P., M. C. Bordage, J. P. Balaguer, and M. Yousfi, 1983, J. Comp. Phys. 50, 116.
Tagashira, H., T. Taniguchi, and Y. Sakai, 1980, J. Phys. D 13, 235.
Thomas, W. R. L., 1969, J. Phys. B 2, 551.
Wagner, E. B., F. J. Davis, and G. S. Hurst, 1967, J. Chem Phys. 47, 3138.
Wedding, A. B., 1984, Thesis, The Flinders University of South Australia.
Wedding, A. B., H. A. Blevin, and J. Fletcher, 1985, J. Phys. D 18, 2361.
Wedding, A. B., and L. J. Kelly, 1989, Aust. J. Phys. 42, 101.

A DESCRIPTION OF THE NON-EQUILIBRIUM BEHAVIOR OF ELECTRONS IN GASES:
MACRO-KINETICS

E. E. Kunhardt

Weber Research Institute
Polytechnic University
Farmingdale, New York 11735

INTRODUCTION

Consider an assembly of (classical) electrons in a background gas
under the influence of an external space-time varying field. The
electron density is assumed to be sufficiently low that their mutual
interaction can be neglected; furthermore, the background gas is assumed
to be in equilibrium with a thermal bath at temperature T, and this
equilibrium is unaffected by the passage of the electrons. Then, every
electron may be treated independently of all others and the behavior of
the assembly may be ascertained from the dynamics of a single electron;
that is, the assembly may be considered as an ensemble of single-
particle systems (note that the number of systems that constitute the
ensemble may change with time due to ionization), characterized by a
stochastic distribution function in phase-space, $\hat{f}(\underset{\sim}{v}, \underset{\sim}{r})$, where $\underset{\sim}{v}$ and $\underset{\sim}{r}$
are velocity and position, respectively. This distribution function and
its equation of evolution provide an exact description of the state
of the assembly and its behavior (Klimontovich, 1986; Balescu, 1975).

Various approximate descriptions of the assembly can be obtained
depending on the space-time "resolution" necessary (or desired) in a
particular situation. In this paper, the focus is on descriptions with
less space-time resolution than that provided by the ensemble-averaged
distribution function, $f(\underset{\sim}{v}, \underset{\sim}{r}, t)$, whose "microscopic" resolution
corresponds to the scales of a two-body collision (Kunhardt, 1988). The
objective is to formulate closed descriptions of the assembly with
"macroscopic" resolution; namely, those corresponding to macroscopic
variables (Kunhardt, 1988). To achieve this, note that in general the
macroscopic variables that characterize the macroscopic dynamics of the

electron assembly depend on velocity averaged properties of the ensemble-averaged distribution over an extended velocity interval; moreover, their dynamical changes occur over space-time scales that are coarser than those of the distribution. This distribution contains more information than necessary to provide a characterization of the assembly in terms of macroscopic variables (Kunhardt, 1988). A distribution function with less space-time resolution (coarser) and equivalent (as far as macroscopic properties) velocity dependence to that of the ensemble-averaged distribution can equally serve to determine macroscopic properties of the assembly and to obtain closed equations of evolution for the macroscopic variables that characterize the dynamics of the assembly (Kunhardt, 1988).

There exist a number of coarser distributions each characterized by a different space-time resolution (Kunhardt, 1988). This resolution is dictated by the characteristic scales of variation of the dynamical macroscopic variables. Because of this, these distributions are collectively referred to as macro-kinetic distributions. A procedure for obtaining macro-kinetic distributions is to expand the ensemble-averaged distribution function in terms of those eigen-functions of the acceleration plus collision operators in the kinetic equation (see next section) whose eigen-values correspond to the desired resolution (Sirovich, 1963). However, for space-time dependent accelerations this may not be possible, and alternate, less compact expansions (for example, in terms of either a local field or eigen-functions of the collision operator only) would have to be used. Moreover, since in general the expansion coefficients have no physical significance, it is desirable to use alternate approaches for obtaining the macro-kinetic distribution.

An alternate, physical, approach is to first identify the macro-scopic variables that describe the dynamics of the assembly. Their equation of evolution contains the characteristic space-time scales that define the resolution of the description (Kunhardt, 1988). These scales are then used to obtain, from the equation of evolution for f, the equation for the macro-kinetic distribution. Since the characteristic scales of the macroscopic equations depend on those of the distribution, the procedure outlined above must be carried out self-consistently. Although not a unique set (rate coefficients can in principle also be used), the velocity moments of the distribution can serve to define the resolution scales for the macro-kinetic distributions. Increasing resolution is obtained by choosing an increasing number of moments in the description, which are selected by ordering the moment equations according to their characteristic space-time scales and keeping those

with less or equal resolution than desired (Kunhardt, 1988). Each finite
set of moments thus selected defines a resolution scale and a corresponding macro-kinetic distribution.

MACRO-KINETIC REGIMES AND DISTRIBUTIONS

The starting point for obtaining the macro-kinetic characterization
of the electron assembly is the distribution function $f(\underset{\sim}{v},\underset{\sim}{r},t)$ (Kunhardt,
1988). It obeys a kinetic equation of the Boltzmann type, namely
(Klimontovich, 1986; Balescu, 1975),

$$\partial_t f + \underset{\sim}{v} \cdot \nabla_{\underset{\sim}{r}} f + \frac{q}{m} \underset{\sim}{E} \cdot \nabla_{\underset{\sim}{v}} f = I(f) \tag{1}$$

where $\underset{\sim}{E} = E(\underset{\sim}{r},t)$ is the electric field (either externally applied or
arising from space-charge) and $I(f)$ is the linear scattering operator
(Sandler and Mason, 1969). No specific form for the operator I need be
assumed at this time. At the macroscopic level, the assembly is
characterized by a "state vector," H_N, whose components are velocity
moments of the distribution and the corresponding macro-kinetic
distribution, $f_M^{(N)}$. That is, $H_N = (S_N, f_M^{(N)})$, where $S_N = (m_j, j =
1, \cdots, N)$ with m_j being a velocity moment of the distribution (a scalar,
vector, or tensor). The dimension of S_N, N, and its components, m_j, are
selected depending on the space-time resolution desired for the
description. Alternatively, S_N and $f_M^{(N)}$ define the scale of resolution
of the macroscopic description. S_N and $f_M^{(N)}$ are obtained as follows.

First, the moment equations [obtained by taking appropriately
weighted integrals (in $\underset{\sim}{v}$ space) of Eq. (1) (Klimontovich, 1986)] are
ordered according to their characteristic scales (Kunhardt, 1988). This
step requires a priori assumptions about the relative magnitude of these
scales, which can be made from physical consideration. In any event, the
ordering used need to be confirmed after a self-consistent description is
obtained. The first three, time-scale ordered moment equations are
(Kunhardt, 1988):

$$\partial_t n + \nabla \cdot (n\underset{\sim}{u}) = \nu n, \tag{2a}$$

$$\partial_t (n\bar{\varepsilon}) + \nabla \cdot \langle \varepsilon \underset{\sim}{v} \rangle - q\underset{\sim}{E} \cdot n\underset{\sim}{u} = - \nu_\varepsilon n\bar{\varepsilon}, \tag{2b}$$

$$\partial t (n\underset{\sim}{u}) + \nabla \cdot \langle \underset{\sim}{v}\underset{\sim}{v} \rangle - \frac{q}{m}\underset{\sim}{E} n = - \nu_m n\underset{\sim}{u}, \tag{2c}$$

where, $n(\underset{\sim}{r},t)$, $\bar{\varepsilon}(\underset{\sim}{r},t)$, and $\underset{\sim}{u}(\underset{\sim}{r},t)$ are the electron density, mean energy,
and average velocity, respectively, the bracket implies an average over
the distribution, $\varepsilon = 1/2\ mv^2$, and ν, ν_ε, and ν_m are the (space-time-
dependent) effective ionization, energy-exchange, and momentum-exchange
frequencies, respectively. These frequencies are defined by:

$$\nu n = \int I(f)d\underset{\sim}{v}, \tag{3a}$$

$$-\nu_\varepsilon n\bar{\varepsilon} = \int \frac{1}{2} mv^2 I(f)d\underset{\sim}{v}, \tag{3b}$$

$$-\nu_m n\underset{\sim}{u} = \int \underset{\sim}{v} I(f)d\underset{\sim}{v}. \tag{3c}$$

An integral without limits implies integration over all space. Since it is difficult to ascribe physical significance to higher-order moments, their equations of evolution are seldom written down. The higher moment equations would also have to be ordered accordingly. It is assumed that their characteristic times are smaller than those defined above. Note that, in general, $\tau_\varepsilon > \tau_m$ for partially ionized gases.

The first three moments, $n, \bar{\varepsilon}$, and $\underset{\sim}{u}$, and their equation of evolution, Eq. (2), can be used to develop three levels of descriptions, each characterized by a space-time resolution scale (Kunhardt, 1988). The most coarse-grained description (i.e., least resolution) has a time scale of the order of $\nu^{-1}(=\tau)$; that is, the scale of the density equation. From Eqs. (2a)-(2c), since $\nu < \nu_\varepsilon < \nu_m$, there is a time for which the mean energy and average momentum of the electrons have relaxed to a state of quasiequilibrium where their subsequent variation is in the scale of τ. In such a scale, the dynamics of the system is determined from Eq. (2a). Consequently, S_N contains one component, n; i.e., $S_1 = [n(\underset{\sim}{r},t)]$. The equation for the corresponding macro-kinetic distribution, $f_M^{(1)}$, is obtained by averaging the Boltzmann equation (BE) over times shorter than τ. An equivalent approach is used in the next section to obtain this equation. By analogy with classical gas kinetics (Sirovich, 1963), the time regime for which this description is valid (namely, the longest time scale) is named the hydrodynamic regime. However, in contrast to gas kinetics, the properties of this state can be derived from a single macroscopic variable (instead of three), the density. This definition of hydrodynamics is less restrictive than that used in the swarm literature which in addition assumes a specific form for the distribution (Kumar et al., 1980).

"Non-hydrodynamic" (higher resolution) descriptions can systematically be obtained by using an additional moment in S_N. Thus, the next less coarse-grained description is in terms of $S_2 = [n(\underset{\sim}{r},t), n\bar{\varepsilon}(\underset{\sim}{r},t)]$ and the corresponding macro-kinetic distribution, $f_M^{(2)}$. This description is valid for times of the order of ν_ε^{-1}. From a practical point of view, the description with most resolution is in terms of $S_3 = [n(\underset{\sim}{r},t), n\bar{\varepsilon}(\underset{\sim}{r},t), nu(\underset{\sim}{r},t)]$ and $f_M^{(3)}$, which is valid for times of the order of ν_m^{-1}.

The approach outlined above for obtaining closed macro-kinetic characterizations is illustrated in the next section where the hydrodynamic description is developed for the case of a quasi-Lorentz gas model.

THE HYDRODYNAMIC REGIME: APPLICATION TO QUASI-LORENTZ GAS MODEL

In this section, the approach outlined in the previous section is used to obtain the macro-kinetic description of the hydrodynamic state for a quasi-Lorentz model (Huxley and Crompton, 1974). This model only takes into account elastic collisions between the electrons and the background gas, and is valid in the limit $m/M \ll 1$ (where M is the mass of an atom/molecule that constitutes the background gas). To obtain analytic solutions, assume that the spatial variation and the field are in the z direction; moreover, the two-term spherical harmonic approximation for f will be used (i.e., $f = f_o^s + f_1^s \cos \theta$). In this case, Eq. (1) becomes (assuming 1-D variation in space)

$$\partial_t f_o^s + \frac{v}{3} \partial_z f_1^s + \frac{1}{3v^2} \partial_v \left(v^2 a f_1^s \right) = I\left(f_o^s \right) \tag{4a}$$

$$f_1^s = - \frac{v}{\nu} \partial_z f_o^s - \frac{a}{\nu} \partial_v f_o^s \tag{4b}$$

where

$$I\left(f_o^s \right) = \frac{1}{v^2} \partial_v \left[\alpha(v) f_o^s + \beta(v) \partial_v f_o^s \right] \tag{4c}$$

$$\nu = N\sigma(v)v$$

$$a = qE/m$$

with,

$$\alpha(v) = \frac{m}{M} v^3 \nu, \quad \beta(v) = \frac{kT}{M} v^2 \nu$$

$\sigma(v)$ is the collision cross-section (only elastic scattering occurs), N, T, and M are the density, temperature, and mass, respectively, of the background gas. In Eq. (4b), it has been assumed that the relative rate of change of f_1 is small compared with the momentum exchange collision frequency. This is consistent with the conditions defining the hydrodynamic regime.

In this regime, the electron assembly is characterized (by definition) by $H_1 = (S_1, f_M^{(1)})$. The evolution of the assembly is determined from Eq. (2a) with the current density given by,

$$n\underset{\sim}{u}(\underset{\sim}{r}, t) = \underset{\sim}{J}(\underset{\sim}{r}, t) = \int \underset{\sim}{v} f_M^{(1)} d\underset{\sim}{v} \tag{5a}$$

and the rate ν by Eq. (3a), with $f = f_M^{(1)}$. Thus, the equation for $f_M^{(1)}$ (Eq. (6) below) and Eq. (2a) form a closed set.

It is convenient in some situations to write the current density in this regime as,

$$\underset{\sim}{J} = n\mu\underset{\sim}{E} - \underset{\sim}{D}\cdot\nabla n + \underset{\sim}{J}_R \tag{5b}$$

where μ is the mobility, D is the diffusion tensor, and $\underset{\sim}{J}_R$ accounts for other contributions to the density that are not proportional to either n or ∇n. (μ and D are in general space-time dependent.) A significant amount of work has been devoted to the theoretical determination of μ, $\underset{\sim}{D}, \underset{\sim}{J}_R$ and Eq. (3a) (Kumar et al., 1980; Huxley and Crompton, 1974; Parker and Lowke, 1969; Thomas, 1969; Lucas, 1970; Tagashira et al., 1977; Kleban and David, 1978; Skullerud and Kuhn, 1983; Penetrante and Bardsley, 1984; Blevin and Fletcher, 1984); equivalently, to the closure of Eq. (2a). These investigations fall into two categories according to the method used; namely, the free path (Huxley and Crompton, 1974) and the perturbed distribution function methods (Kumar et al., 1980; Huxley and Crompton, 1974; Parker and Lowke, 1969). In all approaches, the background (or zeroth order) electron distribution has been taken to be space-independent. The effects of non-equilibrium (arising from density gradients, for example) are then taken into account by introducing the concept of a free path (method (1)) or by perturbing the distribution directly (method (2)). Approaches based on the perturbation method have in general yielded more accurate results. They are characterized by the expansion of either the distribution function in terms of spatial derivatives of the density (Kumar et al., 1980; Huxley and Crompton, 1974; Skullerud and Kuhn, 1983; Penetrante and Bardsley, 1984; Blevin and Fletcher, 1984) or the spatial Fourier transform of the distribution in a power series in the spatial wavenumber (Parker and Lowke, 1969). In all cases, the lowest order solution is spatially uniform so that the expansions are valid in the limit of small density gradients. The results that have been obtained have elucidated a number of phenomena, such as, the properties of the diffusion tensor (Kumar et al., 1980; Parker and Lowke, 1969) and the effect of ionization on electron drift and diffusion (Tagashira et al., 1977). In the approach presented in this paper, an evaluation of μ, $\underset{\sim}{D}$ and $\underset{\sim}{J}_R$ follows from Eq. (5a) after substituting an expression for $f_M^{(1)}$.

Using the spherical harmonic approximation, the current density for the quasi-Lorentz gas model is

$$J_z = - \int \frac{4\pi}{3} \frac{v^2}{v} \partial_z f_o^s \, v^2 dv - \int \frac{4\pi}{3} \frac{v}{v} \, a \partial_v f_o^s v^2 dv \ . \tag{5c}$$

From the previous discussion, in the hydrodynamic regime, the equation for the macro-kinetic distribution of the quasi-Lorentz gas model, $f_M^{(1)}$, is obtained by changing the time scale of the BE (Eq. (1)) to the τ scale. This can be achieved using a technique introduced by Bogoliubov (1962). Mathematically, the change can be accomplished by the following relation:

$$f_o^s(v,z) = f_M^{(1)} \ (v, n(z,t)) = n f_M(v,n) \tag{6}$$

That is, in the τ scale, the space-time dependence of the distribution is implicit through a dependence on the density. Physically, this is equivalent to saying that the space-time dynamics of the assembly is determined by the density. In the swarm literature, the distribution function in the hydrodynamic regime is further restricted to have a density dependence of the form of a (linear) expansion in terms of gradients of the density (Kumar et al., 1980; Huxley and Crompton, 1974). This regime corresponds to a subset of that presented in this paper. Thus, the changes in f can be written as:

$$\partial_t f = \partial_n f_M^1 \partial_t n, \tag{6a}$$

$$\nabla_{\underset{\sim}{r}} f = \partial_n f_M^1 \nabla_{\underset{\sim}{r}} n, \tag{6b}$$

$$\nabla_{\underset{\sim}{v}} f = \nabla_{\underset{\sim}{v}} f_M^1. \tag{6c}$$

(The parenthesis around the superscript has been dropped. This practice will be subsequently continued.) It is convenient to obtain an alternate representation for Eq. (6) by explicitly displaying the nature of the n dependence of f_M. This is done primarily to assist in the physical interpretation of the various approximations to be used for the solution of Eq. (4). Since electron-electron collisions have been neglected, f_M cannot be an explicit function of the density. Its density dependence can only be through normalized derivatives of n; namely, $g_i = \partial_z^i n/n$, $i=1,\cdots,\infty$. That is,

$$f_M(\underset{\sim}{v},n) = f_M(\underset{\sim}{v}, \{g_i\}),$$

where

$$\{g_i\} = (\partial_z n/n, \ \partial_z^2 n/n, \cdots) \ .$$

Then,

$$\partial_n f_M = \underset{i}{\Sigma} \ \partial_{g_i} f_M \partial_n g_i = \underset{i}{\Sigma} \ \partial_{g_i} f_M \frac{\partial_z g_i}{n g_1}$$

89

Using Eqs. (2a), (14), and (6), the (linearized) equation for f_M for this case is found to be:

$$
-\frac{1}{3v^2}\,\partial_v\left[\frac{v^2a^2}{v}\,\partial_v f_M + 2\frac{v^3a}{v}\,g_1 f_M\right] + \left[\frac{d_v(v^3a/v)}{3v^2}\,g_1 - \frac{v^2}{3v}\,g_2\right]f_M
$$

$$
-\frac{1}{3v^2}\,\partial_v\left[2\frac{v^3a}{v}\,\Sigma\,\partial_{g_i}f_M\partial_z g_i\right] + \frac{d_v(v^3a/v)}{3v^2}\,\Sigma\,\partial_{g_i}f_M\partial_z g_i
$$

$$
-\frac{v^2}{3v}\left[\Sigma\partial_{g_i}f_M\left(\frac{\partial_z g_i}{g_1}\{g_2 + g_1^2\} + n\partial_z^2 g_i - n\partial_z g_i\partial_z\ln g_1\right) + \Sigma\partial_{g_i}^2 f_M(\partial_z g_i)^2\right]
$$

$$
= \frac{1}{v^2}\,\partial_v\left[\alpha f_M + \beta\partial_v f_M\right]
\tag{7}
$$

Eq. (7) is the complete working equation from which various approximations to f_M can be obtained, which differ in their range of validity. Some of these solutions are discussed in the subsequent subsections.

An alternate procedure for obtaining an equation for $f_M(v,n)$ (Eq. (7) is to expand $f_M(\underline{v},n)$ in spherical harmonics (instead of $f(\underline{v},\underline{r},t)$), and proceed from Eq. (1) (instead of Eq. (4)). This has not been done in order to start the example from the usual equations for f_o^S and f_1^S, Eqs. (4).

Density Gradient Expansion: Small Spatial Derivatives

This range has been discussed extensively in the literature (Kumar et al., 1980; Huxley and Crompton, 1974; Parker and Lowke, 1969; Thomas, 1969; Lucas, 1970; Tagashira et al., 1977; Kleban and David, 1978; Skullerud and Kuhn, 1983; Penetrante and Bardsley, 1984; Blevin and Fletcher, 1984). It is instructive to obtain the density gradient expansion as a solution to Eq. (7) in the range $\partial_z \sim \delta\partial_z'$, where δ is a small parameter. It is not necessary to specify what δ is at this time (in the end, $\delta\partial_z'$ is again replaced by ∂_z.) δ is strictly used to assist in the ordering of the various terms. Ambiguous ordering of terms has led to a significant amount of misunderstanding in the literature regarding the range of validity of the various solutions obtained and the interpretation of the μ, D, and J_R that appear in Eq. (5b) (Penetrante and Bardsley, 1984). Eq. (7) is rewritten in this range as follows:

$$
\left(\frac{a^2}{3v} + \frac{kT}{M}\,v\right)\partial_v\ln f_M = -\frac{m}{M}\,vv - \frac{2va}{3v}\,g_1'\delta
$$

$$
+ \frac{1}{v^2 f_M}\left\{\int_0^v\left[\frac{d_v(v^3a/v)}{3}\,g_1'\delta - \frac{v^4}{3v}\,g_2'\delta^2\right]f_M dv\right\} - \frac{2va}{3v}\,\Sigma\partial_{g_i'}\ln f_M\partial_z'\,g_i'\delta
$$

$$+ \frac{1}{v^2 f_M} \left\{ \int_0^v \frac{d_v(v^3 a/v)}{3} \sum_i \partial_{g_i'} f_M dv \right\} \partial_z' g_i' \delta + 0\left(\delta^2\right) \tag{8}$$

(subsequently, the primes will be dropped. In the final results, the original variables are used). Expanding $\ln f_M$ in a power series in δ (essentially a Rytov expansion of the solution) (Chernov, 1960),

$$\ln f_M = \sum_i \delta^i S_i = S_0 + S_1 + \delta^2 S_2 + \cdots \tag{9a}$$

$$f_M = f_0 [1 + \delta S_1 + \delta^2 (S_2 + S_1^2/2) + \cdots] \tag{9b}$$

and,

$$f_M^{-1} = f_0^{-1}[1 - \delta S_1 - \delta^2 (S_2 - S_1/2) \cdots] \tag{9c}$$

where

$$f_0 = e^{S_0}$$

The distribution function is normalized such that

$$\int_0^\infty f_0 v^2 dv = 1,$$

whereas all other contributions vanish. Using Eqs. (9) in Eq. (8), and equating coefficients of each power of δ to zero,

$$\delta^0: \qquad \left(\frac{a^2}{3v} + \frac{kT}{M} v\right) \partial_v S_0 = -\frac{m}{M} w$$

with solution,

$$S_0 = -\int_0^v \left(\frac{a^2}{3v} + \frac{kT}{M} v\right)^{-1} \frac{m}{M} wu du \tag{10a}$$

Since S_0 is not a function of g_i, $\partial_{g_i} S_0 = \partial_{g_i} f_0 = 0$. In the next order,

$$\delta^1: \qquad \left(\frac{a^2}{3v} + \frac{kT}{M} v\right) \partial_v S_1 = -\frac{2va}{v} g_1 + \frac{1}{v^2 f_0} \left\{ \int_0^v \left[\frac{d_u(u^3 a/v)}{3} \right] f_0 du \right\} g_1$$

with solution,

$$S_1 = -\int_0^v \left\{ \left[\frac{a^2}{3v} + \frac{kT}{M} v \right]^{-1} \left\{ \frac{2wa}{v} - \frac{1}{w^2 f_0} \int_0^w \frac{d_u(u^3 a/v)}{3} f_0 du \right\} dw g_1 \right\} \tag{10b}$$

91

From this equation, $\partial_{g_i} S_1 = 0$, $i \neq 1$, $\partial_{g_i} S_1 = \dfrac{S_1}{g_1}$, and in next order,

$$\delta^2: \quad \left(\frac{a^2}{3\nu} + \frac{kT}{M}\nu\right)\partial_v S_2 = -\left(\frac{a^2}{3\nu} + \frac{kT}{M}\nu\right)\partial_v S_1^2/2$$

$$-\left[\frac{2va}{3\nu}\frac{S_1}{g_1} + \frac{1}{v^2 f_o}\left\{\int_o^v\left[\frac{u^4}{3\nu} + \frac{d_u(u^3 a/\nu)}{3}\right]\frac{S_1}{g_1}\right] f_o du\right]g_2$$

with solution,

$$S_2 = -S_1^2/2 - \int_o^v \Phi^{(2)}(w)dw g_2 \tag{10c}$$

where,

$$\Phi^{(2)}(w) = \left(\frac{a^2}{3\nu} + \frac{KT}{M}\nu\right)^{-1}\left[\frac{2wa}{3\nu}\frac{S_1}{g_1} + \frac{1}{w^2 f_o}\left\{\int_o^w\left[\frac{u^4}{3\nu} + \frac{d_u(u^3 a/\nu)}{3}\right]\frac{S_i}{g_1}\right] f_o du\right]$$

Using Eqs. (10) and (9b) in Eq. (6) (and returning to the original unprimed variables), the hydrodynamic macro-kinetic distribution in this range is obtained; namely,

$$f_M^1 = nf_o\left[1-[\frac{S_1}{g_1}]g_1 - \left[\int_o^v\Phi^{(2)}(w)dw\right]g_2 + O(\delta^3)\right] \tag{11}$$

This result (obtained from Eq. (9b)) is valid in the regime where each exponential factor ($\delta^i S_i$) in the expansion given in Eq. (9a) is less than one. This requirement imposes on the magnitude of the $g_i's$ conditions defining the range of validity of the density gradient expansion. From this, an approximate condition can be defined, namely, $g_i < (qE/\bar{\epsilon})^i$. The first term in Eq. (10c) cancels out the contribution to the $O(\delta^2)$ term in the expansion of f_M (Eq. (9b)) coming from the square of the $O(\delta)$ term. The resulting solution, given by Eq. (11), is linear in the $g_i's$ and is the density gradient expansion for $f_M^{(1)}$. Eq. (37) is accurate to $O(\delta^3)$.

From Eqs. (11) and (4c), the transport parameters can be obtained. Denoting the coefficients of the $n, \partial_z n$, and $\partial_z^2 n$ terms in Eq. (4c) as drift velocity, v_d, diffusion coefficient, D_o, and curtosis coefficient, D_1, respectively, they are found to be

$$v_d = -\frac{4\pi}{3}\int\frac{u}{\nu}ad_u f_o u^2 du \tag{12a}$$

$$D_o = -\frac{4\pi}{3}\left\{\int \frac{v^2}{v} f_o v^2 dv - \int \frac{v}{v} a\left[\partial_v f_o S_1 + f_o \partial_v S_1\right]v^2 dv\right\} \quad (12b)$$

$$D_1 = -\frac{4\pi}{3}\left\{\int \frac{v^2}{v} f_o S_1 v^2 dv + \int \frac{v}{v} a\left[\partial_v f_o \int_0^v \phi^{(2)}(u)du + f_o \phi^{(2)}(v)\right]v^2 dv\right\} \quad (12c)$$

These results can readily be obtained by a priori assuming an expansion
for f in the form of Eq. (11). The derivation given above indicates that
such expansion is valid in the range where $\partial_z^i n/n$ is of $O(\delta^i)$, i.e., there
is a definite ordering in the expansion. In the range where this
ordering is no longer valid, it is necessary to obtain other solutions to
Eq. (7) for f_M^1.

Consider next the regime where g_1 and g_2 are comparable to each
order and are both of $O(\delta)$. Following the same procedure as used to
obtain Eq. (11) from Eq. (8), the macro-kinetic distribution is found to
be in this regime,

$$f_M^1 = nf_o\left\{1 - S_1^{DGE} + \int_0^v \Psi^{(2)}(u)du\, g_2 + O(\delta^2)\right\} \quad (13)$$

where

$$\Psi^{(2)}(u) = \left[\frac{a^2}{3v} + \frac{kT}{M}v\right]^{-1}\left[\frac{1}{u^2 f_o}\int_0^u \frac{w^4}{3v} f_o dw\right]$$

Among the higher order terms, there are some that are nonlinear in the
g_i's. To $O(\delta^2)$ Eq. (12) is linear in the g_i's and differs from Eq. (11),
to this order, by the term proportional to g_2. Using Eq. (13) in Eq.
(4c), expressions for the transport parameters (to $O(\delta^2)$) can also be
obtained,

$$v_d = v_d^{DGE}$$

$$D_o = D_o^{DGE}$$

$$D_1 = \frac{4\pi}{3}\left\{\int \frac{v^2}{v} f_o \frac{S_1}{g_1} v^2 dv - \int \frac{v}{v} a\left[\partial_v f_o \int_0^v \Psi^{(2)}(u)du + f_o \Psi^{(2)}(v)\right]v^2 dv\right\}$$

where v_d^{DGE} and D_o^{DGE} are given by Eqs. (12a) and (12b), respectively. Eq.
(13) and the expressions for v_d and D_o have been obtained by Penetrante
and Bardsley using a different procedure (1984).

It is important to note that the expressions for the current density
(Eq. (4c)) obtained using Eqs. (13) and (11) are not equal. To $O(\delta^2)$,

Eq. (13) leads to a term proportional to $\partial_z^2 n$, whereas Eq. (11) does
not. In order to obtain terms proportional to $\partial_z^2 n$ using Eq. (11), it is
necessary to go to $O(\delta^3)$; in which case, additional terms also
proportional to $\partial_z^2 n$ than those found in Eq. (13) come into play.
Moreover, going to $O(\delta^3)$ with Eq. (13) brings in terms that are nonlinear
in the g_i's. Such terms are not found in Eq. (11). As far as the
transport parameters, Eqs. (13) and (11) yield the same drift velocity
and diffusion coefficient, but different curtosis, D_1, and higher order
coefficients.

Large Density Gradient

When the density gradient is sufficiently large that terms
proportional to g_1 approach $O(1)$, while terms proportional to g_i, $i \neq 1$,
are of $O(\delta)$, the equation for f_o is found from Eq. (8) to be

$$\partial_v \left[v^2 \left(\frac{a^2}{3v} + \frac{kT}{M} v \right) \partial_v f_o + v^3 \left(\frac{2g_1 a}{3v} + \frac{m}{M} v \right) f_o \right] = - g_1 \frac{d_v (v^3 a / v)}{3} f_o \qquad (14)$$

Solution to this equation for a given velocity dependence of v may be
very difficult. It is however possible to solve this equation
numerically and used to tabulate J as a function of a and g_1 for use with
Eq. (2a). An approximate analytic solution can be obtained by treating
the R.H.S. of Eq. (14) as a perturbation, in which case, the solution
becomes,

$$f_o(v, g_1) = C_o (1 - W_o) e^{-\int_o^v bvdv} \qquad (15a)$$

where

$$b = \frac{2g_1 (a/3v) + (m/M)v}{(a^2/3v) + (kT/M)v} \qquad (15b)$$

$$W_o = - g_1 \int_o^v \left[e^{\int_o^v budu} \right] \frac{\int_o^v \left[d_u (u^3/av) e^{-\int_o^u bwdw} \right] }{3v^2 \left[(a^2/3v) + (kT/M)v \right]} du \; dv \qquad (15c)$$

with the normalization condition $4\pi \int_o^\infty f_o v^2 dv = 1$. Higher order terms
can similarly be included. The resulting expression for the distribution
constitutes an expansion about a non-uniform state, that given by
Eq. (15).

94

Explicit Evaluation of Distribution Function and Transport Parameters:

Constant Collision Frequency, ν_o

In this case (and neglecting the contribution from W_o), Eq. (15a) becomes

$$f_o(v,g_1) = C_o e^{-bv^2/2} \tag{16a}$$

where

$$C_o = (b/\pi)^{3/2} \tag{16b}$$

with b given by Eq. (15b). From Eqs. (15) and (5c), the drift velocity and diffusion coefficient are found to be

$$v_d = a/\nu_o + 0(\partial_z g_1) \tag{17a}$$

and

$$D = D_o/(1 + dg_1) \tag{17b}$$

where,

$$D_o = \frac{2/3(a/\nu_o)^2 + (2kT/m)}{(m/M)\nu_o} \tag{17c}$$

and

$$d = a/(3\nu_o^2 m/M) \tag{17d}$$

Note that, in lowest order, the drift velocity is not affected by the density gradient. The effect of the gradient is to enhance (decrease) the value of the diffusion coefficient in regions with negative (positive) density gradient. Moreover, the diffusion process is found to be nonlinear in g_1. The parameter d has the units of distance, and, from Eq. (17d), it corresponds to the distance covered in an energy exchange time $(m\nu/M)$ by a particle travelling at the drift velocity (a/ν). For $g_1 \ll d^{-1}$, Eq. (17b) can be expanded to yield $D \approx D_o(1-dg_1)$, which when used in the continuity equation (Eq. (2a)) results in a linear equation that contains terms proportional to the third space derivative. For $g_1 d \lesssim 1$, the full expression needs to be retained. This expression for D renders the continuity equation (Eq. (2a)) nonlinear. Eq. (2a) may be written in this case as

$$\partial_t n + v_d \partial_z n - \partial_z (D \partial_z n) = 0 \tag{18}$$

where, v_d and D are given by Eqs. (17a,b) with D_o (Eq. (17c)) being the linear diffusion coefficient, and d (Eq. (17d)) playing the role of a non-linearity parameter. Eq. (18) has a singularity at $d(\partial_z n/n) = -1$ and is valid in the regime $|d(\partial_z n/n)| < 1$. To elucidate the effect of the non-linearity, the time evolution of an initial gaussian density profile, $A\exp(-(z/w)^2)$, has been calculated by numerical solution of Eq. (18). The initial profile and that at t=60 (normalized units), for the case d=0, are shown in Fig. 1a. A reference frame moving with the drift velocity has been chosen so that only diffusion effects are evident. Moreover, the density profiles are plotted as ln vs. $(z/w)^2$, so that in the case of linear diffusion, the plots appear as straight lines whose slope decreases with time. The effect of the non-linearity, $d\neq0$, is shown in Fig. 1b. As expected from the density dependence of D (Eq. (17b)), the leading (trailing) edge of the pulse shows greater (lesser) spreading than for the case d = 0. For a gaussian profile, the effect of the non-linearity is greater towards the tails of the profile where $\partial_z n/n$ is larger. As the profile evolves and diffusion smoothes the gradients, the effect decreases; the tails of the profile, however, maintain their characteristics. As can be seen in Fig. 1b at t=90, the trailing edge (z < 0) is nearly identical to that with d = 0; although, the leading edge remains noticeably different. Physically, the nonlinear diffusion process is due to the fact that the relative exchange of particles between adjacent cells (in space) progressively increases with velocity. This gives rise to a "shear" in particle flow in phase-space. The next effect of the density transport is a relative increase (decrease) in the tail of the velocity distribution in regions with negative (positive) density gradients (in the presence of an electric field) and a diffusion coefficient that depends on the gradient (Eq. (17b)).

CONCLUDING REMARKS

Nonequilibrium descriptions of the dynamics of electrons in gases under the influence of space-time varying fields have been presented. These descriptions are valid in different space-time scales; in particular, the macro-kinetic description is valid for macroscopic space-time scales which are determined from the characteristic scales of the moment equations. The macro-kinetic description has been developed for the case of a quasi-Lorentz gas model in the hydrodynamic regime. With this approach, it has been possible to investigate: 1) the range of validity of the density gradient expansion (DGE), 2) various approximate solutions for the distribution function in the hydrodynamic regime, and 3) the consequences of large density gradients. In particular, the DGE has been shown to be valid over most of the velocity range when the mean electron energy is less than the potential energy in a distance corresponding to the scale of density variation. Moreover, a solution for the distribution function has been found which in lowest order is space-time

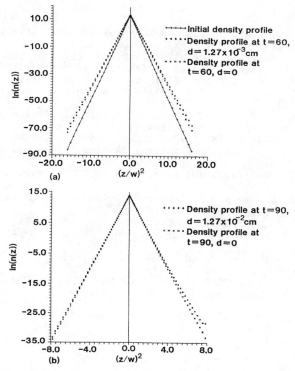

Fig. 1. Effect of non-linear diffusion ($d \neq 0$) on the evolution of an initial density profile. $D_o = 0.1 \mathrm{cm}^2/\mathrm{sec}$. In (a) $d = 5\times10^{-3}\mathrm{cm}$, and in (b) $d = 1.3\times10^{-2}\mathrm{cm}$.

dependent. This distribution has been used to derive a non-linear equation of continuity for density transport and to obtain expressions for the mobility and diffusion coefficient. It has been shown that, in lowest order, the effect of the non-linearity is to enhance (decrease) the diffusion in regions with negative (positive) slopes. These effects are more likely to be experimentally observed at large values of E/N in atomic gases. These conditions result in large values of the non-linearity parameter, d (Eq. (17d)). It is evident that the magnitude of the nonlinear effect depends on the magnitude of the density gradient and d. The values for the nonlinearity parameter and the profile chosen in this paper are such that Eq. (18) is weakly nonlinear.

ACKNOWLEDGMENT

     The author would like to thank M. C. Wang for assistance with the numerical calculations. This work is supported by the Office of Naval Research.

REFERENCES

Balescu, R., 1975, "Equilibrium and Non-equilibrium Statistical
    Mechanics," (Wiley, New York).
Blevin, H. A., and J. Fletcher, 1984, Aust. J. Phys. $\underline{37}$, 593.
Bogoluibov, N. N., 1962, "Studies in Statistical Mechanics," edited by J.
    deBoer and G. G. Uhlenbeck (Interscience, New York).
Chernov, L. A., 1960, "Wave Propagation in a Random Medium,"
    (McGraw-Hill, New York).
Huxley, L. G. H., and R. W. Crompton, 1974, "The Diffusion and Drift of
    Electrons in Gases," (Wiley, New York).
Kleban, P., and H. T. David, 1978, J. Chem. Phys. $\underline{68}$, 2999.
Klimontovich, Y. L., 1986, "Statistical Physics," (Harwood Academic,
    London).
Kumar, K., H. R. Skullerud, and R. E. Robson, 1980, Aust. J. Phys. $\underline{33}$,
    343.
Kunhardt, E. E., C. Wu, and B. Penetrante, 1988, Phys. Rev. $\underline{A37}$, 1654.
Lucas, J., 1970, Int. J. Electron. $\underline{29}$, 465.
Parker, Jr., J. H., and J. J. Lowke, 1969, Phys. Rev. $\underline{181}$, 290.
Penetrante, B. M., and J. N. Bardsley, 1984, J. Phys. $\underline{D}$: Appl. Phys. $\underline{17}$,
    1971.
Sandler, S. I., and E. A. Mason, 1969, Phys. Fluids $\underline{12}$, 71.
Sirovich, L., 1963, Phys. Fluids 6, 218.
Skullerud, H. R., and S. Kuhn, 1983, J. Phys. D: Appl. Phys. $\underline{16}$, 1225.
Tagashira, H., Y. Sakai, and S. Sakamoto, 1977, J. Phys. D: Appl. Phys.
    $\underline{10}$, 1051.
Thomas, W. R. L., 1969, J. Phys. B: Atom Molec. Phys. $\underline{2}$, 551.

NONEQUILIBRIUM EFFECTS IN ELECTRON TRANSPORT AT HIGH E/n

Y. M. Li

GTE Laboratories Incorporated
Waltham, MA 02254

The nonequilibrium effects in electron transport at high E/n are studied using the multibeam model. The reasoning leading to the model is explained. An analytically solvable multibeam model is utilized to illustrate the various features observed in the transient and steady ionization growth at high E/n. Finally, the failure of the concept of the effective field at high E/n for a high-frequency field-driven electron swarm is explained using the multibeam model.

INTRODUCTION

The transport properties of electron swarms in a neutral gas, driven by DC or high-frequency field of high electric-field-to-neutral-gas-density E/n, have raised considerable interest. First, both the break-down studies of gases on the left-hand side of the Paschen curve (Pace and Parker, 1973) and the simulation of self-breakdown in the low-pressure spark gap (Lauer et al., 1981) require detailed knowledge about the transient and steady-state electron impact ionization rate at high E/n. Second, the possibility of having an intense microwave pulse propagating through the atmosphere (Yee et al., 1986), creation of an artificial ionized layer in the atmosphere using crossed microwave beams (Gurevich, 1980), and the possible production of high-pressure nonequi-librium plasma for flue gas treatment using intense microwave power (Guest and Dandl, 1989) all point towards the necessity of having a better understanding of transport properties of electron swarm driven by high-frequency fields of high E/n.

From a fundamental research perspective, extensive experimental and theoretical studies for excitation and breakdown of N_2 and Ar at high E/n have been performed by Phelps and coworkers (Phelps et al., 1987;

Jelenkovic and Phelps, 1987; Phelps and Jelenkovic, 1988). These
studies, based on an electron drift tube with two parallel electrodes,
have shown that the single-beam and multibeam model for spatially
dependent growth of current and excitation rates at very high E/n provide
good descriptions of their experimental results. However, the presence
of electrodes, and the unavoidable secondary emissions, complicate the
interpretation of these results.

Direct determination of the ionization rates for N_2 from the
measurement of temporal growth of electron densities were performed by
Hays and coworkers (Hays et. al., 1987), without the presence of
electrodes. This was accomplished by using microwave heating of a weakly
ionized gas in the presence of a magnetic field. The strength of the
field is tuned to the electron cyclotron resonance, and the electron
swarm driven by the combined microwave field and magnetic field was shown
to be equivalent to the one driven by the DC electric field (Li and
Pitchford, 1989). In the same set-up, transient and steady-state
ionization rates in H_2 at very high E/n are measured and calculated (Hays
et al., 1989). Further discussions are given in DYNAMICS OF IONIZATION
GROWTH.

In principle, direct solution of the collisional Boltzmann equation
for the electron swarm is the only correct way to get distribution
function, excitation, and ionization rates. However, the collisional
Boltzmann equation at high E/n is notoriously difficult to solve, even
numerically, due to the highly anisotropic nature of the distribution and
the existence of the runaway electrons (Phelps and Pitchford, 1985a).
Thus Monte Carlo simulation is the only rigorous and reliable tool for
studying phenomena related to high E/n, for both DC and high-frequency
fields. Recently, Monte Carlo simulation has been treated extensively
(Kunhardt and Tzeng, 1986; Li et al., 1989; Li and Pitchford, 1989) for
high E/n applications and will not be discussed here. Instead, we
concentrate on the multibeam model, first studied by Müller (1962), which
is simple in concept and shows excellent agreement with Monte Carlo
calculations at high E/n (Pitchford and Li, 1988).

In MULTIBEAM MODEL: AN INTUITIVE APPROACH, the physical reasonings
leading to the multibeam model are explained and the relevant equations
are written. In DISTRIBUTION FUNCTION AND ENERGY BALANCE EQUATIONS, the
distribution function associated with the multibeam model is constructed.
Then the averages over the swarm are defined and the energy balance
equation is derived. In DYNAMICS OF IONIZATION GROWTH, various features
of the transient and steady ionization growth are reviewed. An analytically tractable multibeam model is solved, and the qualitative

features are predicted. Finally the ELECTRON SWARM DRIVEN BY HIGH-
FREQUENCY ELECTRIC FIELDS is examined. From the results of Monte Carlo
simulations, the concept of DC effective field is shown to be invalid at
high E/n. The multibeam model is utilized to provide the physical
understandings. A SUMMARY is given. In the Appendix, an attempt is made
to identify the relevant Boltzmann equation which gives the multibeam
model solution. The positive identification has not been made.

MULTIBEAM MODEL: AN INTUITIVE APPROACH

This section gives the equations for the multibeam model. The
assumptions used in arriving at these equations are explained. For
electrons driven by DC electric fields at high E/n, two important
simplifications are possible.

First, at sufficiently high energy, the cross sections for electron-
molecule collision fall with energy. With high E/n, the electrons gain
more energy from the field than dissipate through collisions, and the
runaway regime is attained. In this case, the electron–molecule collision can be considered a perturbation to the free-fall trajectories.
More precisely, the discrete nature of electron-molecule collisions and
the resulting discrete energy and momentum changes are averaged over and
are replaced by the continuous momentum and energy loss terms in the
equations of motion. Borrowing the concepts of "continuous slowing down
approximation" used for high-energy electron degradation calculations,
(Heaps and Green, 1974), the equations for the time evolution of momentum
and energy for a single electron can be written as

$$\frac{dv_o}{dt} = \frac{eE}{m} - \nu_m \, v_o \tag{1a}$$

$$\frac{d\varepsilon_o}{dt} = eE \, v_o - \nu_u \, \varepsilon_o, \tag{1b}$$

where v_o is the component of velocity in the direction of electric field
(or the drift velocity of the primary electron), ε_o is the energy, e and
m are the electron charge and mass, and ν_m and ν_u are the momentum and
energy relaxation frequencies.

The next question is, How do we relate ν_m and ν_u with the microscopic electron-molecule collision cross sections?

The ν_u depends on the total angularly-integrated cross sections Q_k^o
with the threshold energy ε_k for each of the k inelastic processes. The

elastic recoil energy loss, which depends on elastic momentum transfer cross section Q_{el}^m and the ratio of the electron to neutral mass m/M is also included. With $v = (2\varepsilon/m)^{1/2}$, ν_u is given by

$$\nu_u(\varepsilon) = nv \left[\frac{2m}{M} Q_{el}^m(\varepsilon) + \sum_k Q_k^o (\varepsilon)\frac{\varepsilon_k}{\varepsilon} \right]. \tag{2a}$$

The effect of anisotropic electron scattering is manifested in the momentum relaxation frequency. For isotropic scattering,

$$\nu_m(\varepsilon) = nv \left[Q_{el}^m(\varepsilon) + \sum_k Q_k^o \right] \tag{2b}$$

In the limit of forward scattering and in the Born approximation, the contribution from the inelastic collision can be expressed in terms of the energy loss function $L(\varepsilon)$ (Phelps and Pitchford, 1985) for inelastic collisions, and ν_m is given by

$$\nu_m(\varepsilon) = nv \left(Q_{el}^m(\varepsilon) + \frac{L(\varepsilon)}{2\varepsilon} \right) \tag{2c}$$

with

$$L(\varepsilon) = \sum_k Q_k^o(\varepsilon)\varepsilon_k. \tag{2d}$$

The second simplification is in the treatment of the secondary electron production. Secondary electrons are produced via electron impact ionization. Rigorously speaking, the ionization collision is a discrete random event producing a single secondary electron of certain energy with certain probability. In the multibeam model, the secondary electrons are created continuously in time by the primary electron with the rate $\nu_i [\varepsilon_o(t)]$ and $\nu_i (\varepsilon)$ given by

$$\nu_i (\varepsilon) = nv Q_i^o (\varepsilon), \tag{2e}$$

and $Q_i^o (\varepsilon)$ is the ionization collision cross section. In addition, secondary electrons are assumed to be produced with zero energy. In the presence of time-independent spatially uniform electric field, the secondary electrons will follow the trajectory of their parent, i.e.,

$$v_{sec}(t) = v_o(t-t')$$

$$\varepsilon_{sec}(t) = \varepsilon_o(t-t'),$$

where t′ is the time of birth.

This completely specifies the generation process for the secondary electrons, and we can develop an equation describing how the number of electrons grows in time.

Suppose the number of electrons is described by $n_e(t)$, with $n_e(0) = 1$ corresponding to a single primary electron at $t = 0$. The number of secondary electrons produced at time τ to $\tau + d\tau$ is given by $\left(\dfrac{dn_e}{d\tau}\right) d\tau$.

At time t, these electrons will have energy $\varepsilon_0 (t - \tau)$ and ionization rate $\nu_i [\varepsilon_0 (t - \tau)]$. The rate of electron production at t will be the sum of the rates of ionization of the primary and secondary electrons, and is given by

$$\frac{dn_e(t)}{dt} = \nu_i \left[\varepsilon_0(t)\right] + \int_0^t d\tau \left(\frac{dn_e}{d\tau}\right) \nu_i \left[\varepsilon_0(t-\tau)\right]. \tag{3a}$$

This is an integral equation for (dn_e/dt) from which $n_e(t)$ can be calculated.

It is interesting to note that Eq. (3a) can be simplified. Using the procedure of integration by parts, Eq. (3a) becomes

$$\frac{dn_e}{dt} = \int_0^t d\tau \, n_e(\tau) \, \frac{d}{dt} \, \nu_i \left[\varepsilon_0(t-\tau)\right]. \tag{3b}$$

It can be easily shown that the $\dfrac{d}{dt}$ and the integral operator are interchangeable. Further integration with respect to t gives

$$n_e(t) = 1 + \int_0^t d\tau \, n_e(\tau) \, \nu_i \left[\varepsilon_0(t - \tau)\right]. \tag{3c}$$

The set of Eqs. (1a), (1b), and (3c) forms a complete description of the electron swarm in terms of the multibeam model.

The multibeam model presented here differs from the original version of Müller (1962). In our treatment here, the motion of the primary electron is described by both energy and momentum equations, whereas Müller used only the momentum equation for $v_0(t)$, and energy $\varepsilon_0(t)$ is assumed to be $m_e v_0^2/2$. In the multibeam formulation described by Phelps et al. (1987), which was for determining spatial growth of current, only the energy equation was retained.

Since the main purpose of this work is to describe the multibeam model and its various implications, a simplified nitrogen (N_2) cross section from Phelps and Pitchford (1985b) is adopted for all subsequent numerical calculations. Also, only results for isotropic scattering are presented. With this cross section set, the ν_m and ν_u can be determined and are shown in Fig. 1.

DISTRIBUTION FUNCTION AND ENERGY BALANCE EQUATIONS

Once the electron swarm is specified for a given E/N, it is natural to ask for the electron energy distribution function $f(\varepsilon,t)$ for the swarm. From the multibeam model, $f(\varepsilon,t)$ can be constructed explicitly

$$f(\varepsilon,t) = \frac{1}{n_e(t)} \left\{ \int_0^t d\tau \frac{dn_e(\tau)}{d\tau} \delta\left[\varepsilon-\varepsilon_0(t-\tau)\right] + \delta\left[\varepsilon-\varepsilon_0(t)\right] \right\}, \quad (4a)$$

where δ is the Dirac delta function. In Eq. (4a), the first term represents the secondary electron produced in time interval t, and the second term is the primary electron.

The validity of such construction can be understood as follows. First, $f(\varepsilon,t)$ satisfies the normalization condition

$$\int_0^\infty d\varepsilon \; f(\varepsilon,t) = 1. \quad (5)$$

Second, averages over the swarm, such as average energy $\overline{\varepsilon}(t)$, can be defined as usual:

$$\overline{\varepsilon}(t) = \int_0^\infty d\varepsilon \; \varepsilon \; f(\varepsilon,t)$$

$$= \frac{1}{n_e(t)} \left[\int_0^t d\tau \frac{dn_e(\tau)}{d\tau} \varepsilon_0(t-\tau) + \varepsilon_0(t) \right]. \quad (6a)$$

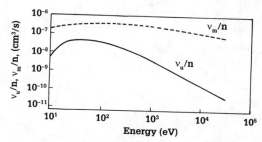

Fig. 1. Density normalized momentum relaxation frequency (ν_m/n) and energy relaxation frequency (ν_u/n) as functions of electron energy.

Equation (6a) has a simple physical interpretation. The average energy $\bar{\varepsilon}$ at time t is the sum of energies of the primary $\varepsilon_o(t)$ and all the secondaries produced at time $\tau < t$ acquiring energies $\varepsilon_o(t - \tau)$, divided by the total number of electrons $n_e(t)$. Similarly, average excitation rates $\bar{\nu}_k$, ionization rate $\bar{\nu}_i$, and drift velocity $\bar{v}_d$ are defined by

$$\bar{\nu}_{k,i}(t) = \frac{1}{n_e(t)} \left[\int_0^t d\tau \, \frac{dn_e(\tau)}{d\tau} \, \nu_{k,i}\left[\varepsilon_o(t - \tau)\right] + \nu_{k,i}\left[\varepsilon_o(t)\right] \right] \qquad (6b)$$

$$\bar{v}_d(t) = \frac{1}{n_e(t)} \left[\int_0^t d\tau \, \frac{dn_e(\tau)}{d\tau} \, v_o(t - \tau) + v_o(t) \right], \qquad (6c)$$

where

$$\nu_{k,i}(\varepsilon) = nv \, Q_{k,i}^o(\varepsilon) \text{ and } v = (2\varepsilon/m)^{1/2}.$$

Now, some of the numerical outputs of the multibeam model, $v_o(t)$, $\varepsilon_o(t)$, $\nu_i[\varepsilon_o(t)]/n$, $\bar{v}_d(t)$, $\bar{\varepsilon}(t)$, and $\bar{\nu}_i(t)/n$ will be examined. In Figs. (2a) to (2c), results are plotted for E/n = 1 kTd, where 1 kTd is 10^{-14} V cm^2. The $v_o(t)$, $\varepsilon_o(t)$, and $\nu_i[\varepsilon_o(t)]/n$ for the primary electron are attaining steady-state values, and the corresponding averages are also shown. The overshoots in $v_o(t)$ and $\bar{v}_d(t)$ are clearly seen. The results for E/n = 10 kTd are given in Figs. (3a) to (3c). The primary electron is running away; $v_o(t)$ and $\varepsilon_o(t)$ are increasing indefinitely with time. The $\nu_i[\varepsilon_o(t)]/n$ for the primary electron attains maximum value at energy about 365 eV, and decreases monotonically with increasing energy (time).

105

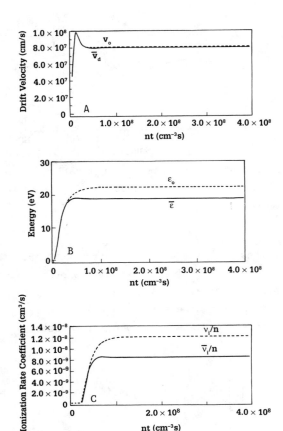

Fig. 2. Numerical results for E/n = 1 kTd as functions of nt.
A. Drift velocity of the primary electron and the average
drift velocity $\bar{v}_d$. B. Energy of the primary electron ε_o
and the average energy $\bar{\varepsilon}$. C. Ionization rate coefficient
of the primary electron v_i/n and the average ionization
rate coefficient $\bar{v}_i$/n.

However, the averages $\bar{\varepsilon}(t)$, $\bar{v}_d(t)$, and $\bar{v}_i(t)/n$ are attaining well-defined
steady-state values. This illustrates the important fact that in the
runaway regime, the steady-state distribution is mainly composed of
transient distributions of younger generations of secondary electrons.
Without the secondary electron production, meaningful steady-state
averages cannot be defined.

   With the averages defined, an energy-balance equation relating $\bar{\varepsilon}$,
$\bar{v}_d$, $\bar{v}_k$, and $\bar{v}_i$ is obtained. An outline of the derivation is given as
follows. To begin with, we have

$$\frac{d}{dt}(n_e\bar{\varepsilon}) = \bar{\varepsilon}\frac{dn_e}{dt} + n_e\frac{d\bar{\varepsilon}}{dt}. \qquad (7a)$$

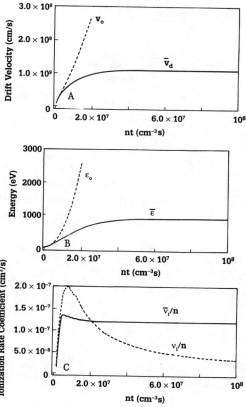

Fig. 3. Numerical results for $E/n = 10$ kTd as functions of nt.
A. Drift velocity of the primary electron v_o and the
average drift velocity $\bar{v}_d$. B. Energy of the primary
electron ε_o and the average energy $\bar{\varepsilon}$. C. Ionization rate
coefficient of the primary electron ν_i/n and the average
ionization rate coefficient $\bar{\nu}_i/n$.

Using the definition for $\bar{\varepsilon}$, we also have

$$\frac{d}{dt}(n_e\bar{\varepsilon}) = \frac{d\varepsilon_o}{dt} + \int_0^t d\tau \frac{dn_e(\tau)}{d\tau} \frac{d}{dt}\varepsilon_o(t - \tau). \tag{7b}$$

Equation (1b) for $d\varepsilon_o/dt$ can be used to simplify Eq. (7b) to give

$$\frac{d}{dt}(n_e\bar{\varepsilon}) = n_e\left[e E \bar{v}_d - \sum_k \varepsilon_k \bar{\nu}_k - \frac{2m}{M} \bar{\nu}_{el} \bar{\varepsilon} \right], \tag{7c}$$

where $\bar{\nu}_{el}(t)$ for elastic collision is defined by

107

$$n_e \bar{\varepsilon}(t) \ \bar{v}_{el}(t) = \varepsilon_o(t) \ v_{el}\left[\varepsilon_o(t)\right]$$

$$+ \int_0^t d\tau \ \frac{dn_e}{d\tau} \ \varepsilon_o \ (t - \tau) \ v_{el}\left[\varepsilon_o(t - \tau)\right] \tag{7d}$$

where

$$v_{el}(\varepsilon) = nv \ Q_m^o(\varepsilon).$$

Notice that the definition for $\bar{v}_{el}(t)$ [Eq. (7d)] differs from the definition for $\bar{v}_k(t)$. Similar situations arise when deriving an energy balance equation from the two-term Boltzmann equation. From Eqs. (3a) and (6b) it is easy to see

$$\frac{dn_e}{dt} = \bar{v}_i(t)n_e. \tag{7e}$$

Thus combining Eqs. (7a), (7c), and (7e), we have

$$\frac{d\bar{\varepsilon}}{dt} = eE \ \bar{v}_d - \sum_k \varepsilon_k \bar{v}_k - \frac{2m}{M} \ \bar{v}_{el} \ \bar{\varepsilon} - \bar{v}_i \ \bar{\varepsilon}. \tag{8a}$$

Equation (8a) is the usual form of energy balance equation, and the summation over k includes the ionization channel. In steady state ($d\bar{\varepsilon}/dt = 0$), we have

$$eE \ \bar{v}_d = \sum_k \varepsilon_k \bar{v}_k + \frac{2m}{M} \ \bar{v}_{el}\bar{\varepsilon} + \bar{v}_i \ \bar{\varepsilon}. \tag{8b}$$

Based on the above discussion, the distribution function $f(\varepsilon, t)$, the power dissipated into various inelastic channels ($\varepsilon_k \ \bar{v}_k$) and power required to bring the secondary electrons to the average energy ($\bar{v}_i \bar{\varepsilon}$) can be computed numerically. It turns out that it is much easier to determine $f(\varepsilon, t)$ using an alternate formula:

$$f(\varepsilon, t) = \frac{1}{n_e(t)} \left\{ \frac{\left(\frac{dn_e}{d\tau}\right) \tau = \tau_o}{\left|\frac{d\varepsilon_o}{dt'}\right| t' = t - \tau_o} + \delta\left[\varepsilon - \varepsilon_o(t)\right] \right\}, \tag{4b}$$

where $\varepsilon = \varepsilon_o(t - \tau_o)$.

Computations based on Eq. (4b) are very convenient once $n_e(t)$ and $\varepsilon_o(t)$ are obtained. The numerical normalization is used as a check for the numerical algorithm and found to be one within 0.2% for all the

108

calculations performed. Some results for $f(\varepsilon,t)$ are given in Figs. (4a) and (4b).

Note that the spiky behavior associated with the δ function distribution for the primary electron is ignored, and $f(\varepsilon,t)$ is determined for three different times. In Fig. (4a), $E/n = 1$ kTd, and the primary electron attains a final energy at around 23 eV. The steady distribution functions computed at three different times are identical. This is not a realistic distribution where electrons are piling up around 22 eV with little at lower energies. Thus the multibeam model fails at such low E/n. In Fig. (4b) the steady-state distributions for $E/n = 10$ kTd at three different times are plotted. The bulk of the distribution functions are identical (and $\bar{\varepsilon}$, ν_i, etc., attain steady values), whereas the primary electron acquires higher energy as time increases. The sudden drop in $f(\varepsilon,t)$ indicates the maximum energies of the primary acquired in three different times. Finally, in Fig. (5), the steady-state ratio of $\bar{\nu}_i\bar{\varepsilon}$ to the power input $e\,E\,\bar{v}_d$ is plotted as a function of E/n. Above 10 kTd, over 96% of the power input is dissipated in bringing the secondary electrons to the average energy.

DYNAMICS OF IONIZATION GROWTH

The multibeam model provides the simplest approach to study the dynamics of ionization growth. We first describe the experimental observations and some of the underlying reasoning, then we will demonstrate that all the observed features can be predicted by an analytically solvable multibeam model.

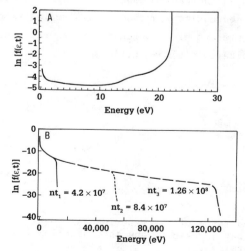

Fig. 4.   Steady-state electron energy distribution function for A. $E/n = 1$ kTd, B. $E/n = 10$ kTd at three different times.

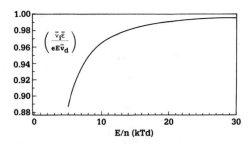

Fig. 5.   Fractions of power dissipated in bringing the secondary
          electrons to the average energy vs. E/n.

In the experimental set-up (Hays et al., 1987) to measure ionization
rate coefficient, the time-dependent electron densities are recorded, and
a schematic of the result is shown in Fig. (6). Two distinct regions are
seen: an exponential growth region at late times, and an early
nonexponential region, which is due to the finite response time needed
for the electron energy distribution function to reach the equilibrium
for a given E/n. Two important parameters are extracted from Fig. (6),
the steady-state ionization rate $\bar{\nu}_i$ and the induction time τ_s, which is a
convenient measure of the finite response time for the electron distribution function. The variations of $\bar{\nu}_i/n$ and $n\tau_s$ with E/n are the main
concern here. Such variations have been calculated numerically using the
multibeam model and plotted in Figs. (8a) and (8b). Recently, measurements for $\bar{\nu}_i/n$ and $n\tau_s$ as functions of E/N in hydrogen were performed,
(Hays et al., 1989), and similar qualitative behavior was observed. From
these figures two distinct features are clearly seen.

First, ν_i/n is an increasing function with E/n at lower E/n and
switches over to a decreasing function. This reflects the fact that as
E/n increases, the average electron energy moves past the energy at which
electron impact ionization is most efficient.

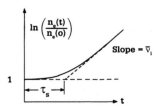

Fig. 6.   A schematic of the experimental results from which the
          ionization rate $\bar{\nu}_i$ and the induction time τ_s are
          determined.

110

Second, $n\,\tau_s$ decreases monotonically with E/n and turns negative at
sufficiently high E/n. The negative induction time is due to a strong
overshoot in the transient ionization rate coefficient, as illustrated in
Figs. (7a) to (7c). Here the ν_i/n, $\bar{\nu}_i/n$, and $\ln\,[n_e(t)]$ calculated using
the multibeam model for E/n = 5 kTd and 25 kTd are shown.

The strong overshoot in ionization rate coefficient $\bar{\nu}_i(t)/n$ for E/n =
25 kTd in Fig. (7b) leads to the negative induction time in Fig. (7c). From
Fig. (7b), the maximum value of the transient ionization rate coefficient
$\bar{\nu}_i(t)/n$ is 1.66×10^{-7} cm^3/s, which is 37% higher than the maximum steady-
state ionization rate coefficient given by 1.21×10^{-7} cm^3/s which occurred
at E/n = 14 kTd. This indicates the strong nonequilibrium effect associated
with the transient distribution function and cannot be described in terms of
the successive evolution of a series of steady-state distributions. Thus,
it is important that such transient distributions can be understood
qualitatively and calculated quantitatively.

Now we are going to construct an analytically solvable multibeam
model which describes the transient and steady-state behavior of the
ionization growth. The model based on Eqs. (1a), (1b), and (3c) is
intractable. To make progress, instead of solving the Eqs. (1a) and (1b)
to get $\varepsilon_o(t)$ and $\nu_i[\varepsilon_o(t)]$, a specific form of $\nu_i[\varepsilon_o(t)]$ is assumed:

$$\nu_i\left[\varepsilon_o(t)\right] = \nu_o(e^{-\alpha_1 t} - e^{-\alpha_2 t}) \tag{5a}$$

with $\alpha_2 > \alpha_1$,

$$\nu_o = \nu_{max}\left/\left\{\left[\frac{\alpha_2}{\alpha_1}\right]^{-\left[\frac{\alpha_1}{\alpha_2-\alpha_1}\right]} - \left[\frac{\alpha_2}{\alpha_1}\right]^{-\left[\frac{\alpha_1}{\alpha_2-\alpha_1}\right]}\right\}\right. \tag{5b}$$

and ν_{max}/n is given by 2×10^{-7} cm^3/s.

According to Eqs. (5a) and (5b), the primary electron ionization
rate rises from zero to maximum value ν_{max} in a short time scale
($\sim 1/\alpha_2$) and decays exponentially in a longer time scale ($\sim 1/\alpha_1$).
Fig. (7a) clearly demonstrates this qualitative behavior, and in
addition, both the rise and decay times decrease with increasing E/n. To
completely specify the model, the following relations for α_1 and α_2 as
functions of E/n are used:

$$\alpha_1/n = \alpha_{10}/n \; (E/n/10)^\beta \tag{6a}$$

$$\alpha_2/n = \alpha_{20}/n \; (E/n/10)^\beta, \tag{6b}$$

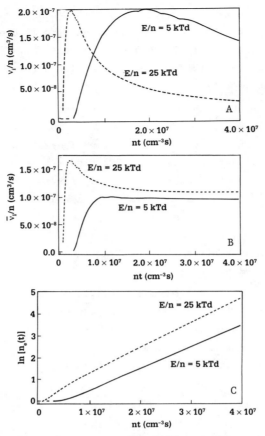

Fig. 7. Numerical results vs. nt for different E/n. A. Ionization
rate coefficient of the primary electron ν_i/n. B. Average
ionization rate coefficient $\bar{\nu}_i/n$. C. Natural logarithm of the
total number of electron $\ln[n_e(t)]$ and negative induction
time obtained for E/n = 25 kTd.

where $\alpha_{10}/n = 4.3 \times 10^{-8}$ cm^3/s, $\alpha_{20}/n = 3 \times 10^{-7}$ cm^3/s and E/n is in
units of kTd. β is considered as a free parameter and in the subsequent
calculations, β is taken as 1, 1.5, and 2. The values of α_{10} and α_{20}
are determined approximately from $\nu_i[\varepsilon_o(t)]$ for E/n = 10 kTd, which is in
turn obtained from the numerical multibeam model.

The integral equation given by Eqs. (3c) and (5a) can be solved
analytically by taking the Laplace transform

$$n_e(s) = \frac{1}{s} + \nu_i(s)\, n_e(s), \tag{7}$$

where

$$f(s) = \int_0^\infty dt \ e^{-st} f(t)$$

$$\nu_i(s) = \nu_0 \left[\frac{1}{(s + \alpha_1)} - \frac{1}{(s + \alpha_2)} \right],$$

and $n_e(s)$ can be determined by

$$n_e(s) = \frac{1}{s\left[1 - \nu_i(s)\right]}. \tag{8}$$

Following the usual procedures of partial fractions, $n_e(s)$ can be simplified and $n_e(t)$ can be solved analytically by taking the inverse Laplace transform. The result is

$$n_e(t) = 1 - D_1(1 - e^{-B_1 t}) - D_2(1 - e^{-B_2 t}), \tag{9a}$$

where

$$D_1 = \frac{2\ \nu_0(\alpha_2 - \alpha_1)}{P_1^2 R\ (R - 1)} \tag{9b}$$

$$D_2 = \frac{2\ \nu_0(\alpha_2 - \alpha_1)}{P_1^2 R\ (R + 1)} \tag{9c}$$

$$P_1 = \alpha_1 + \alpha_2 \tag{9d}$$

$$P_2 = \alpha_2\ (\nu_0 - \alpha_1) - \nu_0 \alpha_1 \tag{9e}$$

$$R = \sqrt{1 + \frac{4\ P_2}{P_1^2}} \tag{9f}$$

$$B_1 = \frac{P_1}{2}\ (1-R) \tag{9g}$$

$$B_2 = \frac{P_2}{2}\ (1+R). \tag{9h}$$

Notice that we have $\alpha_2 \gg \alpha_1$, $P_2 \sim \nu_0\ (\alpha_2 - \alpha_1) > 0$, and $R > 1$; thus $D_1 > 0$, $D_2 > 0$, $B_1 < 0$, and $B_2 > 0$ follows. The exponentially growing term in Eq. (9a) is represented by $(D_1\ e^{-B_1 t})$, and the average ion rate $\bar{\nu}_i$ in steady state is given by

$$\bar{\nu}_i = -\ B_1. \tag{10a}$$

Suppose the $n_e(t)$ is represented by $D_1 e^{-\bar{\nu}_i(t-\tau_s)}$ for late times. The extrapolation back to $t = 0$ gives

$$D_1 e^{\bar{\nu}_i \tau_s} = 1,$$

and τ_s is solved to be

$$\tau_s = \frac{1}{\bar{\nu}_i} \ln \left(\frac{1}{D_1} \right). \tag{10b}$$

Thus the steady-state ionization rate $\bar{\nu}_i$ and the induction time τ_s are determined in terms of α_1, α_2, and ν_{max}. The subsequent E/n variations of $\bar{\nu}_i/n$ and $n\tau_s$ reflect those of α_1 and α_2 given by Eqs. (6a) and (6b), and the results are given in Fig. (8a) and (8b) for $\beta = 1$, 1.5, and 2.

Corresponding results based on the numerical solutions of the multibeam model [Eqs. (1a), (1b), and (3c)] are also given.

The general behavior of the $\bar{\nu}_i/n$ and $n\tau_s$ with E/n agrees quite well with the full numerical multibeam model results. However, the functional form of ν_i given by Eq. (5a) cannot accurately mimic the actual ν_i, thus some quantitative discrepancies between the analytic model and the numerical results for $\bar{\nu}_i/n$ are not surprising. However, the good agreements for $n\tau_s$ at higher E/n with $\beta = 1$ may indicate that the rise and decay times for $\nu_i(t)$ are very much inversely proportional to E/n.

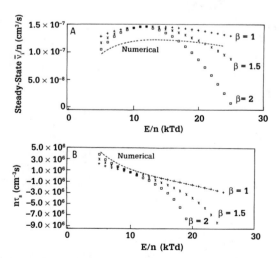

Fig. 8. Comparison between the analytic models of different β and the numerical results. A. Steady-state average ionization rate coefficient $\bar{\nu}_i/n$. B. Density-normalized induction time $n\tau_s$ as functions of E/n.

ELECTRON SWARM DRIVEN BY HIGH-FREQUENCY ELECTRIC FIELDS

For an electron swarm driven by a high-frequency electric field (E = E_o sin ωt) of low E/n, the distribution function can be adequately determined from the two-term expansion of the time-dependent Boltzmann equation (Allis, 1956). Transport properties can be calculated accordingly. If the frequency (ω) of the field is much larger than the energy relaxation frequency (ν_u), a time average over the period of the field ($2\pi/\omega$) can be performed, and a two-term approximated Boltzmann equation with the DC effective field (E_{eff}) is obtained. The E_{eff} is given by

$$E_{eff} = \frac{E_o}{\sqrt{2}} \frac{\nu_m(\varepsilon)}{\left[\omega^2 + \nu_m^2(\varepsilon)\right]^{1/2}}. \tag{11}$$

In summary, the calculation of $\bar{\varepsilon}$, $\bar{\nu}_k$, $\bar{\nu}_i$ for an electron swarm driven by high-frequency electric field can be converted to the one driven by a DC effective field E_{eff} given in Eq. 11. For detailed derivations of the E_{eff}, in the presence or absence of a static magnetic field and arbitrary polarization of the electric field, see the work of Allis (1956), Hays et al. (1987) and Li and Pitchford (1989). However, the validity of the concept of effective field at high E/n remains an open question.

Using Monte Carlo simulation, the effective field was first shown to be invalid for an electron swarm driven by a high-frequency electric field at high E/n (Li and Pitchford, 1989). The transport properties $\bar{\varepsilon}$, $\bar{\nu}_i/n$, and Θ/n (Θ is the time average absorbed power per electron) were determined for a swarm driven by a high-frequency field (HF) and by DC effective field (DC). The percentage deviation of the quantity ψ is given by

$$\Delta\psi = \frac{\psi(HF) - \psi(DC)}{\psi(HF)} \times 100\%. \tag{12}$$

The results of the percentage deviations are shown in Fig. 9. It is clear that $\bar{\varepsilon}$, $\bar{\nu}_i/n$, and Θ/n are higher in the HF case than in the corresponding DC effective field case, and the percentage deviation increases with increasing E_{eff}/n.

For the swarm driven by a high-frequency field, the secondary electrons produced at time τ will see an electric field E_o sin $\omega\tau$ and will not follow the trajectory of the primary electron. Thus the multibeam model cannot be applied directly. If $\omega \gg \nu_m$, such that there are a number of field oscillations between two momentum transfer collisions, the initial phase of the secondary electron relative to the field can be averaged over, so that both the primary and the secondary electrons follow the same initial-phase-averaged (or cycle-averaged) trajectory.

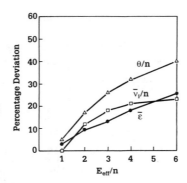

Fig. 9.   The percentage deviations of average energy $\bar{\varepsilon}$, average
ionization rate coefficient $\bar{\nu}_i/n$ and density normalized
absorbed power θ/n vs. E_{eff}/n.

Then the multibeam model is again applicable and utilized to explain the
difference between the high-frequency-driven swarm and the DC effective
field-driven swarm. To carry the analysis further, let us consider
motion of the primary electron as given by Eqs. (1a) and (1b) with E
given either by (1) $E = E_o \sin \omega t$ or (2) $E = E_{eff}$. If ν_m is constant,
the equation for v_o can be solved with $v_o(0) = 0$. For the HF case:

$$v_o(t) = \frac{e\, E_o}{m\left(\nu_m^2 + \omega^2\right)} \left(\nu_m \sin \omega t - \omega \cos \omega t + \omega\, e^{-\nu_m t}\right). \qquad (13a)$$

For the DC case:

$$v_o(t) = \frac{e\, E_{eff}}{m\, \nu_m}\left(1 - e^{-\nu_m t}\right). \qquad (13b)$$

Substituting Eqs. (13a) and (13b) into Eq. (1b), the corresponding
energy equations for HF and DC are

$$\frac{d\,\varepsilon_o}{dt} = \frac{e\, E_o^2 \sin \omega t}{m\left(\omega^2 + \nu_m^2\right)}\left[\nu_m \sin \omega t - \omega \cos \omega t + \omega\, e^{-\nu_m t}\right] - \nu_u \varepsilon_o \qquad (14a)$$

$$\frac{d\,\varepsilon_o}{dt} = \frac{e\, E_{eff}^2}{m\nu_m}\left(1 - e^{-\nu_m t}\right) - \nu_u \varepsilon_o. \qquad (14b)$$

Suppose $\omega \gg \nu_m$ and Eq. (14a) is averaged over one cycle. It
simplifies to

$$\frac{d\langle\varepsilon_o\rangle}{dt} = \frac{e\, E_{eff}^2}{m\nu_m}\left[1 + \frac{1}{\pi}\left(\frac{\omega}{\nu_m}\right)\left(\frac{\omega^2}{\omega^2+\nu_m^2}\right)\left(1 - e^{-\frac{2\pi\nu_m}{\omega}}\right)e^{-\nu_m t} - \nu_u\langle\varepsilon_o\rangle\right], \quad (15)$$

where $\langle\varepsilon_o\rangle$ is the cycle averaged ε_o.

116

By comparing between Eqs. (15) and (14a), and in the runaway

regime $\left(\dfrac{eE_{eff}^2}{m\nu_m} > \nu_u \varepsilon_o \right)$, three interesting facts are quite apparent:

1. For $t \gg 1/\nu_m$, $d\varepsilon_o/dt$ is the same for both HF and DC. The above analysis is essentially the derivation of effective field formula [Eq. (11)], using the single electron picture.

2. The transient term associated with $e^{-\nu_m t}$ enhances the rate of gaining energy for the HF case [Eq. (15)], but reduces the rate for the DC case [Eq. (14a)].

3. For $t \sim 1/\nu_m$, the electron in the HF case will gain more energy than in the DC case, thus $\langle\varepsilon_o(t)\rangle$ for HF is higher than $\varepsilon_o(t)$ for DC. The most pronounced differences in energies occur in this transient regime.

The physical reason for the failure of the DC effective field at high E/n can be explained as follows. As shown in Figs. (3a) and (3b), the majority of the population of the steady-state distribution is always consisting of secondary electrons of the "younger" generations, such that the energies they had acquired after their birth are close to the steady-state average energy $\bar{\varepsilon}$. Since the electron driven by a high-frequency field gains much higher transient energy, it will naturally give rise to a steady state distribution of higher average energy and ionization rate.

Furthermore, the transient term in Eq. (15) is an increasing function of ω/ν_m. For the same effective field, the swarm driven by the electric field of higher frequency (higher ω/ν_m) will have higher average energy and ionization rate. This conclusion has also been verified using Monte Carlo simulation (Li and Pitchford, 1989).

SUMMARY

In this work, the multibeam model is examined and applied to two interesting cases. First, the transient and steady-state ionization growth at high E/n is studied by a tractable multibeam model. Second, the concept of the DC effective field for swarms driven by high frequency electric field of high E/n is demonstrated to be invalid.

ACKNOWLEDGEMENTS

Much of the work presented here was obtained in collaboration with L.C. Pitchford, and her constant support in the past few years is gratefully acknowledged. The author would also like to acknowledge helpful discussions with G.B. Hays, J.B. Gerardo, A.V. Phelps, and J.T. Verdeyen.

APPENDIX

This is not a successful attempt, but the effort recorded here may stimulate further work. The question we try to answer is whether one can find a Boltzmann-like kinetic equation whose solution is the distribution function generated by the multibeam model. A reasonable kinetic equation is given as follows:

$$\frac{\partial f}{\partial t} + \frac{d\varepsilon_o}{dt} \frac{\partial f}{\partial \varepsilon} = \delta(\varepsilon) \int_o^\infty d\varepsilon' \nu_i(\varepsilon') f(\varepsilon',t), \tag{A1}$$

where $\varepsilon_o(t)$ satisfies the following equations:

$$\frac{d\varepsilon_o(t)}{dt} = eE\, v_o(t) - \nu_u \varepsilon_o(t) \tag{A2}$$

and

$$\frac{dv_o(t)}{dt} = \frac{eE}{m} - \nu_m v_o(t) \tag{A3}$$

Let us now construct the solution for Eq. (A1). The homogeneous solution (where the ionization term of the right-hand side is set to zero) with the initial condition $f(\varepsilon,0) = \delta(\varepsilon)$ corresponds to the evolution of a single primary electron with zero energy is given by

$$f_p(\varepsilon,\ t) = \delta\left[\varepsilon - \varepsilon_o(t)\right]. \tag{A4}$$

The particular solution corresponding to the evolution of the continuously produced secondary electrons is assumed to be in form of

$$f_s(\varepsilon,\ t) = \Theta(\varepsilon)\Theta(\xi)\, A(\xi), \tag{A5}$$

where Θ is the step function and

$$\xi = \varepsilon_o(t) - \varepsilon, \tag{A6}$$

where function A has to be specified.

Substituting f_s back into Eq. (A1), we have

$$\delta(\varepsilon)\Theta(\xi)A(\xi)\frac{d\varepsilon_o}{dt} = \delta(\varepsilon) \int_o^\infty d\varepsilon' \nu_i(\varepsilon') f(\varepsilon',t),$$

which implies

$$A\left[\varepsilon_o(t)\right]\frac{d\varepsilon_o}{dt} = \int_o^\infty d\varepsilon' \nu_i(\varepsilon') f(\varepsilon',t). \tag{A7}$$

Since $f(\varepsilon, t) = f_p(\varepsilon, t) + f_s(\varepsilon, t)$, and f_s is given by Eq. (A5). Eq. (A7) becomes an integral equation for $A[\varepsilon_o(t)]$:

$$A[\varepsilon_o(t)] \, \frac{d\varepsilon_o}{dt} = \nu_i[\varepsilon_o(t)] + \int_o^\infty d\varepsilon' \, \nu_i(\varepsilon') \theta(\varepsilon_o(t) - \varepsilon') A(\varepsilon_o(t) - \varepsilon'). \quad (A8)$$

Let $\varepsilon_o(t) - \varepsilon' = \varepsilon_o(\tau)$, Eq. (A8) becomes

$$A[\varepsilon_o(t)] \, \frac{d\varepsilon_o}{dt} = \nu_i[\varepsilon_o(t)] + \int_o^t d\tau \, \frac{d\varepsilon_o}{d\tau} A[\varepsilon_o(\tau)] \nu_i[\varepsilon_o(t) - \varepsilon_o(\tau)]. \quad (A9)$$

Using the definition

$$n_e(t) = \int_o^\infty d\varepsilon \, f(\varepsilon, t), \quad (A10)$$

and integrating Eq. (A1) with respect to ε, we have

$$\frac{dn_e}{dt} = \int_o^\infty d\varepsilon' \, \nu_i(\varepsilon') f(\varepsilon', t). \quad (A11)$$

Comparing Eq. (A11) and Eq. (A7), the following identification can be made:

$$\frac{dn_e}{dt} = A[\varepsilon_o(t)] \, \frac{d\varepsilon_o}{dt}. \quad (A12)$$

Eq. (A9) can be rewritten as

$$\frac{dn_e}{dt} = \nu_i[\varepsilon_o(t)] + \int_o^t d\tau \left(\frac{dn_e}{d\tau}\right) \nu_i[\varepsilon_o(t) - \varepsilon_o(\tau)] \quad (A13)$$

Notice that Eq. (A13) is very similar to Eq. (3a). However, the kernals of the integral equations, $\nu_i[\varepsilon_o(t) - \varepsilon_o(\tau)]$ and $\nu_i[\varepsilon_o(t - \tau)]$ are different. Thus a positive identification of the kinetic equation which the multibeam model obeys has not been found yet.

REFERENCES

Allis, W. P., 1956, "Handbuch der Physik," Flugge, S., (Ed.), Springer, Berlin, 21, 383-444.

Guest, G. E., and R. A. Dandl, 1989, Plasma Chem. Plasma Process. 9, 55S.

Gurevich, A. V., 1980, Sov. Phys. Usp. 23, 862.

Hays, G. N., J. B. Gerardo, L. C. Pitchford, Y. M. Li, and J. T. Verdeyen, 1989, private communication.

Hays, G. N., L. C. Pitchford, J. B. Gerardo, J. T. Verdeyen, and Y. M. Li, 1987, Phys. Rev. A 36, 2031.

Heaps, M. G., and A. E. S. Green, 1974, J. Appl. Phys. 45, 3183.

Jelenkovic, B. M., and A. V. Phelps, 1987, Phys. Rev. A 36, 5310.

Kunhardt, E. E., and Y. Tzeng, 1986, J. Comput. Phys. 67, 279.

Lauer, E. J., S. S. Yu, and D. M. Cox, 1981, Phys. Rev. 23, 2250.

Li, Y. M., and L. C. Pitchford, 1989, Bull. Am. Phys. Soc. 34, 306.

Li, Y. M., L. C. Pitchford, and J. T. Moratz, 1989, Appl. Phys. Lett. 54, 1403.

Müller, K. G., 1962, Z. Phys. 169, 432.

Pace, J. D., and A. B. Parker, 1973, J. Phys. D. 6, 1525.

Phelps, A. V., and B. M. Jelenkovic, 1988, Phys. Rev. A 38, 2975.

Phelps, A. V., B. M. Jelenkovic, and L. C. Pitchford, 1987, Phys. Rev. A 36, 5327.

Phelps, A. V., and L. C. Pitchford, 1985a, Phys. Rev. A 31, 2932.

Phelps, A. V., and L. C. Pitchford, 1985b, JILA Information Report No. 26, University of Colorado, Boulder, CO.

Pitchford, L. C., and Y. M. Li, 1988, Bull. Am. Phys. Soc. 33, 136.

Yee, J. H., R. A. Alvarez, D. J. Mayhall, D. P. Bryne, and J. DeGroot, 1986, Phys. Fluids 29, 1238.

ELECTRON COLLISION CROSS SECTIONS FOR PROCESSING PLASMA GASES FROM SWARM
AND DISCHARGE DATA

Larry E. Kline

Westinghouse Science and Technology Center
Pittsburgh, PA 15235

INTRODUCTION

   Plasma etching and deposition are key processes in integrated circuit
manufacturing. The important microscopic physical processes in plasma processing discharges include 1) electron impact dissociation and ionization,
2) gas phase chemical reactions by ions and neutrals and 3) surface chemical
reactions including ion reactions. Process modeling is widely used in the
design of both integrated circuits themselves and in the design of integrated circuit manufacturing processes, and empirical models of plasma
processes are currently in use. However, there has been limited use of
plasma process simulation models which are based on the solution of the
fundamental equations which describe the underlying, physical processes.
Kline and Kushner (1989) describe the current state of the art in the
formulation and validation of physically based, first principles, models for
plasma etching deposition processes. They discuss modeling of the electron,
heavy particle and surface kinetics. Boeuf and Belenguer (to be published)
have recently reviewed discharge kinetic models. The limited use of first
principle models is due, in part, to the limited availability of the data
needed for such models. The needed data include electron cross sections,
electron and ion transport data and data for heavy particle gas phase and
surface chemical reaction rates.

   This chapter describes recent progress toward the determination of the
electron collision cross section and electron transport data for some of the
gases which are used in plasma processing. The rest of the chapter is
organized as follows: The next section describes some typical experimental
data for plasma processing discharges. This experimental data serves to
establish the range of electric field-to-gas density (E/N) values which are
experimentally observed and give an indication of the range of cross section

Nonequilibrium Effects in Ion and Electron Transport
Edited by J. W. Gallagher et al., Plenum Press, New York, 1990

data needed to model plasma processing discharges. The third section describes two basic approaches which have been used to calculate electron energy distributions (EED's) and discusses several aspects of the calculated EED's. Next, some of the various approaches which have been used to model plasma processing discharges are briefly described. This description of plasma processing discharge models provides additional information about the type of data needed. The last section describes some of the techniques being used to measure electron cross section and transport data for plasma processing gases and briefly reviews some of the available data.

DISCHARGE EXPERIMENTAL DATA

Many different types of discharges are used for plasma processing. The important experimental parameters include the gas pressure, mixture and flow rate, the applied electrical power, frequency and voltage and the reactor geometry, including the means of coupling the electrical power to the gas and the gas flow pattern. When electrodes are used to couple power into the plasma, the electrode spacing and electrode area ratio are reactor design parameters. Electrodes are not needed for some types of inductively coupled radio frequency (rf) discharges and are not usually used for microwave (μw) discharges. Most of the numerous reactor types which are used operate at low pressures, in a discharge regime which is on the "left hand side" of the Paschen breakdown curve. Operation in this regime leads to very high E/N values which are discussed first. Then the consequences of these very high E/N values are discussed, including the mechanisms for power deposition in the discharge. The low operating pressures also lead to sheath regions which occupy a substantial part of the reactor volume. The discharge electrons are not in equilibrium with the local electric field in these sheath regions, and the energy which the electrons gain in the sheaths is often deposited in the glow region outside the sheaths. Experimental data on sheath fields versus position and the spatial distribution of the electron energy deposition is presented to illustrate these points. The discharge ions may also play an important role in the discharge energy balance. Recent experimental data which emphasizes the importance of the discharge ions are briefly discussed.

Breakdown and Sustaining Fields

Both the voltage required to produce a discharge, i.e., the breakdown voltage, and the voltage required to sustain a discharge are functions of the parameter Nd (= gas density x gap spacing). Figure 1 shows curves of measured values (Perrin et al., 1988; Den Hartog et al., 1988; Viadud et al., 1988; Scheller et al., 1988; Horwitz, 1983; Thompson and Sawin, 1986) of the E/N needed to sustain a discharge versus nd for several gases. The values plotted are average electric field to gas density ratio, V/Nd = E/N.

122

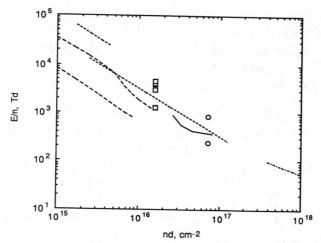

Fig. 1. Operating or breakdown electric field-to-gas density ratio
versus gas density X distance for SiH_4 (− − −), (Perrin et
al., 1988); He (O), (Den Hartog, 1988); N_2 (− · −), (Viadud
et al., 1988); BCl_3 (□) (Scheller et al., 1988); Ar
(-----), (Horwitz, 1983); and SF_6 (———), (Thompson and
Sawin, 1986).

All of the data are for parallel plane gaps. E/N values are shown for a dc
discharge in He and radio frequency (rf) discharges in Ar, N_2, BCl_3, SiH_4,
SF_6 and an Ar + BCl_3 mixture. The rf voltages are rms values. All of the
E/N values shown are discharge-sustaining voltages except for the SF_6 data
which is rf breakdown data.

Several features are apparent from the data shown in Fig. 1. First,
for the range of conditions shown in the figure, the E/N values rapidly
increase as nd decreases for all of the gases shown and for the entire range
of frequencies. A similar variation of E_s/N versus Nd is observed at μw
frequencies, where E_s is the breakdown field. Also, the E/N values are
similar for all of the gases shown. Although the data are not shown in
Fig. 1, the breakdown and sustaining voltages are similar for each gas. The
increase in E/N as Nd decreases which is shown in Fig. 1 occurs because the
number of ionizing collisions per cm of electron travel decreases as the gas
density decreases. As a result, the applied field required to accelerate
the electrons and produce enough ionization to balance electron losses
increases as Nd decreases.

These low pressure breakdown and discharge maintenance E/N values are
one to two orders of magnitude higher than the breakdown and discharge
maintenance E/N values which are observed for uniform field gaps of a few
centimeters at atmospheric pressure. The range of E/N values over which a
discharge can be sustained is also large at low pressures, and the

experimental data suggest that two modes of discharge operation are possible (Boeuf and Belenguer, to be published; Perrin et al., 1988; Viadud et al., 1988; Popov and Godyak, 1985). In the first mode, secondary electrons which are produced at the cathode by ion impact help to sustain the discharge. In the second mode, secondary electrons are not important.

Electrode Sheath Characteristics

Although the average E/N values for 1) producing a discharge, i.e., breakdown of the gas and 2) sustaining a discharge are similar, there is an important difference between these two experimental situations. During the breakdown process, the charged particle densities are small. Consequently, space charge effects are also small and the applied fields in a parallel plane gap are spatially uniform, at least for dc and rf applied voltages. Once a discharge has been established, electrode sheath regions form because the electrons are much more mobile than the positive ions. The field in these sheath regions decreases the electron loss rate and increases the ion loss rate so that the electron and ion loss rates to the electrodes and chamber walls are equal in an electro-positive gas. This loss rate is termed the "ambipolar" diffusion loss rate (Allis and Brown, 1951). In electronegative gases the negative ions can also play an important role in determining the properties of these sheath regions (Boeuf and Belenguer, to be published; Gaebe et al., 1987).

The electric fields in the sheath regions in dc discharges in helium have recently been measured spectroscopically by Den Hartog et al. (1988). They also measured population densities of the He 2^1s and 2^3s metastable states. Their measured fields versus position for five different current densities are shown in Fig. 2. The average E/N values corresponding to these conditions are shown in Fig. 1. The corresponding dc breakdown voltage for their Nd value (7.16 x 10^{16} cm^{-2}) lies in the middle of their measured range of discharge operating voltages. Their predicted He 2^3s metastable production rates are shown in Fig. 3 for one of the current densities in Fig. 2. Note that the high field region, where the electrons gain energy, is located close to the cathode. In contrast, the region of maximum electron energy deposition, as indicated by the peak in the metastable production rate, is in the negative glow region where the field is approximately zero. These results clearly show that the electron energy distribution is not in local equilibrium in these helium discharges. Den Hartog et al. show that the metastable production rates in Fig. 3 are consistent with their measured metastable densities and confirm the lack of local equilibrium for the electrons. Their model uses their measured field values. Modeling of discharges in helium is feasible because of the extensive data base of measured and calculated cross sections and metastable diffusion and quenching rates for helium.

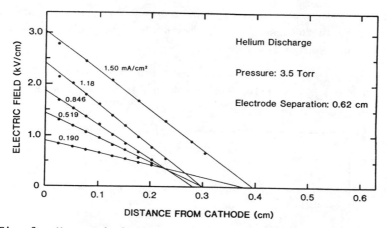

Fig. 2. Measured electric fields versus position for dc glow discharges in helium (Den Hartog et al., 1988).

Space and time resolved electric fields in rf discharges have been measured spectroscopically by Gottscho and his colleagues (1987; 1989) in BCl_3. Their results for 50 KHz discharges are shown in Fig. 4 for several different electrode area ratios. Similar measurements have been made by Gottscho at higher frequencies. Note that the field in the center of the discharge in Fig. 4 is negligible compared with the field in the sheath regions. In fact, the field in the center of the discharge is below the threshold for quantitative field measurements in Gottscho's papers (1987; 1989). These measurements also show that the sheath field and sheath thickness both increase as frequency decreases as the electrode area ratio increases. In addition, they show that almost all of the applied voltage is dropped within the sheath regions for frequencies from 50 KHz to 10 MHz in BCl_3 and in BCl_3-rare gas mixtures. Since the sheath regions occupy only a

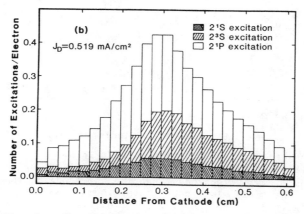

Fig. 3. Predicted excited state production rates versus position for a dc glow discharge in helium (Den Hartog et al., 1988).

fraction of the interelectrode gap, and since essentially all of the applied voltage drop is in the sheaths, the sheath field values are several times higher than the average field values.

Gottscho and his colleagues have also measured space- and time-resolved Ar metastable and BCl profiles for discharges in BCl_3 and $Ar-BCl_3$, $Ar-Cl_2$

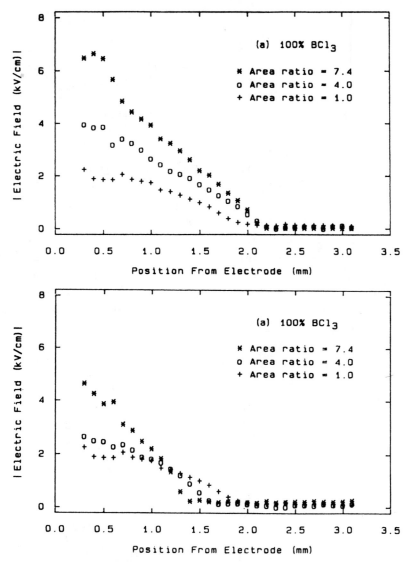

Fig. 4. Measured cathodic electric fields versus position for different electrode area ratios in rf discharges in BCl_3. Top: small electrode (when the electrode voltage relative to the large electrode is a maximum). Bottom: large electrode (when the electrode voltage relative to the small electrode is a minimum). From Gottscho et al., 1989.

and Ar-SF$_6$ mixtures (Scheller et al., 1988). These results, together with the measured space- and time-resolved fields just discussed, again provide clear indications that the electrons are not in local equilibrium in these rf discharges.

Many of the features which have been observed in the space and time resolved field measurements of Gottscho and coworkers have been qualitatively predicted by Boeuf and Belenguer (to be published). However, these efforts to theoretically understand the microscopic physics of rf discharges in BCl$_3$ and BCl$_3$-rare gas mixtures have been hampered by a lack of electron and ion collision cross section and transport data.

ELECTRON ENERGY DISTRIBUTIONS AND EXCITATION RATES

The discharge data just described clearly show that the electrons are not in equilibrium with the local electric field in both dc and rf low-pressure discharges. Therefore, the electron kinetics in these discharges can be completely understood only by calculating the space- and time-dependent behavior of the EED. Non-equilibrium electron kinetics calculations of this kind are very difficult to perform, as discussed in the next section. Fortunately, much simpler dc EED calculations provide considerable insight into the electron kinetics and electron energy deposition pathways in these discharges. Some dc EED results for CH$_4$ are described in this section to illustrate several aspects of election behavior at high E/N values. First, the Boltzmann equation for electrons is described in order to provide the necessary background for the discussion.

The Boltzmann Energy Balance Equation for Electrons

The steady state and time dependent behavior of the electron energy distribution can be described by the Boltzmann equation (Holstein, 1946). In its most general form, the Boltzmann equation is a continuity equation for the electron density $n_e(\vec{r}, \vec{v})$:

$$\frac{\partial n_e}{\partial t} + \vec{v} \cdot \nabla_r n_e + \vec{a} \cdot \nabla_v n_e =$$

$(\partial n_e / \partial t)$ elastic collisions

$+ (\partial n_e / \partial t)$ inelastic collisions

$+ (\partial n_e / \partial t)$ attaching collisions

$+ (\partial n_e / \partial t)$ ionizing collisions   (1)

where the terms on the left hand side express the continuity of the electron density in a six dimensional phase space with three position coordinates and three velocity coordinates. The collision terms on the right hand side account for the instantaneous redistribution of electrons in velocity space due to elastic and inelastic collisions, the loss of electrons in attaching collisions and the gain of new electrons in ionizing collisions. These terms are described in detail by Holstein (1946). The accelerating force, $\vec{a}$, includes contributions from the time- and space-dependent applied electric and magnetic fields and the effects of space charge distortion of the applied electric field.

There are two basic approaches which have been used to calculate dc EED's. In the first approach, a representation is chosen for the electron energy distributions and substituted into the Boltzmann equation, Eq. (1), to obtain a system of equations which is then solved numerically. In the second approach, Monte Carlo simulation techniques are used. Both of these approaches require a complete set of electron collision cross sections, for each gas or gas mixture to be studied, as the input data. The study of mixtures is straightforward because cross sections are used as the input data. The cross sections for each component of a mixture are weighted by the fractional concentration for that mixture component.

The results presented here were calculated by performing Monte Carlo simulations for electrons in CH_4 using an anisotropic cross section set. The dc EED's were calculated by following a group of electrons as they move in a uniform field. Boundaries were not included in the simulation. The EED develops during an initial transient period which is ignored in estimating the properties of the dc EED. Figure 5 shows calculated rate coefficient values versus E/N for elastic collisions, vibrational excitation, neutral dissociation and ionization in CH_4. Figure 6 shows the corresponding calculated fractional electron energy deposition versus E/N. Note that most of

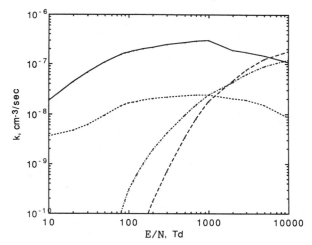

Fig. 5. Calculated excitation rate coefficients for elastic collisions
(———), vibrational excitation (-----), neutral dissociation
(— · —) and ionization (— — —) versus E/N in CH_4.

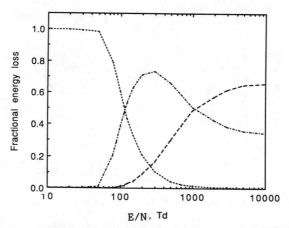

Fig. 6. Calculated fractional electron energy loss to vibrational excitation (-----), dissociation (- · -) and ionization (- - -) versus E/N in CH_4.

the electron energy is lost to dissociation and ionization for E/N values > 300 Td. Furthermore, the fractional energy losses to dissociation and ionization are comparable for E/N values > 500 Td. These results for CH_4 show that all of the electron energy is lost to the processes which drive the plasma chemistry even at E/N values which are relatively low compared with the average discharge E/N values shown in Fig. 1.

The dc EED results can also be used to estimate the electron energy relaxation time versus E/N. The time constant for electron energy relaxation in an applied dc field is:

$$(\tau_u N)/\eta = u_0/[e(E/N)W]$$

where u_0 is the steady state value of the electron energy, u, W is the electron drift velocity and η is the fractional electron energy loss per collision. If u_0 is approximated by the measurable quantity $(3/2)D/\mu$, where $\mu=W/E$ is the electron mobility and D is the electron diffusion coefficient, then $(\tau_u N)/\eta$ can be approximated as:

$$(\tau_u N)/\eta = (3/2)(DN/W^2)$$

The quantity $(\tau_u N)/\eta$ is a function of E/N since DN, η and W are functions of E/N. Calculated dc values of $(\tau_u N)/\eta$ and u_0 for CH_4 are plotted in Fig. 7 as functions of E/N. Note that the mean energy is a rapidly increasing function of E/N, and the time constant for electron energy relaxation is a strongly decreasing function of E/N. These dependencies result from the variation of the electron energy losses with E/N which are shown in Fig. 6. At the low end of the E/N range shown in Figs. 5, 6 and 7, the dominant collisional loss is vibrational excitation with corresponding energy losses of 0.161 eV and 0.361 eV. As E/N increases, the dominant energy loss processes become dissociation, ionization and dissociative ionization with

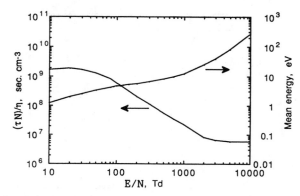

Fig. 7. Calculated values of (electron energy relaxation time x gas
density)/fractional energy loss, $(\tau N)/\eta$, and calculated
electron mean energy versus E/N in CH_4.

minimum energy losses of 9, 12.6 and 14.3 eV, respectively. Electron energy
relaxation is much faster at high E/N values where dissociation, ionization
and dissociative ionization are the dominant electron energy losses.

These results also imply that low energy electrons will relax slowly
even at high E/N values, because they can lose energy only by vibrational
excitation and elastic collisions. Thus the high energy part of the EED can
follow much stronger time and space variations in the applied electric
field, compared with the low energy part of the EED below the dissociation
threshold at 9 eV. The high energy part of the EED will be in local equilibrium with the field when 1) τ_u is short compared with one rf cycle and
2) the electron mean free path is short compared with the characteristic
distance over which E/N changes. Note that both the mean free path and the
electron energy relaxation time increase as the pressure decreases. These
results provide insight into the applicability of the rf discharge models
which are discussed next.

MODELS OF rf DISCHARGES

Many different approaches have been used to predict and understand the
behavior of electrons in rf discharges. Four different approaches will be
discussed here along with the data needed to apply each approach.

Space and Time Averaged Electron Energy Distributions

Many aspects of the behavior of plasma processing discharges can be
predicted by assuming that the time and space averaged behavior of discharge
electrons can be approximated by the dc behavior which would be obtained at
the dc E/N value corresponding to the time- and space-averaged field in the

discharge. The resulting constant electron rate coefficient values can then be used in a chemical kinetic model. The data needed in this approach is either calculated or measured electron collisional rate coefficients. In fact, the only rate coefficients which are needed are those which affect the discharge chemistry. This approach is attractive because it is very simple. However, it cannot predict any of the non-equilibrium or time- and space-dependent behavior of the type described in the Discharge Experimental Data section.

Electron Energy Distributions in Spatially Uniform rf Fields

A second approach to rf discharge modeling is to neglect spatial variations in E/N while including time variations. Winkler et al. (1987) and Capitelli et al. (1988) have performed an extensive series of calculations of this kind. They assumed a spatially uniform field and studied the effect of varying ω/N, at fixed values of E/N, for several different gases. ω is the applied rf frequency. The effects of electron production and loss were neglected in most of these calculations. However, a recent study for SF_6 (Winkler et al., 1987) included electron production by ionization and electron loss by attachment. In the SF_6 calculations steady state occurs, i.e., the electron production and loss rates are equal, at an E/N close to the dc limiting E/N value over the range $(\pi/2)9x10^7 < \omega/N < \pi\ 9x10^8\ sec^{-1}$ $Torr^{-1}$. Similar calculations have been performed by Makabe and Goto (1988) for CH_4.

These calculations require a complete set of electron collision cross sections, for each gas or gas mixture to be studied, as the input data. As in the dc EED calculations, the study of mixtures is straight forward because cross sections are used as the input data.

The E/N values assumed in most of these calculations were well below the E/N values shown in Fig. 1. The assumed E/N values may be representative of the E/N values in the central, low-field glow region of rf discharges. This approach is valid for predicting the behavior of rf discharges at relatively high pressure ($\geq$ 1 Torr) where the bulk electrons in the glow region are in local equilibrium. However, it cannot be used to model the sheath regions or to model much lower pressure discharges where there is no local equilibrium. Models which can be used for both the sheath and glow regions are discussed next.

Self Consistent Continum Models of rf Discharges

Self-consistent models of rf discharges have recently been reviewed by Boeuf and Belenguer (to be published). These models simultaneously solve continuity equations together with Poisson's equation. The general model formulation used in these studies includes a continuity equation for each charged species considered, Poisson's equation to relate the magnitude of the space charge distorted applied field to the net local space charge density, and in some cases, momentum and energy equation for the electrons. Appropriate boundary conditions are also required for the densities and the fields and the electron energy and momentum. The following formulation was used by Graves and Jensen (1986) for a plasma containing electrons and one positive ion species

$$\frac{\partial n_e}{\partial t} + \nabla \cdot j_e = r_i \tag{2}$$

$$\frac{\partial n_+}{\partial t} + \nabla \cdot j_+ = r_i \tag{3}$$

$$\frac{\partial}{\partial t} \left(\frac{3}{5} n_e h_e \right) + \nabla \cdot q_e = j_e \cdot eE + r_i u_i \tag{4}$$

$$\nabla^2 V = \frac{e}{\varepsilon_o} (n_e - n_+) \tag{5}$$

where j_e is the electron flux, r_i is the ionization rate, j_+ is the ion flux, V is voltage, e is the electron charge, ε_o is the permittivity of free space, h_e is the electron enthalpy, q_e is the enthalpy flux, E is the electric field and u_i is the ionization energy; Graves and Jensen also give the needed constitutive equations and boundary conditions. The data needed in order to apply Eqs. (2)-(5) is a set of electron transport coefficients for each gas or gas mixture to be studied. These transport coefficients can be measured or calculated in dc EED calculations of the type described in the last section.

The model described by Eqs. (2)-(5) is capable of predicting many of the observed properties the sheath and glow regions in low pressure rf discharges, even though it is essentially a local equilibrium model which describes the properties of the "bulk" electrons, i.e., electrons which are in local equilibrium with the local electric field. For example, the results of Graves (1987) and Richards et al. (1987) predict local maxima in the light emission near the discharge electrodes in qualitative agreement with the experimental observations of several workers. Graves also shows that the time behavior of the light emission which is predicted by his model is in qualitative agreement with the experimental results. Boeuf presents results which show that the model he used can predict the frequency dependence of the phase relationship between the discharge voltage and current and the magnitude of the sheath voltage. His model also correctly predicts some of the sheath effects observed by Gottscho and coworkers (1987; 1989) in comparing pure BCl_3 discharges with discharges in BCl_3-rare gas mixtures. All of these models correctly predict the qualitative space time behavior of the space-charge distorted electric field which has been observed experimentally by Gottscho. Specifically, the predicted fields are small in the central regions of the discharges at all times during the rf cycle, increase in the direction toward the electrodes within the sheath regions, have maxima at each electrode when that electrode is the instantaneous cathode, and have minima at each electrode when that electrode is the instantaneous anode. Also, most of the models predict a repulsive sheath at both electrodes at all times during the rf cycle except when the electrode is the instantaneous anode.

The domain of applicability of this type of model can be considerably broadened by adding "beam" electrons to the model. These are high energy, forward-directed electrons which are produced by secondary emission at the instantaneous cathode and accelerated through the sheath regions. Boeuf and Belenguer (to be published) describe a number of calculated results which were obtained using continuity equation models which include both beam and bulk electrons. The beam part of the model uses electron cross sections as its input data. These models are attractive, compared with detailed Monte Carlo simulations, because they require much less computation. However, the Monte Carlo simulation approach provides a more accurate description of the electron kinetics and can include a detailed treatment of anisotropic scattering. Anisotropic scattering is potentially important because it affects the range of the beam electrons. Monte Carlo simulation has been used in a number of rf discharge modeling studies. The Monte Carlo approach is described next.

Monte Carlo Simulations

In a Monte Carlo simulation the trajectories of a large number of electrons are followed as they are accelerated by the local electric field and are scattered and lose energy in elastic and inelastic collisions. Thus, the simulation is capable of predicting the time- and space-dependent electron-molecule collision rates for processes such as dissociation and ionization whether or not the electrons are in equilibrium with the local electric field in the discharge. A complete set of electron collision cross section data is required as the input data for a Monte Carlo simulation. Detailed treatment of anisotropic scattering and electron reflection at the discharge boundaries can easily be included in Monte Carlo simulations, as long as the appropriate electron scattering and reflection data are available.

Most of the Monte Carlo simulations which have been performed for rf discharges have assumed a space-time variation for the local electric field based on the experimental studies of Gottscho (1987) and on measurements of ion energy distributions arriving at the electrodes in rf discharges (Thompson, et al., 1986). Monte Carlo calculations of this kind have been performed by Kushner, (1983; 1986; 1988) and Kline et al. (1989). These treatments are similar except for the treatment of secondary electrons and the details of the assumed space-time variation of the electric field.

Kline et al. (1989) added the secondary electrons produced in ionizing collisions and adjusted the number of simulation particles as in the Monte Carlo simulation model of Kline and Siambis (1972). Kushner added secondary electrons at a rate equal to the diffusion loss rate. The calculations of Kushner (1983) and Kline et al. (1989) assumed stationary sheath boundaries. In more recent calculations Kushner (1986; 1988) has assumed a moving sheath boundary.

The Monte Carlo simulations described above are not completely self consistent due to their use of parametrically described, assumed electric fields. Kitamori et al. (1989), Boswell and Morey (1987), Hoffman and Hitchon (1989), and Surendera and Graves (1989) have reported self consistent simulations for parallel plane rf discharges. These Monte Carlo simulations can provide a very detailed description of the electron kinetics in rf discharges. However, they are very computationally intensive. Therefore, they have been used mainly as checks on the simpler models described above.

Discharge Model Input Data Needs

The input data needs of each of the discharge models described in this section are summarized in Table 1. Note that electron impact cross sections are needed, or can be used, as the input data for all of the models, since electron transport data can be calculated from cross sections. Calculation of electron transport data from cross sections is discussed in the next section. The types of electron molecule and atom cross sections which are needed and the methods used to determine these cross sections are also discussed in the next section and sources for cross section data are briefly listed.

Table 1. Input data needed by various rf discharge models

Model	Input data needed
Space-time averaged EED	Electron transport data
Spatially uniform time varying EED	Electron cross sections
Self consistent, local equilibrium models	Electron and ion transport data
Self consistent, two- electron group models (beam/bulk models)	Electron cross sections, electron and ion transport data
Monte Carlo models	Electron cross sections

ELECTRON IMPACT CROSS SECTIONS

A complete set of electron impact cross sections for each gas in the gas mixture to be studied is needed in order to use the discharge models just discussed. In principle, cross sections for dissociation fragments and excited states are also needed in order to completely describe plasma processing discharges, because significant dissociation and excitation occurs under many plasma process conditions. However, the description of these species is incomplete in present plasma processing models because there is a very limited amount of cross section data available for short-lived, chemically reactive and excited species. Some ionization cross section data for chemically reactive species are available and are discussed below. The types of electron collision processes which must be included are elastic scattering, vibrational excitation, electronic excitation, dissociation, attachment, ionization and dissociative ionization. The availability of cross section data for each of these types of electron collision processes is dependent on the ease of detecting either collision products or the presence or absence of electrons which have participated in a specific type of collision. The methods which are used to measure these various types of cross sections are briefly discussed below.

Some of the gases which are used in plasma etching and plasma enhanced chemical vapor deposition are listed in Table 2. All of the electron impact ionization is dissociative for many of the polyatomic molecules listed in Table 2. Molecules for which the parent ion is not observed include CF_4, C_2F_x and C_3F_x species, SF_6, SiF_4, CCl_xF_y, CCl_4, $SiCl_4$, CF_3H, CBr_xF_y, SiH_4, Si_2H_6, TEOS, B_2H_6, $TiCl_4$, WF_6, MoF_6 and $MoCl_5$. Furthermore, spectroscopic data shows that these molecules do not have stable electronically excited states. Instead, electronic excitation leads to dissociation and the available data suggests that a large fraction of the dissociation products are in their ground electronic state. This conclusion appears to be valid, for example, for CH_4, CF_4 and SiH_4 where reasonably complete cross section data are available. The dominance of the neutral dissociation channel limits the availability of cross section data for these molecules because neutral fragments in their electron ground state are difficult to detect. This point is discussed further below when the data presently available for the gases listed in Table 2 are briefly described.

Cross Sections from Beam and Swarm Data

Most of the cross sections needed as input data to rf discharge models are measured in beam experiments. In beam experiments a monoenergetic beam of electrons is reacted with a volume filled with gas or with a beam of gas molecules or atoms. Conditions are adjusted so that the probability of more than one collision per electron is negligible. The methods which are used to measure the various types of cross sections listed above are reviewed by Christophorou (1983-84).

135

Table 2. Gases used in plasma etching and deposition

Etching Gases (for Si, GaAs, Si oxide, Si nitride, metals)

CF_4, C_xF_y, SF_6, SiF_4, NF_3, CCl_xF_y, Cl_2, CCl_4, BCl_3, HCl, PCl_3, $SiCl_4$, H_2, O_2, CF_3H, Br_2, CBr_xF_y, Rare gases, plus numerous stable and intermediate product species

Deposition Gases (For Si, Si oxides, Ti oxides, Si,B nitride, metals)

SiH_4, Si_2H_6, TEOS, O_2, N_2O, CO_2, NH_3, N_2, B_2H_6, BCl_3, BBr_3, $TiCl_4$, WF_6, MoF_6, $MoCl_5$, H_2, Rare gases, plus numerous stable and intermediate product species

---

The fact that charged products are detected facilitates the measurement of ionization and attachment cross sections. As a result, ionization cross sections have been measured for almost all of the gases listed in Table 2 and attachment cross sections have been measured for most of the gases which form negative ions. Since ionization cross sections are usually measured by detecting the positive ion products, the differential cross section data for ionization is limited.

Total cross sections are usually measured by measuring the attenuation of the electron beam. Total cross section measurements are difficult at low energies because it is difficult to produce a monoenergetic electron beam at low energies. Thus, total cross sections are often not available from beam experiments at low energies. Low energy momentum transfer cross sections can be determined from swarm experiments, and this approach has been used to study some of the gases listed in Table 2.

Vibrational cross sections are usually measured by detecting electrons which have lost one or more vibrational qunata of energy. Detection of electrons which have lost a specific amount of energy is more difficult as the energy loss becomes smaller. In addition, vibrational spacings decrease as the size of the molecule increases. Consequently, vibrational cross section measurements become more difficult as the size of the molecule increases. Cross sections for vibrational excitation have been measured for most of the small molecules listed in Table 2 but not for the larger molecules. Swarm data can also be used to determine vibrational cross sections. Swarm data has been used to estimate vibrational cross sections for some of the gases listed in Table 2.

Cross sections for electronic excitation have been measured for those molecules listed in Table 2 which have bound electronically excited states. However, as discussed above, many of the larger molecules listed in Table 2 do not have bound, electronically excited states. Therefore, the cross sections needed for these molecules are dissociation cross sections. Measurements of these dissociation cross sections is discussed next.

Dissociation cross sections where one (or more) of the dissociation fragments is in an electronically-excited radiating or metastable state have been measured for many molecules. Becker (to be published) reviews some of the recent measurements of these dissociative excitation cross sections for plasma processing gases.

Dissociative excitation cross section data for processes where an excited fragment is produced are available for many gases. However, the data for gases where a complete cross section set is available suggests that many dissociative excitation processes produce two (or more) fragments which are all in their electronic ground states. The cross sections for these neutral dissociation processes are difficult to measure because the products are difficult to detect. In fact, direct detection of neutral dissociation products has been performed for only a few molecules (Melton and Rudolph, 1967; Cosby and Helm, 1989).

Another approach to measuring neutral dissociation cross sections is to measure total dissociation and dissociative ionization. The neutral dissociation cross section is then the difference between the total dissociation cross section and the dissociative ionization cross section. Total dissociation cross sections have been measured for a number of plasma processing gases by Winters and Inokuti (1982).

Cross section data for chemically reactive and excited species is limited, as discussed above. However, Freund and his colleagues (Freund, 1985; Baiocchi et al., 1984; Hayes et al., 1987; 1988) have recently measured ionization cross sections for a number of reactive species which are relevant to plasma processing discharges.

When a "complete" set of electron cross sections has been assembled, the consistency of the cross section set can be checked by using the cross sections to predict swarm data including the electron drift velocity and diffusion coefficients and ionization and attachment coefficients. Comparison of the predicted swarm data with measured swarm data provides an indication of the consistency of the cross section set and can also be used to make adjustments to the cross section set. In fact, this approach can be used to determine an unknown cross section in some cases where information about all other cross sections is available. The use of swarm data to determine cross sections has been reviewed by Phelps (1968) and Crompton (1983). The procedure used is shown schematically in Fig. 8. Recently, Hayashi (1985; 1987) has used this approach to develop cross section sets for many plasma processing gases.

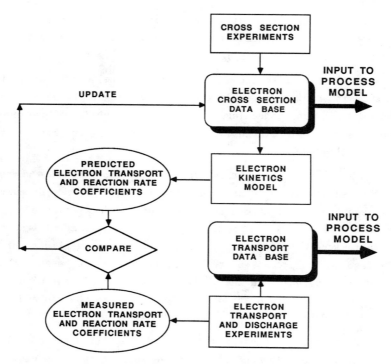

Fig. 8.   Schematic diagram showing the computational procedure used
to determine electron cross sections from electron beam
data, electron swarm data and discharge experiments.

Cross Sections from Discharge Data

Since the data for neutral dissociation cross sections are very limited,
it is useful to consider other approaches to determining dissociation rate
coefficients and/or cross sections. One approach is to measure the products
which are formed in discharge dissociation of a plasma processing gas. Ryan
and Plumb have used this approach to study several gases including CF_4, CF_4 +
O_2 mixtures, SF_6 and SF_6 + O_2 mixtures. They used mass spectrometric
composition measurements in conjunction with a chemical kinetics model to
determine rate coefficients. In principle, it is possible to "work
backwards" from this kind of experimental result to determine the neutral
dissociation cross section. This approach can be used where all of the other
cross sections are known. The needed data are known, at least approximately,
for many of the gases listed in Table 2. In these cases, the shape of the
neutral dissociation cross section can be estimated and used, together with
the other cross sections, to calculate electron energy distributions (EED's)
and rate coefficients. For example, the shape of the neutral dissociation

Table 3. Gases of Interest to Plasma Processing For Which Comprehensive
          Electron Impact Cross Sections are Available

Etching Systems:

CF_4 Winters and Inokuti, 1982; Hayashi et al., 1985; Stephan
              et al., 1985; Masek et al., 1987; Curtis et al., 1988;
              Spyrou et al., 1983

CCl_2F_2 Hayashi et al., 1985; Novak and Frechette, 1985; Okabe and
              Kuono, 1986; Leiter and Mark, 1985

F_2 Hayashi and Nimura, 1983; Deutsch et al., 1986; Stevie and
              Vasile, 1981

Cl_2 Stevie and Vasile, 1981; Rogoff et al., 1986

C_2F_6 Winters and Inokuti, 1982; Spyrou et al., 1983

CCl_4 Hayashi et al., 1985; Leiter et al., 1984

HCl   Hayashi et al., 1985; Davies, 1982

SF_6 Deutsch et al., 1986; Phelps and VanBrunt, 1988; Hayashi
              and Nimura, 1984

Deposition Systems:

SiH_4 Hayashi et al., 1985; Ohmori et al., 1986; Perrin et al.,
              1982; Garscadden et al., 1983; Chatham et al., 1984

CH_4 Hayashi et al., 1985; Chatham et al., 1984; Ohmori et al.,
              1986; Davies et al., 1989

N_2O Hayashi et al., 1987

Si_2H_6 Hayashi et al., 1985; Perrin et al., 1982; Chatham et al.,
              1984

C_2H_6 Hayashi et al., 1987; Chatham et al., 1984

Diluent Gases:

He   Hartog et al., 1988

Ar   Tachibana, 1986; Nakamura and Kurachi, 1988

---

cross section can be estimated by using the ionization cross section shape for the same gas and shifting the threshold or by analogy with known neutral dissociation cross section shapes for similar gases. Since dissociative ionization cross sections must be known to use this approach, the ionization contribution to dissociation will be one of the outputs of the EED calculations. Therefore, the magnitude of the neutral dissociation cross section can be estimated by comparing the predicted and measured neutral dissociation rates. Although this approach is approximate, it should allow plasma processing modeling over a much wider range of discharge conditions than would be possible by simply using the rate coefficients determined for one specific set of discharge conditions.

Cross Section Data Sources

The available cross section data for various plasma processing gases is summarized in Table 3. The references listed in this table include both complete cross section compilations for these gases and individual cross sections. Complete compilations of cross sections or cross section references are given Den Hartog et al. (1988), Hayashi et al. (1985), Hayashi et al. (1987), Masek et al. (1987), Novak and Frechette (1985), Okabe and Kuono (1986), Hayashi and Nimura (1983), Davies (1982), Rogoff et al. (1986), Phelps and VanBrunt (1988), Hayashi and Nimura (1984), Ohmori et al. (1986), Garscadden et al. (1983), Ohmori et al. (1986), Davies et al. (1989), and Tachibana (1986). The remaining references which are listed in Table 3 give either individual cross sections or cross section sets which are useful only for low energies. There are many other sources of specific cross sections for these gases which are referenced in the papers listed in the bibliography at the end of this chapter. Cross sections for a number of free radicals and excited states have been measured by Freund and his colleagues (Freund, 1985; Baiocchi et al., 1984; Hayes et al., 1987; Hayes et al., 1988). In addition there are two ongoing projects to develop cross section data compilations for plasma processing gases. One of these is being performed under the sponsorship of the International Union of Pure and Applied Chemistry (IUPAC) Subcommittee on Plasma Chemistry. This effort is being chaired by L. E. Kline. The second effort is being carried out by W. L. Morgan of the Joint Institute for Laboratory Astrophysics in Boulder, Colorado.

REFERENCES

Allis, W. P., and S. C. Brown, 1951, Phys. Rev. 84, 519.
Baiocchi, F. A., R. C. Wetzel, and R. S. Freund, 1984, Phys. Rev. Lett. 53, 771.
Becker, K. H., "Electron Molecule Collision Cross Sections for Etching Gases," in Non Equilibrium Processes in Partially Ionized Gases, NATO ASI Series, M. Capitelli, and J. N. Bardsley, Eds. (to be published).

Boeuf, J. P., and P. Belenguer, "Fundamental Properties of r.f. Glow Discharges, an Approach Based on Self Consistent Numerical Models," in Non Equilibrium Processes in Partially Ionized Gases, NATO ASI Series, M. Capitelli, and J. N. Bardsley, Eds. (to be published).

Boswell, R. W., and I. Morey, 1987, Appl. Phys. Lett. 52, 21.

Chatham, H., D. Hils, R. Robertson, and A. Gallagher, 1984, J. Chem. Phys. 81 1770.

Christophorou, L. G., 1983-84, Ed., Electron-Molecule Interactions and Their Applications, Vols. 1 and 2, Academic, New York.

Cosby, P. C., and H. Helm, 1989, Bull. Am. Phys. Soc. 34, 325.

Crompton, R. W., 1983, Ptoc. 16th International Conf. on Phenomena in Ionized Gases, Duesseldoef, Invited Papers, p. 58.

Curtis, M. G., I. C. Walker, and K. J. Mathieson, 1988, J. Phys. D 21, 1271.

Davies, D. K., August 1982, "Measurements of swarm parameters in chlorine-bearing molecules," Report No. AFWAL-TR-82-2083, Air Force Wright Aeronautical Lab., Wright-Patterson Air Force Base, Ohio.

Davies, D. K., L. E. Kline, and W. E. Bies, 1989, J. Appl. Phys. 65, 3311.

Den Hartog, E. A., D. A. Doughty, and J. E. Lawler, 1988, Phys. Rev. A 38, 2471.

Deutsch, H., P. Scheier, and T. D. Mark, 1986, Int. J. Mass Spect. 74, 81.

Freund, R. S., 1985, "Electron Impact Ionization Cross Sections for Atoms, Radicals and Metastables", in Swarm Studies and Inelasitc Electron Molecule Collisions, L. C. Pitchford, B. V. McKoy, A. Chutjian, and S. Trajmar, p. 167-187, Springer-Verlag, New York.

Gaebe, C. E., T. R. Hayes, and R. A. Gottscho, 1987, Phys. Rev. A35, 2993.

Garscadden, A., G. L. Duke, and W. F. Bailey, 1983, Appl. Phys. Lett. 43, 1012.

Gottscho, R. A., 1987, Phys. Rev. A36, 2223.

Gottscho, R. A., G. R. Scheller, D. Stoneback, and T. Intrator, 1989, J. Appl. Phys. 66, 492.

Graves, D. B., 1987, J. Appl. Phys. 62, 88.

Graves, D. B., and K. F. Jensen, 1986, IEEE Trans. Plasma Science, PS-14, 78.

Hayashi, M., and T. Nimura, 1983, J. Appl. Phys. 54, 4879.

Hayashi, M., and T. Nimura, 1984, J. Phys. D 17, 2215.

Hayashi, M., 1985, "Electron Collision Cross Sections for Molecules Determined From Beam and Swarm Data," in Swarm Studies and Inelasitc Electron Molecule Collisions, L. C. Pitchford, B. V. McKoy, A. Chutjian, and S. Trajmar, Springer-Verlag, New York, p. 167-187.

Hayashi, M., 1987, "Electron collision cross sections for C_2F_6 and N_2O," in Gaseous Dielectrics V, L. Christophorou, and D. Boulden, Eds., Pergamon, New York.

Hayes, T. R., R. C. Wetzel, and R. S. Freund, 1987, Phys. Rev. A 35, 578.

Hayes, T. R., R. C. Wetzel, F. A. Baiocchi, and R. S. Freund, 1988, J. Chem. Phys. 88, 823.

Hofman, G. J., and W. N. Hitchon, 1989, Bull. Am. Phys. Soc. 34, 296.

Holstein, T., 1946, Phys. Rev. 70, 367.

Horwitz, C. M., 1983, J. Vac. Sci. Technol. A1, 60.

Kline, L. E., and J. G. Siambis, 1972, Phys. Rev. A5, 794.

Kline, L. E., and M. J. Kushner, 1989, "Computer Simulation of Materials Processing Plasma Discharges," in CRC Critical Reviews of Solid State and Materials Sciences, 1-35.

Kline, L. E., W. D. Partlow, and W. E. Bies, 1989, J. Appl. Phys. 65, 70.

Kohler, K., J. W. Coburn, D. E. Horne, and E. Kay, 1985, J. Appl. Phys. 57, 59; K. Kohler, D. E. Horne, and J. W. Coburn, 1985, J. Appl. Phys. 58, 3350.

Kushner, M. J., 1983, J. Appl. Phys. 54, 4958.

Kushner, M. J., 1986, IEEE Trans. Plasma Science PS-14, 188.

Kushner, M. J., 1988, J. Appl. Phys. 63, 2532.

Leiter, K., and T. D. Mark, 1985, "Electron Ionization Cross Sections for CF_2Cl_2," in Proc. 7th Int. Symp. Plasma Chem., Eindhoven.

Leiter, K., K. Stephan, E. Mark, and T. D. Mark, 1984, Plasma Chem. Plasma Proc. 4, 235.

Makabe, T., and N. Goto, 1988, J. Phys. D 21, 887.

Masek, K., L. Laska, R d'Agostino, and F. Cramarossa, 1987, Contrib. Plasma Phys. 27, 15.

Melton, C. E., and P. S. Rudolph, 1967, J. Chem. Phys. 47, 1771.

Nakamura, Y., and M. Kurachi, 1988, J. Phys. D 21, 718.

Novak, J. P., and M. F. Frechette, 1985, J. Appl. Phys. 57, 4368.

Ohmori, Y., K. Kitamori, M. Shimozuma, and H. Tagashira, 1986, J. Phys. D 19, 437.

Ohmori, Y., M. Shimozuma, and H. Tagashira, 1986, J. Phys. D 19, 1029.

Okabe, S., T. Kuono, 1986, Japanese J. Appl. Phys. 24, 1335.

Perrin, J., J. P. M. Schmitt, G. De Rosny, B. Drevillon, J. Huc, and A. Lloreyt, 1982, Chem. Phys. 73, 383.

Perrin, J., P. R i Cabarrocas, B. Allain, and J-M. Freidt, 1988, Japanese J. Appl. Phys. 27, 2041.

Phelps, A. V., 1968, Rev. Mod. Phys. 40, 399.

Phelps, A. V., and R. J. VanBrunt, 1988, J. Appl. Phys. 64, 4269.

Popov, O. A., and V. A. Godyak, 1985, J. Appl. Phys. 57, 53; 1986, 59, 1759; V. A. Godyak and A. S. Khanneh, 1986, IEEE Trans. Plasma Science, PS-14, 112 and the references therein.

Richards, A. D., B. E. Thompson, and H. H. Sawin, 1987, Appl. Phys. Lett. 50, 492.

Rogoff, G. L., J. M. Kramer, and R. B. Piejak, 1986, IEEE Trans. Plasma Sci. PS-14, 103.

Scheller, G. R., R. A. Gottscho, and T. Intrator, 1988, J. Appl. Phys. 64, 4384.

Spyrou, S. M., I. Sauers, and L. G. Christophorou, 1983, J. Chem. Phys. 78, 7200.

Stephan, K., H. Deutsch, and T. D. Mark, 1985, J. Chem. Phys. 83, 5712.

Stevie, F. A., and M. J. Vasile, 1981, J. Chem. Phys. 74, 5106.

Surendra, M., D. B. Graves, and I. J. Morey, August 1989, submitted to Appl. Phys. Lett.

Tachibana, K. 1986, Phys. Rev. A34, 1007.

Tagashira, H., A. Date, and K. Kitamori, 1989, Proc. 9th Int'l. Symp. on Plasma Chem., Pugnochiuso, Italy, p. L75.

Thompson, B. E., K. D. Allen, A. D. Richards, and H. H. Sawin, 1986, J. Appl. Phys. 59, 1890.

Thompson, B. E., and H. H. Sawin, 1986, J. Appl. Phys. 60, 89.

Viadud, P., S. M. A. Durrani, and D. R. Hall, 1988, J. Phys. D 21, 57.

Winkler, R., M. Dilnardo, M. Capetelli, and J. Wilhelm, 1987, Plas. Chem. and Plas. Proc. 7, 245.

Winkler, R., M. Dilnardo, M. Capetelli, and J. Wilhelm, 1987, Plas. Chem. and Plas. Proc. 7, 125; M. Capitelli, C. Gourse, R. Winkler, and J. Wilhelm, 1988, Plas. Chem. and Plas. Proc. 8, 399 and the references therein.

Winters, H. F., and M. Inokuti, 1982, Phys. Rev. A 25, 1420.

WHEN CAN SWARM DATA BE USED TO MODEL GAS DISCHARGES?

Michael J. McCaughey and Mark J. Kushner

University of Illinois
Department of Electrical and Computer Engineering
Gaseous Electronics Laboratory
607 E. Healey
Champaign, IL 61820 USA

ABSTRACT

Electron swarm data (e.g., first Townsend coefficient, diffusion coefficient, drift velocity) have at least two important uses. The first is that swarm data may be used to derive electron impact cross sections. These cross sections may then be used, in conjunction with solving Boltzmann's equation, to obtain reaction rate coefficients to model gas discharges. The second use is to directly model gas discharges using the coefficients obtained from swarm experiments. The use of swarm data in this manner presupposes that the conditions under which the swarm parameters were measured resemble those of the gas discharges in question. With this in mind, one must consider whether differences in current density, nonequilibrium aspects of the electron energy distribution, and cleanliness of the system affect the validity of using swarm data to directly model gas discharges in this manner. In this paper, we discuss the parameter space in which swarm data may be directly used to model gas discharges, while paying attention to these parameters. We will also suggest that swarm data may be extended beyond the accepted parameter space by using scaling laws.

INTRODUCTION

Electron swarm parameters are transport coefficients and reaction rate coefficients obtained directly from experiments and which are typically catalogued as functions of electric field/gas number density (E/N) (Dutton, 1975; Wedding et al., 1985; Hunter et al., 1986). In the typical experiment, a small burst of electrons (containing a few as 10^4 electrons) is generated at the cathode and accelerated towards the anode in a uniform E/N (Hunter et al., 1986). The time of flight, dispersion and multiplication of the electron cloud are recorded. These values are used to derive the electron drift velocity, diffusion coefficient, and ionization coefficient. The parameters so obtained are averages over the electron energy distribution (EED).

The further use of these swarm parameters falls into two categories. The first is that the swarm data may be "deconvolved" to yield the fundamental electron impact cross sections (Phelps, 1987; Hayashi, 1987). This procedure begins by proposing a set of cross sections for, say, momentum transfer, vibrational excitation, and ionization. The EED is then calculated by solving Boltzmann's equation, and the distribution averaged transport coefficients are computed as a function of E/N. These quantities are then compared to the experimentally derived swarm parameters, and adjustments are iteratively made to the proposed set of cross sections as needed. The cross sections so obtained are not necessarily unique. However, they are fundamental values which may then be used to model gas discharges under rather arbitrary conditions by first calculating the EED for those conditions.

The second use of the swarm data allows one to directly model gas discharges as a function of E/N (Wu and Kunhardt, 1988; Morrow, 1987). For example, the electron continuity equation for density n_e may include the terms

$$\frac{\delta n_e}{\delta t} = v_d \cdot \alpha \cdot n_e + \cdots - \frac{D n_e \cdot}{\Lambda^2} \tag{1}$$

In Eq. 1, v_d is the electron drift velocity (cm/s), α is the ionization coefficient (cm^{-1}) and D is the diffusion coefficient (cm^2/s), all of which are functions of E/N and can be obtained directly from swarm experiments. The use of swarm data in this manner implicitly assumes that the conditions of the discharge resemble those of the swarm experiment. For example, in a properly performed swarm experiment, the EED is in equilibrium with the applied E/N so that the Local Field Approximation (LFA) is valid. Additionally, the current density in the swarm experiment is sufficiently low that collisions with excited states of the gas make no important contribution to ionization. Finally, the system is extremely clean. If any of these conditions are violated, then the direct use of swarm data to modeling gas discharges is questionable.

New modeling techniques are currently being developed which are fully capable of describing the nonequilibrium aspects of electric discharges (Sommerer et al., 1989; Boeuf and Marode, 1982; Kushner, 1987). This is accomplished by using the fundamental cross sections derived from swarm data, and calculating the time and spatially dependent EED. One may then ask whether it is necessary to directly use swarm data to model gas

144

discharges, given these capabilities. In the foreseeable future, there will inevitably be instances where the complexity of solving Boltzmann's equation in, for example, many dimensions and as a function of time is beyond the scope of the problem, or the fundamental electron impact cross sections are not available. In these cases, we must rely upon the use of swarm data to model gas discharges.

The question we wish to address in this paper is "In what parameter space may swarm data be directly used to model gas discharges?" The intent of this work is to both begin to answer that question and to motivate the investigation of new scaling parameters. These new parameters will enable swarm data to be extended beyond what we will call the "intrinsic" regime. By extending the parameter space in which swarm data may be used, it becomes a more utilitarian tool in modeling gas discharges.

NONEQUILIBRIUM EFFECTS

Implicit in the use of swarm data is the assumption that the electron energy distribution is in equilibrium with the applied electric field or more accurately, the local E/N. This restriction is quite limiting in two regards: swarm data may not always be used during transients or when there are gradients in the electric field which may produce nonequilibrium effects (Moratz, 1988). A classic example of this behavior is in the cathode fall of a glow discharge. Under these conditions a "beam" component of the EED may exist (Sommerer, et al., 1989). The "beam component of the EED most dramatically affects the rate constants for high threshold processes, such as ionization.

To investigate the conditions for which the beam component of the EED significantly contributes to the rate of high threshold inelastic processes and, hence, when the use of swarm data is not valid, the following computational experiment was performed. We considered a "box" in which a current is flowing, driven by a uniform E/N. An electron beam of fixed energy, V_b, is injected into the box and allowed to slow down below the inelastic threshold energies. The beam is specified to carry a given fraction of the total current, $\delta = j_b/j$. The remainder of the current is carried by the "bulk" portion of the EED which is in equilibrium with the applied E/N. As a measure of the importance of the beam component of the distribution, we calculate the rate constant for ionization, k_I ($cm^3 s^{-1}$), as a function of V_b, δ and E/N. We call this experiment the Beam-In-a-Box (BIB).

The electron energy distribution for the BIB was obtained by first calculating the slowing of the injected beam using a Monte Carlo simulation. From this calculation, the influx of electrons to the bulk EED was obtained and used to solve Boltzmann's equation using a conventional 2-term expansion

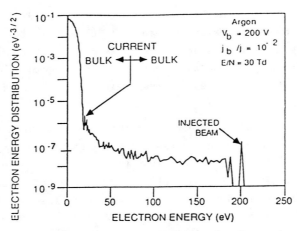

Fig. 1.   Electron energy distribution for "Beam-in-a-Box" experiment.

(Kushner, 1989). A typical EED for a BIB sustained in argon is shown in Fig. 1. The energy at which we split the EED between the beam current and bulk current is shown and is at the intersection of the bulk distribution with the slowing beam component. This division between the beam and bulk currents causes a non-intuitive systematic "slant" to the resulting rate coefficients as a function of, for example, E/N. The reason for this choice of normalization, however, will become clear when we propose the use of a new scaling parameter.

The rate constant for ionization of argon as a function of E/N is shown in Fig. 2 for different values of δ, the fraction of current in the beam. For E/N values of < 30 Td (1 Td = 10^{-17} V-cm^{-2}), k_I with the beam is higher than the value for bulk alone, provided that $\delta \geq 10^{-4}$. However, for E/N > 30 Td the value of k_I for only the bulk is nearly the same as the combined beam-bulk values provided that $\delta \leq$ 2-3 x 10^{-3}. In this region, swarm data are "valid" in spite of the EED being nonequilibrium. For values of δ > 10^{-2}, the ionization coefficient of the bulk never "catches up" to the beam-bulk value, and swarm data are not valid.

The enhancement in k_I, that is the increase in k_I for the BIB compared to its value for the bulk distribution only, is a function of E/N, δ, and type of gas. The enhancement in k_I for V_b = 200 V and $\delta = 10^{-2}$ is shown in Fig. 3 as a function of E/N for discharges in Ar and N_2. As one would expect, the enhancement can be many orders of magnitude at low values of E/N. The enhancement is larger for molecular gases, at a given value of E/N, than for atomic gases.

The cited choice in the division of the current between the beam and the bulk is quite convenient when defining the enhancement factor. Given

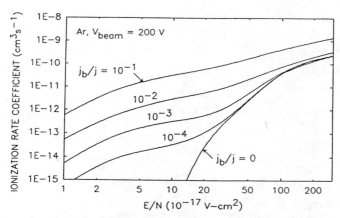

Fig. 2. Rate coefficient for ionization in the BIB experiment for
various fractional beam currents.

this normalization, the enhancement factor is independent of beam voltage
for values of $V_b > 100$ V and depends only on the type of gas, δ and E/N.
This effect is shown in Fig. 4 where k_I is plotted as a function of E/N for
V_b = 100, 200 and 300 V. since the ratios of the cross sections for
ionization compared to non-ionizing electron energy-loss processes do not
significantly change for energies between 50-60 eV and many hundreds of eV,
the initial energy of the beam is not important in determining its relative
contribution to ionization. The only important parameter is the amount of
current which is in the beam.

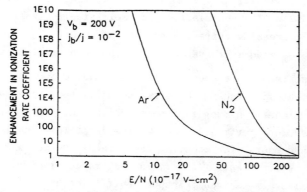

Fig. 3. Enhancement in the rate coefficient for ionization for Ar and
N_2 in the BIB experiment.

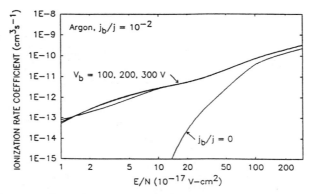

Fig. 4. Rate coefficient for electron impact ionization for various
beam voltages in the BIB experiment, showing invariance
with our definition of beam current.

These results for the contribution of the beam component of the EED to
ionization bracket the region of E/N and δ for which swarm data may be used.
The region is fairly narrow; however, the deviation from swarm data is
fairly well behaved. Given this behavior we ask: Can we extend existing
swarm data to include nonequilibrium effects by using an enhancement factor?
This factor would increase the swarm derived ionization coefficient according to the values of E/N, δ and V_b. This enhancement factor could be used
in modeling gas discharges provided the user could specify these parameters
as a function of position and time.

CURRENT DENSITY EFFECTS

It is well known that electron impact ionization in gas discharges can
occur from excited states as well as from the ground states of atoms and
molecules. Due to the lower threshold energy and larger cross section for
this process, multistep ionization (that is, ionization of excited states)
can contribute a significant, if not dominant, fraction of the ionization
(Friedland, 1989). The contributions of multistep ionization should
increase with increasing current density since the relative number density
of excited states also increases with increasing current density.

Swarm experiments are specifically designed to operate at low current
densities so that space-charge effects and multistep processes are not
important. The question must then be asked whether transport coefficients
derived from swarm experiments, such as the first Townsend ionization
coefficient, are useful for modeling real gas discharges where multistep
processes may be important? If not, in what range of current densities can
one expect that swarm-derived transport coefficients are valid for modeling?
Finally, is there a "fix" one can apply so that swarm derived transport
coefficients can be used at higher current densities?

To try to answer these questions we constructed a simple model of a positive column gas discharge in argon. The following processes were included in the model.

$$e + Ar \rightleftarrows Ar^* + e \qquad\qquad e + Ar^* \rightleftarrows Ar^+ + e + e \qquad (2a)$$

$$\rightarrow Ar^+ + e + e \qquad\qquad Ar + Ar^* + Ar \rightarrow Ar_2^* + Ar \qquad (2b)$$

$$Ar + Ar^+ + Ar \rightarrow Ar_2^+ + Ar \qquad Ar^+, Ar_2^+ \xrightarrow{D} (2)Ar \qquad (2c)$$

$$e + Ar \rightarrow Ar + e \qquad\qquad Ar^*, Ar_2^* \xrightarrow{D} (2)Ar \qquad (2d)$$
$$\text{(momentum transfer)}$$

where Ar^* represents the density of argon excited states and D denotes loss by diffusion to the wall. Rate coefficients for these electron impact processes were obtained by solving Boltzmann's equation. The rate coefficients were catalogued as functions of E/N, $[Ar^*]/[Ar]$, and $[e]/[Ar]$, as the EED depends upon all of these factors. The rate coefficients were then used in the model invoking the Local Field Approximation. DC values of the densities of the species were obtained by integrating their respective rate equations until the steady state was reached.

Results from this simple model are shown in Fig. 5 where the effective ionization coefficient, α/N (cm^2), is plotted as a function of current density, j (A/cm^2), in an Ar positive column. The scaling parameter is pR, where p is the gas pressure and R is the radius of the discharge tube. In a simple positive column α/N should be independent of j because losses are dominated by diffusion. At low values of j, α/N is nearly a constant and essentially equal to the value one would use if modeling the discharge strictly using swarm data. At an intermediate value of j, the density of Ar^* begins to become significant, and α/N increases due to the multistep ionization process. Finally, at larger values of j, α/N is again nearly a constant value. At these higher values of j the ratio $[Ar^*]/[Ar]$ reaches a nearly constant value as the excitation of Ar^* is balanced by depletion of that level due to super-elastic and ionizing collisions (see Fig. 6). Note that the current density above which the swarm-derived values of α/N are not valid is only 10 $\mu A\text{-}cm^{-2}$ for pR = 1 cm-Torr, but as large as 1 $mA\text{-}cm^{-2}$ for pR = 10 cm-Torr.

This exercise is illuminating because we have isolated two operating regimes. The first is what we call the "intrinsic" regime where the effective ionization coefficient is equal to the value derived from swarm data. In this regime, the positive column may be directly modeled using swarm data. The second is what we call the "saturated" regime in which the ratios of species densities have saturated (see Fig. 6) and in which a new "equilibrium" has been achieved. Under these conditions, α/N is

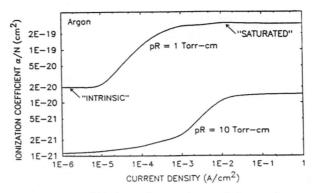

Fig. 5. Ionization coefficient in an Ar positive column as a
function of current density showing intrinsic and saturated
regimes.

again a constant and a fixed value greater than the intrinsic regime. The
degree of enhancement uniquely depends upon the gas mixture and the product
pR at pressures below which diffusion losses dominate and the three body
reactions are not important. The dependence is less straight-forward at
higher pressures. The question is then whether there is a second operating
regime for which "corrected" swarm data may be used? This second operating
regime includes the higher current densities for which some degree of
equlibrium has been achieved in the excited state manifold and in which α/N
is a constant and independent of j. The fact that α/N is independent of j
in the saturated regime qualifies it as a valid transport coefficient for
modeling. It appears that operation between the "intrinsic" and "saturated"
current densities is not well modeled using transport coefficients because,
for example, α/n is a function of j.

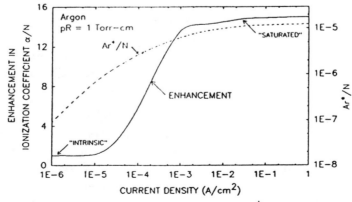

Fig. 6. Enhancement in α/N and ratio of Ar^*/N as functions of
current density showing saturation effect.

150

CLEANLINESS EFFECTS

An underlying premise in using swarm data to model gas discharges, and for that matter in using any transport coefficients or cross sections, is that the identities of all species are known. Furthermore, these species are the same as those in the experiment in which the transport data was measured. This is an obvious, "zeroth" order assumption with one exception. Real gas discharge devices (e.g., lasers, lamps, plasma processing reactors) are unavoidably contaminated by impurities, both microscopic (trace gases) and macroscopic (dust or particulates) in size. Because of this contamination the operational gas mixtures will always be different from those of the swarm experiments. The question is, then, how contaminated can the real gas mixture be before swarm data measured in the "pristine" mixture are no longer valid? The effect of trace gas impurities on transport coefficients has been addressed by many other workers. We will not repeat their work here other than to remind the reader that minute amounts of impurities, especially molecular gases in atomic gases, can significantly change the transport coefficients. For example, in Fig. 7 we show the change in the Townsend ionization coefficient for a helium discharge contaminated with water at three values of E/N. At low values of E/N where negligibly small amounts of ionization might be expected, a water impurity of < 0.01% can result in significant ionization.

The topic we will address here concerns the fact that real gas discharges are unavoidably contaminated with large particulate matter, or dust, resulting from sputtering of electrode and wall materials or gas phase polymerization (Dorrin and Khapov, 1986; Roth et al., 1985). If sufficiently large [diameter $\geq \lambda_D$ (the Debye length)], the "dust" appears to be a macroscopic, electrically floating surface in the plasma and negatively charges in the same manner as a dieletric wall of the discharge tube. In doing so, a sheath forms in the vicinity of the surface of the dust. The dust then appears as a massively large negative ion with a shielding volume many λ_D thick. These repulsive centers can be expected to detrimentally effect electron transport. The question we wish to answer is how clean does the discharge need to be in order for swarm data, measured in pristine plasmas, to be valid? That is, what is the dust density which changes electron transport coefficients?

To answer this question we have developed a hybrid Monte Carlo-Molecular Dynamics simulation to calculate the electron energy distribution in contaminated plasmas. In this model, schematically illustrated in Fig. 8, conventional Monte Carlo techniques are used to advance pseudo-electrons in a uniform applied electric field far from a dust particle. As the pseudo-electron nears the sheath of the particle, molecular dynamics techniques are used to advance the electron while allowing collisions to

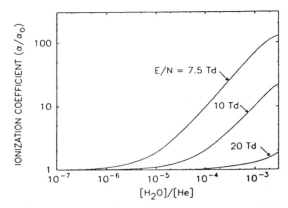

Fig. 7. Increase in ionization coefficient in a He discharge contaminated by water. The values of α/N at $[H_2O]/[He]$ = 10^{-7} are 1.79×10^{-22}, 3.6×10^{-21}, and 1.8×10^{-19} cm^2 at 7.5, 10 and 20 Td, respectively.

occur in the sheath. Electrons striking the particle are removed from the simulation as having been "collected" by the dust. When the pseudo-electron passes the particle and is far from the sheath, conventional Monte Carlo techniques are again adopted.

The floating sheath potential in a Maxwellian plasma is given by

$$\phi_s = - \left(\frac{kT_e}{2q} \right) \cdot \ln \left(\frac{T_e M_I}{T_I m_e} \right) \tag{3}$$

where T_e and T_I are the electron and ion temperatures; m_e and M_I are the electron and ion masses (Mitchner and Kruger, 1973). Since, however, the EED in our plasma is not Maxwellian, this value for ϕ_s is not necessarily valid. The model accounts for the non-Maxwellian nature of the plasma by calculating ϕ_s by balancing the flux of electrons and ions to the surface of the dust and updating the sheath potential over time to reflect changes in the electron energy distribution. ϕ_s is obtained from

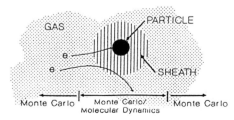

Fig. 8. Schematic of the hybrid Monte Carlo-Molecular Dynamics simulation for obtaining the EED in dusty plasmas.

$$\Gamma_e = \frac{n_e}{4} \int_{-e\phi_s}^{\infty} f(\varepsilon) \left(\frac{2\varepsilon}{m_e}\right)^{1/2} d\varepsilon = \frac{n_I V_I}{4} = \Gamma_I \qquad (4)$$

where Γ is the electron or ion flux, V_I is the ion thermal velocity, and $f(\varepsilon)$ is the EED. This expression accounts for the fact that only electrons with energy greater than ϕ_s can reach the surface of the dust.

The electric potential in the shielding volume around the dust was approximated as a $\frac{1}{r}$ Coulomb potential with Debye shielding

$$\phi(r) = \phi_s \left(\frac{d}{r}\right) \exp(-(r-d)/l) \qquad (5)$$

where d is the radius of the dust particle and r is the radial distance of the electron from the center of the particle. The characteristic shielding volume radius, 1, is the minimum of λ_D and the distance required for the shielding electric field to decay to 10% of the applied electric field in half the interparticle spacing. This latter constraint is necessary only at high dust densities in order to avoid the unphysical condition of the Debye spheres of adjacent dust particles overlapping. The force due this potential is then added to the applied electric field and used in the molecular dynamics portion of the model.

The electron energy distribution in an Ar discharge calculated by the model for pristine and dusty plasmas is shown in Fig. 9. The conditions are E/N = 7 Td, p = 0.1 Torr, and T_{gas} = 300 K. The dust density is $\rho = 10^5$ cm^{-3} and electron density is 5.0 x 10^{11} cm^{-3}. These parameters were chosen to simulate a gas laser or magnetically assisted plasma processing reactor. The scattering of electrons from the Debye shield around the dust causes a reduction in the high-energy component of

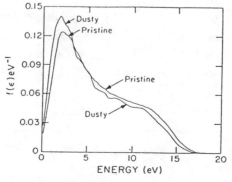

Fig. 9. Electron energy distributions in pristine and dusty argon plasmas (E/N = 7 Td, $\rho = 10^5$ cm^{-3}.

the EED and an increase in the low-energy component. The ionization rate coefficient for pure argon and in dusty plasmas ($\rho = 10^5$ cm^{-3}) is shown as a function of applied E/N in Fig. 10. The depression of the high energy component of the EED results in a decrease in the ionization rate coefficient, k_I. The difference between ionization coefficients in the pristine and dusty plasmas decreases as E/N, and average electron energy, increase. This behavior is consistent with the $1/\varepsilon$ energy scaling of the Coulomb-like "cross section" of the Debye-shielded dust particles. Ionization rate coefficients of Ar as a function of dust density are shown in Fig. 11 for two E/N values. These coefficients have been normalized by their pristine values to emphasize the effect of the dust. Increased dust densities result in lower overall rates of ionization as a result of the increasing reduction of the high energy component of the EED. The threshold dust density for which this effect occurs, 10^3-10^5 cm^{-3}, decreases with decreasing E/N. Below this value of ρ, transport coefficients obtained from "pristine" swarm experiments may be used. Above this value, swarm data is no longer applicable. Ionization coefficients also decrease with increasing diameter of the dust particles.

We see, then, that electron transport in plasmas contaminated by particulate matter may not necessarily be well represented by swarm data obtained under pristine conditions. Due to the many variables upon which transport coefficients may depend under these conditions (e.g., λ_D, ρ, dust radius, ϕ_s), it is not clear that there is a simple scaling that may be applied to pristine swarm data to extend it into the contaminated regime.

CONCLUDING REMARKS

In this article, we have briefly examined the conditions under which swarm data may be used to model electric discharges. The parameter space (with respect to equilibrium of the EED, current density and cleanliness) in which the "intrinsic" or "pristine" swarm data may be used is fairly small. There do, however, appear to be new sets of scaling parameters which may be used to adapt intrinsic swarm data to model gas discharges under more relevant conditions. Realization of these scaling parameters will permit more engineering oriented models to be constructed.

ACKNOWLEDGMENTS

This work was supported by the National Science Foundation with grants ECS 88-15781 directed by Dr. L. Goldberg and CBT 88-03170 directed by Dr. W. Grosshandler.

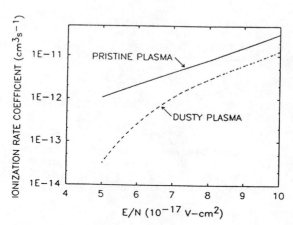

Fig. 10. Ionization rate coefficient in pristine and dusty argon plasmas ($\rho = 10^5$ cm^{-3}).

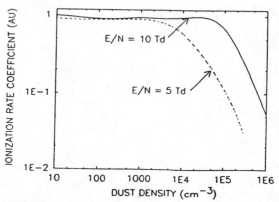

Fig. 11. Ionization rate coefficient for Ar plasmas as a function of dust density. The coefficients have been normalized by their pristine values.

REFERENCES

Boeuf, J. P., and E. Marode, 1982, J. Phys. D 15, 2169.
Dorrin, V. I., and Yu. I. Khapov, 1986, Sov. J. Quant. Electron 16, 1034.
Dutton, J., 1975, J. Phys. Chem., Ref. Data 4, 577.
Friedland, L., J. H. Jacob, and J. A. Mangano, 1989, J. Appl. Phys. 65, 3790.
Hayashi, M., 1987, "Swarm Studies and Inelastic Electron-Molecule Collisions," edited by L. C. Pitchford, B. V. McKoy, A. Chutjian, and S. Trajmar, (Springer-Verlag, New York), pp. 167-187.
Hunter, S. R., J. G. Carter, and L. G. Christophorou, 1986, J. Appl. Phys. 60, 24.
Kushner, M. J., 1987, J. Appl. Phys. 61, 2784.
Kushner, M. J., 1989, J. Appl. Phys. 66, 2297.
Mitchner, M., and C. H. Kruger, 1973, "Partially Ionized Plasmas," (Wiley, New York), pp. 126-134.
Moratz, T. J., 1988, J. Appl. Phys. 63, 2558.
Morrow, R., 1987, Phys. Rev. A 35, 1778.
Phelps, A. V., 1987, "Swarm Studies and Inelastic Electron-Molecule Collisions," edited by L. C. Pitchford, B. V. McKoy, A. Chutjian, and S. Trajmar, (Springer-Verlag, New York), pp. 127-142.
Roth, R. M., K. G. Spears, G. D. Stein, and G. Wong, 1985, Appl. Phys. Lett. 46, 253.
Sommerer, T. J., W. N. G. Hitchon, and J. E. Lawler, 1989, Phys. Rev. A 39, 6356.
Wedding, A. B., H. A. Blevin, and J. Fletcher, 1985, J. Phys. D 18, 2361.
Wu, C., and E. E. Kunhardt, 1988, Phys. Rev. A 37, 4396.

NON-EQUILIBRIUM EFFECTS IN DC AND RF GLOW DISCHARGES

David B. Graves and M. Surendra

Department of Chemical Engineering
University of California, Berkeley
Berkeley, CA 94720

INTRODUCTION

Glow discharges are used in a variety of applications, including electronic materials processing, lasers, discharge switches and lighting sources. In the microelectronics industry, considerable interest has developed within the last decade in plasma film etching and deposition. A major advantage of glow discharges in microelectronics material processing is that the gas can remain at or near room temperature while considerable physical and chemical energy is supplied to the surface. Indeed, the enormous range of conditions that can be achieved in low pressure plasmas by varying operating conditions and reactor design makes a plasma environment an extraordinarily versatile surface processing tool. The major processes occurring in the plasma that result in surface modification are (1) electron-impact dissociation of molecules to create chemically active molecular fragments; and (2) ion bombardment of surfaces bounding the plasma. The first effect is chemical: the original parent species is often not reactive and the fragment resulting from electron-impact dissociation is often very reactive at the surface. The second effect is primarily physical in that ions with tens to hundreds of electron volts of energy can dramatically influence virtually all surface processes from bond creation and breaking to desorption, lattice disruption, etc. The structure of the plasma determines the nature of these surface effects: electron density and energy distribution change dissociation rates and plasma potentials, ionization rate profiles and other parameters influence ion energies incident at surfaces.

Although low pressure, weakly ionized plasmas have proven remarkably useful, the complexities in chemically reacting discharges makes process development inefficient. As a result, a number of investigators have attempted to develop models and simulations that will help unravel the

couplings and complexities in this technology. In this paper, attention
will be focused on recent work in simulations of chemically simple discharges, and in particular on treating the non-equilibrium aspects of dc
(direct current) and rf (radiofrequency) glows.

The discharges that will be discussed here are characterized by a
number of parameters. The plasma densities are usually between 10^8 and 10^{12}
cm^{-3}; average electron energies range from a few tenths of an electron volt
to tens of electron volts (however, some electrons may have energies well in
excess of this); ions have energies that are on the same order as the room
temperature neutral gas (except in sheaths); degrees of ionization range
from 10^{-3} to 10^{-6}; neutral gas pressures typically range from 10^{-2} Torr to
10 Torr; and applied voltages vary from 50 to 500 volts, typically. These
discharges are often struck between parallel plate electrodes separated by a
few to perhaps tens of centimeters, and attention in this paper will be
restricted to this configuration. In direct current glows, electron emission from the cathode due to ion, photon, metastable etc. bombardment
sustains the discharge. In rf glows, electron emission may or may not be an
important discharge mechanism. We will not consider discharges with
appreciable quantities of negative ions.

In swarm experiments, electric fields tend to be uniform and static,
spatial gradients are usually mild and charged particle density is low
enough that space charge can be ignored. However, in discharges none of
these simplifying conditions hold, in general. Quantities vary as a function of position (especially near bounding surfaces) and time (in rf
discharges), and space charge is a central issue coupling electron and ion
motion. Furthermore, the strong gradients that form naturally near walls
not only influence the rest of the glow, but also are of great importance in
applications that involve surface processing. These gradients, and in rf
discharges time variations of these gradients, can introduce non-local and
non-equilibrium effects in discharge structure.

DISCHARGE MODELS

We know that weakly ionized plasmas are intrinsically non-equilibrium
in the sense that electrons are not at local thermal equilibrium with other
species in the discharge. The form of the electron velocity distribution
is, in general, not Maxwellian (and may not even be close to isotropic) and
depends in a complex way on the gas, the electric field profile and
gradients in the discharge. Ions tend to be close to the neutral gas
temperature, and it is commonly assumed that ions are at the local gas
temperature. However, ions in the sheaths often show large deviations from
the local gas temperature; indeed, the highly directed nature of ion motion

in the high field sheath regions is not well described by an isotropic
velocity distribution. Space charge is important and must be accounted for
in modeling the discharge. As a result of these considerations, one would
like to be able to formulate a description of the plasma that can account
for both electron and ion non-equilibrium as well as predicting the self-
consistent electric field profile. However, rigorously accounting for
electron and ion non-equilibrium in a self-consistent model has proven to be
relatively expensive and difficult, so discharge modelers have explored
simpler alternatives involving moment equations, also known as the equations
of change (Self and Ewald, 1966; Ward, 1962; Graves and Jensen, 1986;
Richards et al., 1987; Barnes et al., 1987; Boeuf, 1987). Sometimes, these
equations are referred to as 'fluid' equations. These equations have been
very useful in helping to understand the complexities of discharge structure
(Boeuf and Belenguer, 1989).

A technique that has been applied to problems in fusion and space plasma
physics, namely the particle-in-cell method, is now being adapted to weakly
ionized, collisional glow discharge plasmas. This technique, while limited in
some respects, can be used to obtain a self-consistent, fully kinetic description of electrons and ions in discharges. The particle-in-cell method and
some recent results will be described later in the paper (Birdsall and
Langdon, 1985; Boswell and Morey, 1988; Surendra et al., 1990b).

A complete macroscopic picture of a discharge involves the number
density, flux and energy of all charged species in addition to the self-
consistent electric field, all as a function of position and time. The
charged species will be assumed to be immersed in some neutral gas at a
given density. Due to the weakly ionized nature of the gas and with the
assumption of no chemistry, no equations need be solved for neutral species.
In some discharges, this is a good approximation. Of course, in chemically
active plasmas the local neutral gas composition can vary, and especially if
molecular dissociation products are present (for which cross sections are
rarely known), the situation is complicated substantially. There is a
serious need for cross section data for unstable molecular fragments in
models of plasma chemical processes.

Local Field Assumption

The simplest self-consistent model consists of continuity equations for
electrons and ions coupled with Poisson's equation and the 'local field'
assumption. The latter assumption associates the instantaneous and local
electric field with values for rate and transport coefficients that go into
continuity equations. Since rate and transport coefficients from swarm
experiments are correlated with E/N, swarm data can be used directly in this

159

model. A typical set of local field equations are the following, assuming quantities vary with only a single dimension, x:

$$\partial n_e / \partial t + \partial j_e / \partial x = R_{ion} \tag{1}$$

$$\partial n_+ / \partial t + \partial j_+ / \partial x = R_{ion} \tag{2}$$

$$\partial E / \partial x = e/\varepsilon_0 (n_+ - n_e). \tag{3}$$

In (1), n_e is electron density, j_e is electron flux (the product of number density and directed velocity) and R_{ion} is ionization rate. Corresponding quantities for positive ions are in (2). Equation (3) is Poisson's equation, which provides the electric field E. (1)-(3) are incomplete as written since no expressions have yet been provided for j_e, j_+ or R_{ion}. In a local field model, one commonly assumes that electron and ion motion are collisionally dominated so that momentum balance equations reduce to expressions involving only drift and diffusion:

$$j_e = -\mu_e n_e E - D_e \partial n_e / \partial x \tag{4}$$

$$j_+ = \mu_+ n_+ E - D_+ \partial n_+ / \partial x. \tag{5}$$

In (4) and (5), μ is mobility and D is diffusivity for the respective species. As mentioned above, (4) and (5) result when one begins with the momentum balance equation for each species, then discards all terms but the force due to the applied field (mobility term) and that due to pressure or density gradient (diffusivity term). This neglects the acceleration term (time rate of change of net velocity) and the inertial term (involving the product of net velocity and the gradient in net velocity). Unless the electron velocity distribution function is highly anisotropic (as will be seen later for high energy electrons), these assumptions are generally acceptable for electrons. For ions in rf discharges, it is in principle necessary to include the acceleration term because the ion—neutral momentum transfer collision frequency is not large with respect to applied frequencies in common use (Richards et al. 1987).

In local field models the quantities normally correlated with electric field are ion mobility and ionization rate. Boeuf (1987) used the expression reported by Ward (1962) for electrons and helium ions in a helium background gas:

$$\alpha /p = A \exp[-B(p/E)^{1/2}] \tag{6}$$

$$\mu_+ = C_1/p \ (1 - C_2 E/p) \qquad\qquad E/p < W_1 \tag{7a}$$

$$\mu_+ = C_3/(Ep)^{1/2}(1 - C_4 E^{-3/2} p^{3/2}) \qquad E/p > W_1. \tag{7a}$$

The Townsend first ionization coefficient α is related to $R_{ion} = \alpha j_e$. A, B, C_1, C_2, C_3, C_4 and W_1 are constants, determined by fits to experimental data. Electron and ion diffusivities and electron mobility were taken to be constants. Values used by Ward (1962) are listed in Table 1.

The above set of equations have been used extensively in the field of semiconductor device simulations, where of course the positive species are holes rather than ions (Barnes et al., 1987; Boeuf, 1987). Furthermore, numerical techniques to solve the equations have been well developed, and much progress has been made in extending these calculations to two spatial dimensions. The local field approximation is the simplest and computationally most efficient self-consistent discharge simulation technique. As a result, it would be useful to have a careful analysis of its strengths and weaknesses so that it can be exploited when possible. Preliminary results suggest that the approach is at least qualitatively correct for applied (radio) frequencies above the ion plasma frequency and at pressures above several hundred millitorr. At lower frequencies, directed electrons that originate in sheaths can dominate plasma structure and the local field approach can be qualitatively in error. In such cases, the use of separate moment equations for high energy 'beam' electrons can be successful, at least qualitatively, in representing discharge structure. At pressures below several hundred millitorr, mean free paths become too large for the local field, drift-diffusion assumption to be satisfactory. At lower pressures, other techniques must be used, and one possibility (the particle-in-cell method) is discussed later.

Table 1. Local Field Model Parameters for Ionization and Transport Coefficients in Helium Discharges (Ward, 1962; Boeuf, 1987)

A	$4.4 \text{ cm}^{-1} \text{ Torr}^{-1}$
B	$14.0 \text{ V}^{1/2} \text{ (cm Torr)}^{-1/2}$
C_1	$8.0 \times 10^3 \text{ cm}^2 \text{ Torr (V sec)}^{-1}$
C_2	$8.0 \times 10^{-3} \text{ Torr cm V}^{-1}$
C_3	$4.1 \times 10^4 \text{ cm}^{3/2} \text{ Torr}^{1/2} \text{ V}^{-1/2} \text{ sec}^{-1}$
C_4	$27.44 \text{ (V cm}^{-1} \text{ Torr}^{-1})^{1/2}$
W_1	$25 \text{ V cm}^{-1} \text{ Torr}^{-1}$
μ_e	$10^6 \text{ cm}^2 \text{ Torr V}^{-1} \text{ sec}^{-1}$
D_e	$10^6 \text{ cm}^2 \text{ Torr sec}^{-1}$

Beam Electron Models

In low frequency rf discharges and dc discharges, it is well known that electrons created at electrode surfaces via secondary emission processes and subsequently accelerated into the plasma play the key role in sustaining the discharge (Boeuf and Belenguer, 1989). Under these conditions, the so-called beam electrons (motion strongly directed with a small random component) are primarily responsible for important electron-impact inelastic collisions such as ionization, dissociation, excitation etc. However, the number density of these fast electrons is typically orders of magnitude smaller than the relatively cold bulk electrons sitting in the plasma or quasineutral region. An obvious strategy is to write separate equations for the fast or beam electron (Boeuf and Belenguer, 1989; Phelps et al., 1987; Sommerer et al., 1988; Gottscho et al., 1988; Surendra et al., 1990a) and bulk electrons and ions. The beam electrons in such a model would be responsible for all ionization in the discharge, so the local field assumption is relaxed for ionization. However, the electric field in the discharge is dominated by space charge from bulk electrons and ions, and acceleration of beam electrons depends on this field profile; hence, the bulk and beam equations are coupled.

In a <u>single</u> beam model, the assumption is made that all (fast or beam) electrons have the same purely directed velocity at every point in space at each point in time. Then the beam electrons are completely characterized by their number density n_{be} and velocity v_{be}; the product of density and velocity is beam flux j_{be}. Number density and velocity are determined by continuity and energy balance equations, respectively. The energy of the directed electrons ε_{be} is of course $1/2 m_e v_{be}^2$, where m_e is electron mass. Two issues complicate this simple model in a discharge: first, beam electrons can be created in the high field cathode fall region, but not in the low field plasma region; and second, some method of transferring electrons from the beam group to the bulk group must be devised for beam electrons that suffer enough inelastic collisions before they reach the anode. Both issues introduce ad-hoc features into the model. The equations for beam electrons in a single beam model (ignoring time variations) are the following (Surendra et al., 1990a):

$$dj_{be}/dx = (1-\alpha)R_{ion}, \text{ for } \varepsilon_{be} > \varepsilon_{ion}, \text{ and} \qquad (8a)$$

$$dj_{be}/dx = -\alpha R_d, \text{ for } \varepsilon_{be} < \varepsilon_{ion}, \text{ where} \qquad (8b)$$

$$R_{ion} = \sigma_{ion}(\varepsilon_{be}) v_{be} n_n n_{be}, \text{ and} \qquad (9a)$$

$$R_d = \sigma_{en} v_{be} n_n n_{be}. \qquad (9b)$$

162

In equation (8a), α is 0 when in the cathode fall, so beam multiplication occurs, and is 1 in the bulk or plasma region. In this region, ionization creates bulk, not beam, electrons. Equation (8a) is for electron energies above the ionization threshold (ε_{ion}), so the flux either increases or remains constant. If beam electron energy is below this threshold, equation (8b) states that beam electrons are lost at some rate R_d, which depends on the total collision cross section, σ_{en}. R_{ion} and R_d depend on the beam velocity or energy, which is calculated via the beam energy equations:

$$d(j_{be}\varepsilon_{be})/dx = ej_{be}\, dV/dx - \left\{R_{ion}\varepsilon_{ion} + R_e\varepsilon_{ext}\right\}, \text{ for } \varepsilon_{be} > \varepsilon_{ion}, \text{ and} \quad (10a)$$

$$d(j_{be}\varepsilon_{be})/dx = ej_{be}\, dV/dx - \left\{R_d\varepsilon_{be} + R_e\varepsilon_{ext}\right\}, \text{ for } \varepsilon_{be} < \varepsilon_{ion}, \text{ and} \quad (10b)$$

$$R_e = \sigma_{ext}(\varepsilon_{be})v_{be}n_n n_{be}. \quad (11)$$

In (10a), the divergence of beam energy flux is balanced by the energy gained by acceleration in the field (which can be an energy gain or loss depending on the sign of the field), and the collisional losses, which here we have restricted to ionization and electronic excitation (R_e). For molecular gases, other terms would be included.

Bulk electron and ion transport coefficients used are listed in Table 2; cross sections for beam electrons are plotted in Fig. 1. It is intended that these transport coefficients and cross sections approximate electron-argon collisions. Typical results (all taken from Surendra et al., 1990a) for these single beam equations are shown in Fig. 2. The cathode fall extends about 2 mm from the cathode (at position x = 0.0), and in this region positive ion density is nearly uniform, resulting in a nearly linear variation in electric field profile. Although it is not possible to see on the scale of the figure, the electric field goes through zero very near the maximum in the plasma density, at a position about 5 mm from the cathode. Ions fall to the cathode if they are created on the cathode side of the density maximum and to the anode if they are created on the anode side of the density peak. This is because ion transport is completely controlled by field-induced drift; therefore, ions always move in the direction of the local electric field. Electrons move from cathode to anode, and carry most of the discharge current in the region to the anode side of the density maximum. Our results for dc discharges invariably show the zero crossing for the electric field close to the maximum in plasma density and therefore that ions created on the cathode side of the density maximum fall to the cathode.

Although the single beam model is adequate to qualitatively represent the non-local nature of fast electron transport, it clearly has limitations. Even if one assumes that fast electron motion is completely forward directed

Table 2.   Conditions and Discharge Parameters for Single Beam –
           Fluid Model of DC Glow Discharge in Ar (Surendra et al.,
           1990a)

Quantity	Value
Discharge Voltage	300 V
Electrode Gap	2.0 cm
Gas Pressure	0.6 Torr
Electron Diffusivity	2.0×10^5 cm^2 Torr sec^{-1}
Electron Mobility	2.0×10^5 cm^2 Torr V^{-1} sec^{-1}
Ion Diffusivity	4.0×10^2 cm^2 Torr sec^{-1}
Ion Mobility	$\mu_+ = \mu_{+0} [1+a|E/n_n|]^{-1/2}$
μ_{+0}	1.42×10^3 cm^2 Torr V^{-1} sec^{-1}
a	7.36×10^{14} cm^{-2} V^{-1}
Secondary Electron Emission Coefficient, γ	0.033

(that is, no angular scattering), if electrons are created at different
positions in the cathode fall or if they lose energy via inelastic collisions at different locations, the distribution function will not be a delta
function in velocity space: there will be a distribution of fast electron
velocities.   We have therefore explored a multibeam model which can predict

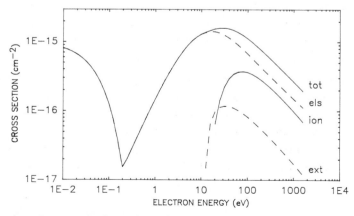

Fig. 1.   Cross sections (total, elastic, ionization, composite
          excitation) for electron-neutral collisions in argon.

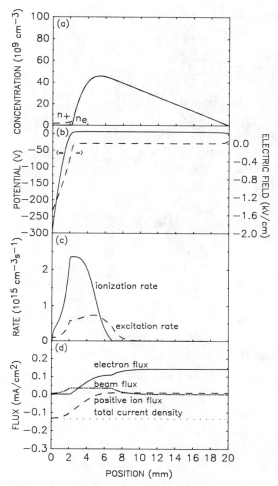

Fig. 2.   Solution to the single beam model (300V, 0.6 torr).   (a) Slow
electron and ion densities; (b) potential and electric field;
(c) inelastic rates; (d) charged particle fluxes and total
current density.

a distribution of fast electron velocities.   The approach we take is based
on the ideas of Carman and Maitland (1987) which can be shown to be equivalent to a finite difference solution to the Boltzmann equation for fast
electrons assuming completely forward scattering.   Details can be found in
Surendra et al. (1990a).   A typical fast electron flux distribution in the
cathode fall region is shown in Fig. 3.   Note the existence of the spikes at
high energy.   These electrons have either experienced no inelastic
collisions or only a single inelastic collision, and therefore have either
exactly the cathode fall potential or just less than that quantity.   Another

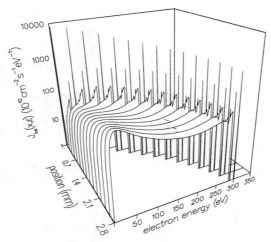

Fig. 3. Energy distribution of fast electron flux as a function of
         position at 300 V and 0.6 torr from the multibeam model
         solution.

important point is the relatively flat distribution from about 50 eV to the
cathode fall potential. This distribution is significantly different from a
single velocity approach.

    The last example of treatment of fast electrons in dc discharges is the
use of a hybrid fluid-particle model (Surendra et al., 1990a). In order to
take account of angular scattering of fast electrons, we have used a Monte
Carlo simulation of fast electrons and have made the treatment self-
consistent by iterating with the bulk fluid equations in a manner identical
to the one used with fluid beam electron models. In this approach, electrons in the cathode fall are all assumed to be fast electrons, but in the
low-field region, only electrons with energy above the first inelastic
threshold (about 12 eV for argon) remain in the fast group. Slower electrons drop into the bulk electron group and contribute to space charge but
do not experience inelastic collisions.

    The assumed form for angular dependence of the electron-neutral
collision cross sections is given in Fig. 4. This form was chosen partly
for convenience and because it qualitatively represents the gross features
of angular scattering as a function of electron energy: at higher energies,
electrons tend to forward scatter, whereas at lower energies, angular
scattering is important. Figure 5 summarizes the results of various
scattering assumptions for fast electrons when treated using the particle
simulation approach. In Fig. 5a, fast electrons are assumed to scatter
isotropically; in Fig. 5b, scattering follows the angular form in Fig. 4;
and Fig. 5c assumes completely forward scattering. The latter assumption

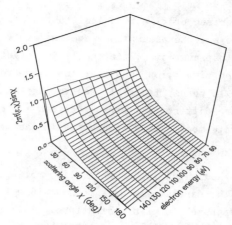

Fig. 4.   Normalized differential cross sections used in particle-
          fluid hybrid simulation of dc glow: $f(\varepsilon,\chi) = \varepsilon/\{4\pi[1 + \varepsilon$
          $\sin^2(\chi/2)]\ln(1 + \varepsilon)\}$.

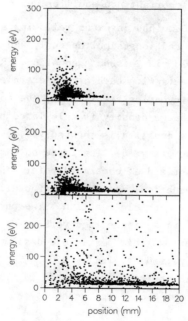

Fig. 5.   Snap shot of energy and position of electrons for different
          scattering assumptions. $E(x)$ and electron beam flux from
          the cathode are from the multibeam model solution at 275 V
          and 0.6 torr: (a) isotropic scattering; (b) anisotropic
          scattering; (c) full forward scattering.

results in a fast electron profile identical to that from the multibeam model, as it should since the assumptions are the same. The major point from this figure is that angular scattering is important under the conditions used in the simulation, and that the anisotropic form assumed gives results closer to purely isotropic scattering than purely forward scattering.* **Accurate differential cross sections are essential for quantitatively accurate models of discharges with fast electrons.**

Particle-In-Cell Technique

A simulation method which is, in principle, more rigorous than fluid or moment equations is the particle-in-cell method since it is a self-consistent kinetic approach to plasma structure. This technique (Birdsall and Langdon, 1985; Hockney and Eastwood, 1981) has been developed primarily for application in collisionless fusion and space plasmas, but is beginning to attract the interest of investigators in plasma materials processing (Boswell and Morey, 1988). A brief description of the method will be given here, but details can be found in the many articles and monographs that are available (Tajima, 1989; Kruer, 1988; Dawson, 1983). We then present some preliminary results from PIC simulations of rf discharges between parallel plate electrodes. The primary motivation for these simulations is to understand some of the important non-equilibrium effects in rf discharges.

In essence, the PIC method involves following a few thousand to perhaps a few million electrons and ions simultaneously and solving for the self-consistent electric field via Poisson's equation. Rather than try to follow how each particle interacts with each other particle (as would be done, for example, in a molecular dynamics simulation), the contribution of each particle to the local charge density is calculated by interpolation to a spatial grid. Poisson's equation is then solved, typically via finite difference, on this discrete grid. The force on each particle is determined by finding the electric field at each grid point (this is the numerical solution to Poisson's equation) and then interpolating from the nearest grid points to the individual particle position. The particle's position and velocity is advanced using a discretized version of Newton's second law. The time step taken (for explicit particle codes) must be a small fraction of the electron plasma period, typically 0.05. The normal cycle is shown in Fig. 6. Collisions with neutrals are treated in a straightforward manner, using a standard Monte Carlo approach, such as those employed with Monte

---

* In making this comparison between scattering laws, we have kept the total cross section constant, thereby changing the momentum transfer cross section. Some authors have suggested keeping the momentum transfer cross section constant.

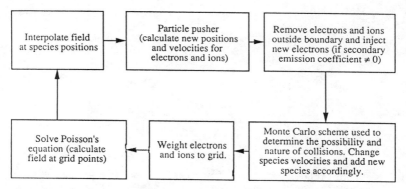

Fig. 6. PIC computation cycle, calculating the positions and
velocities of electrons and ions self-consistently with the
electric field. Collisions with neutrals are treated using
a Monte Carlo approach.

Carlo swarm simulations. Since the PIC time step is much less than
electron-neutral or ion-neutral collision times, it is more convenient to
use the constant time step technique than the null collision cross section
Monte Carlo approach. Typically, about 4% of electrons have a collision per
time step.

The key idea behind the particle-in-cell method is that in a plasma one
is most often interested in the collective behavior of large numbers of
charged particles interacting over relatively large distances. This is a
consequence of the fact that charged particles interact primarily via relatively long range Coulomb forces. As a result, one can employ a particle
simulation that uses 'finite size particles' to smooth out the microfields
associated with individual local particle encounters. The fact that
particles are interpolated to a spatial grid in PIC codes acts as the
important smoothing mechanism that allows meaningful results to be obtained
with relatively few particles (Dawson, 1983).

The electron-neutral collision cross sections we have used in this
simulation are shown in Fig. 7. Only elastic and ionization collisions are
included in the preliminary simulation. The values chosen are meant to
approximate helium cross sections. Ion collisions are modeled assuming a
constant collision frequency, so the cross section varies inversely with ion
velocity:

$$\sigma_{ion} = K_{ce}/v_+ \qquad\qquad (12)$$

Operating conditions and numerical values for the simulation are listed
in Table 3. We chose an applied frequency of 30 MHz because ion mass in
helium is relatively low and it was the intent to be well above the applied
frequency at which ions can follow the time varying fields. (Future

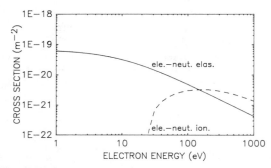

Fig. 7. Model cross sections for elastic and ionization electron-
neutral collisions used in PIC simulation.

simulations will address lower frequencies.) Results are averaged over at
least 10 cycles in plotted results.

Figure 8a is a plot of ion density as a function of position and rf
phase (as a parameter). At 30 MHz, the ion density profile doesn't modulate
significantly during the period, which is expected. Figure 8b is the corresponding electron density profile, and the sheath modeulation is marked.
Note that the discharge is symmetric but 180 degrees (π radians) out of

Table 3. Conditions and Discharge Parameters for RF PIC Simulation

Discharge Voltage	$800\ V_{RF}$
Applied Frequency	2π x 30 MHz
Electrode Gap	4.0 cm
Gas Pressure	0.1 Torr
Secondary Electron Emission Coefficient, γ	0.0
Electron Reflection Coefficient	0.25
Ratio of Ion Mass to Electron Mass	4.0 x 1836 (4 amu)
Number of Grid Points	100
Time Step	1.7×10^{-10} sec
Total Number of Particles at Steady State Electrons Ions	6.5×10^{3} 7.0×10^{3}
Number of Periods Integrated	350

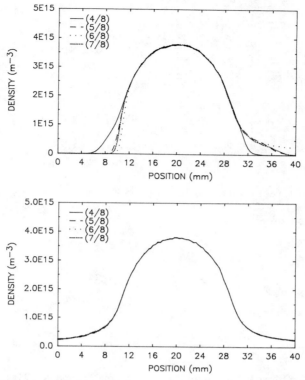

Fig. 8.   Solution to PIC model of rf discharge at 0.1 torr, 30 MHz
and 800 V$_{rf}$ applied voltage: (a) ion density profiles at
four times in the rf period; (b) corresponding electron
density profiles, showing sheath modulation.

phase.   For example, all quantities at position x = 12 mm at 6/8 through the
period are identical to those at position x = 28 mm at 2/8 through the
period.   (The exception is for vector quantities which must change sign at
the midplane, such as electric field; in this case, however, it is the
absolute value that is symmetric).   Thus results are similar to profiles
from local field fluid simulations (Boeuf, 1987) at a lower applied
frequency of 10 MHz.   The voltage and electric field profiles are shown in
Figs. 9a and 9b, respectively.   Note that as expected, the fields are
largely confined to the sheaths.

A major advantage of the PIC method is that velocity distributions are
generated self-consistently.   In particular, the electron velocity (or
energy) distribution function is of interest because ionization (and other
important inelastic collision) rates depend sensitively on the high energy
tail of the distribution.   Boltzmann calculations in (spatially uniform) dc
and rf fields for electrons in atomic gases predict that the tail of the
distribution is depressed relative to low energy electrons (Winkler et al.,

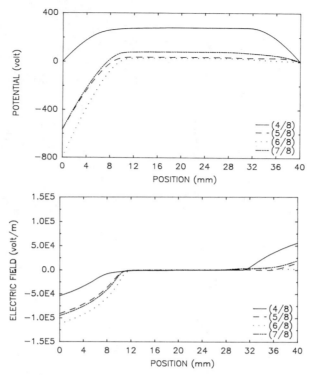

Fig. 9. PIC solutions for the same conditions as Fig. 8:
(a) potential profiles; (b) electric field profiles.

1984; Vriens, 1973, Postma, 1969). This is because of the existence of
inelastic collisional losses above some threshold energy, typically 10-20
eV. One can often approximately represent the electron energy distribution
function – (eedf) in rare gases as two electron temperatures: one
temperature for the 'bulk' group and one for the 'tail' group. Inelastic
collisions cause the tail temperature to be less than the bulk temperature.
However, in PIC simulations of 30 MHz rf discharges, we have found evidence
of a higher tail temperature compared to the bulk temperature rather than
the reverse. This is demonstrated in Fig. 10. One can clearly see the
difference between the eedf from the PIC simulation (in the center of the
discharge) and a swarm simulation which utilizes a uniform electric field
with about the same mean electron energy. In the former, the tail has a
higher equivalent temperature (which is modulated in the rf field) than the
bulk, and in the latter, the tail has a lower equivalent temperature than
the bulk group.

We believe this 'tail enhancement' is a direct result of heating at the
plasma-sheath boundaries during the time the sheath expands into the plasma.
As Boeuf and Belenguer (1989) have outlined, one can distinguish three types

172

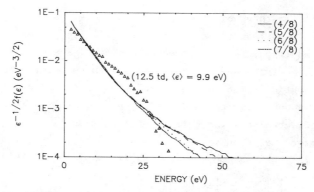

Fig. 10. Electron energy distribution function, averaged from the
central 8 mm of the discharge as a function of time in the
rf period. Also shown is eedf from a local field simulation at the same mean electron energy as the PIC results.

of electron heating in an rf discharge: (1) secondary electrons (and
progeny) that accelerate through sheaths completely analogously to the dc
case; (2) bulk Joule heating due to electric fields and conduction current
in the body of the plasma; and (3) sheath heating which is due to the
oscillating motion of the sheaths (Godyak, 1972; Liebermann, 1988; Kushner,
1986). Note that in our PIC simulations, the secondary electron emission
coefficient has been set to zero. The observed tail enhancement is
therefore not due to secondary electrons that fall through the rf sheaths.

    This effect of sheath heating and tail enhancement appears to be unique
to rf discharges since it relies on both time- and space-varying electric
fields at the boundaries of the quasineutral plasma (i.e. the plasma-sheath
boundaries). It is therefore difficult to directly compare this case to
studies that incorporate either solely time-varying (Kunhardt et al., 1988)
or solely space-varying (Moratz et al., 1987) electric fields. Godyak and
Oks (1979) have measured electron energy distribution functions in 40.58 MHz
Hg discharges at pressure-gap (pL) products below about 0.02 Torr-cm and
find that, indeed, the distribution is enriched with fast electrons as
compared to a Maxwellian. It was suggested (Godyak and Oks, 1979) that this
is because only energetic electrons are capable of overcoming the ambipolar
electric fields that tend to confine lower energy electrons to the center of
the discharge. Godyak (1972) has also demonstrated that if one considers a
group of Maxwelliam electrons interacting with a sinusoidally varying sheath
velocity, the distribution function is enhanced at high electron energies
after a single 'collision' with the sheath.

    Note that pL in our simulation is 0.4 Torr-cm, above the limit observed
by Godyak and Oks (1979) for fast electron enhancement. We attribute the
difference to the fact that in the model gas we use in the simulation, the

inelastic collision threshold energy is about 24 eV, whereas mercury (the gas used in the experiments) has inelastic collisions at much lower electron energies. In addition, fast electron enhancement is probably a function of parameters such as frequency, rf voltage, sheath velocity, etc., all of which are no doubt different in our simulations as compared to the experiments. Since we have eliminated secondary electrons from the simulation, the only other mechanism for electron heating is Joule heating in the bulk of the discharge. We cannot rigorously exclude bulk Joule heating as the source of the tail enhancement, since the fields in the bulk do vary with both position and time in the period, and are therefore not identical to local field conditions (which invariably show tail depression). However, the small magnitude of the bulk fields (about 1 volt/cm Torr) suggests that this is not the mechanism responsible for enhancing the tail of the electron energy distribution. More simulations and careful comparison to experiment will be necessary to completely resolve this issue.

APPLICATION OF SWARM DATA TO NON-EQUILIBRIUM DISCHARGES

Most swarm experiments are conducted under conditions in which perturbations due to spatial gradients and time-varying quantities are minimized. One obtains transport and rate coefficients under well defined conditions, parameterized by the value of E/N and the gas composition. These data can be used directly in discharge simulations that employ either the local field approximation or those that parameterize electron and ion transport through a mean energy, obtained through a simultaneous solution of an energy balance equation. Any discharge simulation that attempts to predict velocity or energy distributions requires cross sections, and perhaps differential cross sections. The latter are especially important when directed electrons are present, for example in dc and low frequency rf discharges. Cross sections can be inferred from swarm data by iterating on trial cross section sets until rate and transport coefficients predicted from the trial cross sections agree with measured coefficients. It hardly needs to be stressed that without accurate cross sections, quantitative discharge simulations are not possible. Other phenomena, such as ion-molecule or fast neutral-molecule ionization and excitation collisions, have not been discussed here. Phelps and coworkers (1987) have demonstrated the importance of these collisions at very high E/N, and these collisions may be significant in the high-field sheath regions of glow discharges. Unpublished results from our laboratory suggest that the simple dc discharge models we have presented fail at moderate to high values of the cathode fall voltage (above about a few hundred volts). This appears to be due to complications involving emission (and probably ionization) near the cathode, unexplained ion emission in the center of the discharge, and the roles of contaminant impurities (such as H_2 absorbed in the chamber walls) and materials sputtered from the cathode (such as Fe and Al, depending on electrode materials used). Quantitative models of discharges which include these complications have yet to be fully developed and tested.

Even when it is desirable to use less sophisticated (and less expensive) methods involving the solution of moment equations, it is important to be able to evaluate the accuracy of the simpler approaches with more complete models (Boeuf and Belenguer, 1989; Moratz et al., 1987). Then the simpler and less expensive techniques can be used when multiple spatial dimensions are to be included in simulations, and especially when detailed chemistry is to be incorporated into the model. Both additions will be included in future models of plasma processes, so there is a strong motivation for developing acceptably accurate approximate models. An important challenge is to find ways to use swarm data meaningfully, and also to expand the swarm data base to include transient, chemically active species. Also, data for fast neutral particles and ions are needed. A summary of discharge models and their utilization of swarm data is provided in Table 4.

Table 4. Discharge Models and Utilization of Swarm Data

Equations	Data	Strengths	Weaknesses
First two moment eqns. (continuity, momentum) for electrons & ions.	Use swarm data directly: $\mu(E/N)$; $D(E/N)$; $\alpha\ (E/N)$, etc.	Simple; inexpensive to solve.	Inaccurate for non-local, non-equilibrium phenomena.
First three moment eqns. for electrons (add energy balance); first two moment eqns. for ions.	Use swarm data indirectly: $\mu(\varepsilon)$; $D(\varepsilon)$; $\alpha\ (\varepsilon)$, etc.; $\varepsilon=\varepsilon(E/N)$.	Can treat electron energy with temperature gradients and bulk electron motion.	Inaccurate for strongly non-local and non-equilibrium phenomena such as beam electrons. Also adds significantly to computational cost.
Add moment eqns. for fast electrons. Local field approx. for cold electrons and ions.	Cross sections for beam; swarm data for cold electrons and ions.	Handles strongly non-local effects due to highly forward scattered electrons.	Ad-hoc treatment for two separate groups of electrons. Ignores angular scattering & therefore over-predicts fast electron penetration.
Monte Carlo for fast electrons, including angular scattering. Local field approx. for cold electrons and ions.	Differential cross sections for particle electrons; swarm data for cold electrons and ions.	Handles highly non-local effects but is more accurate than beam models.	More costly than beam models. Still requires ad-hoc two electron groups.
All species treated as particles.	Differential cross sections.	Ion and electron velocity distributions obtained in self-consistent field. Coding is simple. Treats non-local, non-equilibrium effects with no ad-hoc assumptions.	Relatively expensive computationally. Explicit particle codes are limited to relatively short time- and space-scales. Implicit codes not yet well developed for bounded plasmas.

REFERENCES

Barnes, M. S., T. J. Cotler, and M. E. Elta, 1987, J. Appl. Phys. 61, 81.

Birdsall, C. K., and A. B. Langdon, 1985, "Plasma Physics via Computer Simulation," McGraw-Hill, New York.

Boeuf, J. P., 1987, Phys. Rev. A 36, 2782.

Boeuf, J. P., and Ph. Belenguer, 1989, to appear in "Non Equilibrium Processes in Partially Ionized Gases," NATO ASI Series, M. Capitelli, and J. N. Bardsley, (Eds.).

Boswell, R. W., and I. J. Morey, 1988, Appl. Phys. Lett. 52, 21.

Carman, R. J., and A. Maitland, 1987, J. Phys. D 20, 1021.

Dawson, J. M., 1983, Rev. Mod. Phys. 55, 403.

Godyak, V. A., 1972, Sov. Phys. Tech. Phys. 16, 1073.

Godyak, V. A., and S. N. Oks, 1979, Sov. Phys. Tech. Phys. 24, 1255.

Gottscho, R. A., A. Mitchell, G. R. Scheller, N. L. Schryer, D. B. Graves, and J. P. Boeuf, 1988, Proc. Seventh Symp. on Plasma Processing, G. S. Mathad, G. C. Schwartz, and D. W. Hess, (Eds.), 88-22, The Electrochem. Soc., Pennington, N.J.

Graves, D. B., and K. F. Jensen, 1986, IEEE Trans. Plasma Sci. PS-14, 78.

Hockney, R. W., and J. W. Eastwood, 1981, "Computer Simulation Using Particles," McGraw-Hill, New York.

Jelenkovic, B. M., and A. V. Phelps, 1987, Phys. Rev. A 36, 5310.

Kruer, W. L., 1988, "The Physics of Laser-Plasma Interaction," Addison-Wesley, Redwood City, California.

Kunhardt, E. E., J. Wu, and B. Penetrante, 1988, Phys. Rev. A 37, 1654.

Kushner, M. J., 1986, IEEE Trans. Plasma Sci. PS-14, 188.

Liebermann, M. A., 1988, IEEE Trans. Plasma Sci. PS-16, 638.

Moratz, T. J., L. C. Pitchford, and J. N. Bardsley, 1987, J. Appl. Phys. 61, 2146.

Phelps, A. V., B. M. Jelenkovic, and L. C. Pitchford, 1987, Phys. Rev. A 36, 5327.

Postma, A. J., 1969, Physica 43, 581.

Richards, A. D., B. E. Thompson, and H. H. Sawin, 1987, Appl. Phys. Lett. 50, 492.

Scott, D. A., and A. V. Phelps, 1987, Bull. Am. Phys. Soc. 32, 1158.

Self, S. A., and H. N. Ewald, 1966, Physics of Fluids 9, 2486.

Sommerer, T. J., J. E. Lawler, and W. N. G. Hitchon, 1988, J. Appl. Phys. 64, 1775.

Surendra, M., D. B. Graves, and G. M. Jellum, 1990a, Appl. Phys. Lett., 56, 1022.

Surendra, M., D. B. Graves and I. J. Morey, 1990b, Phys. Rev. A, 41, 1112.

Tajima, T., 1989, "Computational Plasma Physics: With Applications to Fusion and Astrophysics," Adison-Wesley, Redwood City, California.

Vriens, L., 1973, J. Appl. Phys. 44, 3980.

Ward, A. L., 1962, J. Appl. Phys. 33, 2789.

Winkler, R., A. Deutsch, J. Wilhelm, and Ch. Wilke, 1984, Beitr. Plasmaphys. 24, 285.

MEASUREMENTS OF ATTACHMENT COEFFICIENTS IN THE PRESENCE OF IONIZATION

D. Kenneth Davies

Westinghouse Science and Technology Center
Pittsburgh, Pennsylvania 15235, USA

ABSTRACT

A critical comparison of available methods for the determination of
the attachment coefficient of an electronegative gas at values of E/N
(the ratio of electric field to gas density) above the onset of
ionization is presented. It is concluded that the shielded-collector
pulsed drift tube method leads to the most direct and unambiguous
determination of the attachment coefficient under these conditions.
Examples of data determined using this method are given for dry and
humid air, CO_2, HCl, and CCl_4.

INTRODUCTION

Many discharge devices operate in a region of E/N (the ratio of electric field to gas density) where a significant number of electrons have
sufficient energy to ionize the gas medium. Thus, the determination of
swarm parameters in this region of E/N provides a particularly desirable
contribution to the basic database necessary for obtaining a complete
understanding of device behavior. In this paper, we focus on the determination of electron attachment coefficients as a function of E/N in
electronegative gases above the onset of ionization; in this region of E/N
attachment processes are usually dissociative. Thus, in the context of this
paper, low-energy attachment processes are precluded, since the measurement
of attachment coefficients below the onset of ionization are more straightforward and not subject to ambiguous interpretation.

The methods that have been traditionally used for measurements of
attachment coefficients in the presence of ionization are briefly reviewed,
and their strengths and shortcomings discussed. An alternative technique is

Nonequilibrium Effects in Ion and Electron Transport
Edited by J. W. Gallagher *et al.*, Plenum Press, New York, 1990

described which provides a more direct approach to the measurement of attachment coefficients in the presence of ionization. Examples of attachment data as a function of E/N obtained using this technique are described for dry and moist air, CO_2, HCl, and CCl_4.

IONIZATION GROWTH IN THE PRESENCE OF ELECTRON ATTACHMENT

The determination of swarm parameters from experimental data obtained from the different experimental techniques all rely on the appropriate solutions to the continuity equations (Huxley and Crompton, 1974) for electrons and ions:

$$\frac{\partial n_e}{\partial t} = n_e w_e (\alpha - \eta) + D_L \frac{\partial^2 n_e}{\partial z^2} - w_e \frac{\partial n_e}{\partial z}, \tag{1}$$

$$\frac{\partial n_+}{\partial t} = n_e w_e \alpha + w_+ \frac{\partial n_+}{\partial z}, \tag{2}$$

$$\frac{\partial n_-}{\partial t} = n_e w_e \eta - w_- \frac{\partial n_-}{\partial z}, \tag{3}$$

where n_e, n_+, and n_- are the densities of electrons, positive ions, and negative ions, respectively; w_e, w_+, and w_- are the drift velocities of electrons, positive ions, and negative ions, respectively; D_L is the electron longitudinal diffusion coefficient; and α and η are the ionization and attachment coefficients, respectively.

Most of the experimental data have been analyzed neglecting diffusion effects. While this is generally a justifiable assumption for ions, the effects of electron diffusion become important at high values of E/N where the ratio $D_L(\alpha - \eta)/w_e$ is appreciable compared to unity. In cases where electron detachment is important, detachment terms must be included in the continuity equations and solutions to the equations have been obtained for such conditions (Raether, 1964; Llewellyn-Jones, 1967).

Under steady-state conditions where a continuous stream of electrons is released from the cathode, and neglecting the effects of electron diffusion, the solution of the continuity equations [Eqs. (1)-(3) with the right hand sides set equal to zero] leads to the following expressions for the electron and negative-ion density distributions:

$$n_e(z) = n_o \exp[(\alpha - \eta)z], \tag{4}$$

$$n_-(z) = n_o \frac{\eta}{(\alpha - \eta)} \frac{w_e}{w_-} \{\exp[(\alpha - \eta)z] - 1\}, \tag{5}$$

where n_o is the electron density at the cathode. The total current density at the anode is then given by

$$j(d) = j_0 \left\{ \frac{\alpha}{(\alpha - \eta)} \exp[(\alpha - \eta)d] - \frac{\eta}{(\alpha - \eta)} \right\}, \tag{6}$$

where $j_0 = e n_0 w_e$ is the current density at the cathode.

Under conditions where a pulse of electrons of duration δt is released from the cathode, such that $\delta t \ll d/w_e$, the solution of Eq. (1) leads to the expression for the time-dependent electron density in the field direction given by:

$$n_e(z,t) = \frac{n_0 \, \delta t \, w_e}{(4 \pi D_L t)^{3/2}} \exp\left[-\frac{(z - w_e t)^2}{4 D_L t} + (\alpha - \eta)w_e t \right], \tag{7}$$

for $0 \leq t \leq d/w_e$, where n_0 is the initial net electron charge released from the cathode. Eq. (7) reduces to

$$n_e(z,t) = n_0 \delta t \, w_e \exp[(\alpha - \eta)w_e t], \tag{8}$$

when $D_L(\alpha - \eta)w_e \ll 1$.

In the case of attachment/ionization studies, since the drift velocity of electrons is much greater than that of the ions, the initial electron pulse δt (in the pulsed methods) is usually made sufficiently long so that the electrons may be considered to have established a steady-state distribution throughout the drift space and yet any ions which are formed will have moved an insignificant distance in that time. Thus, neglecting the effects of electron diffusion, the electron density spatial distribution is given by Eq. (4). The ion density distributions at time $t = 0$ (which are identified with those at time δt) are then determined from solutions to Eqs. (2) and (3), omitting the drift terms and substituting for n_e from Eq. (4); i.e.,

$$n_+(z,0) = n_+(z,\delta t) = \alpha \, w_e \, n_0 \, \delta t \exp[(\alpha - \eta)z], \tag{9}$$

$$n_-(z,0) = n_-(z,\delta t) = \eta \, w_e \, n_0 \, \delta t \exp[(\alpha - \eta)z]. \tag{10}$$

The ion densities as a function of time at later times are then determined by solving Eqs. (2) and (3) with the source terms set equal to zero. The resulting expressions for the time dependent positive- and negative-ion densities are given by:

$$n_+(z,t) = \alpha \, w_e \, n_0 \, \delta t \exp[(\alpha - \eta)(z + w_+ t)], \tag{11}$$

$$n_-(z,t) = \eta \, w_e \, n_0 \, \delta t \exp[(\alpha - \eta)(z - w_- t)], \tag{12}$$

for $0 \leq t \leq d/w_+$ and $0 \leq t \leq d/w_-$, respectively.

The analyses of the different experimental methods are based on these solutions and depend on whether the method is a steady-state or time-resolved one; in the case of the latter, the analysis further depends on the experimental arrangement, i.e., on whether the drifting charged species are detected continuously by the current induced in the external circuit or are sampled at some specific drift position. More complex solutions taking into account electron diffusion effects have been derived (Blevin et al., 1981; Purdie and Fletcher, 1989). However, for the purposes of the present comparison, it is convenient to restrict the discussion to the simpler solutions.

METHODS USED FOR MEASUREMENTS OF ATTACHMENT COEFFICIENTS ABOVE THE ONSET OF IONIZATION

Three basic methods have been used to determine attachment coefficients at values of E/N above the onset of ionization: (1) steady-state current growth, (2) pulsed Townsend, and (3) shielded-collector pulsed drift tube. We now discuss each of these techniques in more detail.

Steady-State Current Growth Method

This method is an extension of the classical Townsend approach (Townsend, 1925) for measuring ionization coefficients introduced initially by Harrison and Geballe (1953) and illustrated in Fig. 1. The apparatus usually consists of two plane-parallel electrodes, whose separation may be varied, contained in a chamber so that the ambient gas and gas pressure can be controlled. The electric field between the anode and cathode electrodes is arranged to be as uniform as possible either through the use of guard rings or by suitably profiling the electrode surfaces. Measurements of the amplification of a dc current injected into the uniform-field region (e.g., by irradiating the cathode with uv light) are recorded as a function of electrode separation but for a constant value of the electron mean energy parameter E/N.

The appropriate expression for the analysis of these data is given by Eq. (6). If the currents, I_d, are measured at constant incremental electrode separations, Δd, then Eq. (6) may be rewritten to give (Davies, 1976)

$$I_{d+\Delta d} = I_d \exp \left[(\alpha - \eta)\Delta d \right] + \left[I_0 \eta / (\alpha - \eta) \right] \{ \exp[(\alpha - \eta)\Delta d] - 1 \}. \quad (13)$$

Thus, a plot of $I_{d+\Delta d}$ versus I_d is linear of slope M_1 and intercept N_1, from which

$$(\alpha - \eta) = (1/\Delta d) \ln M_1, \quad (14)$$

$$I_0 \eta / (\alpha - \eta) = N_1 / (M_1 - 1). \quad (15)$$

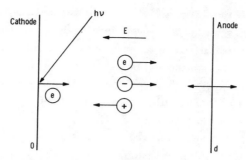

Fig. 1. Schematic diagram of the steady-state current growth method.

If the expression for I_d is now rewritten in the form (Davies, 1976)

$$\ln [I_d + N_1/(M_1 - 1)] = (\alpha - \eta)d + \ln [I_0\alpha/(\alpha - \eta)], \qquad (16)$$

a plot of $\ln [I_d + N_1/(M_1 - 1)]$ versus d is linear of slope M_2 and intercept N_2 from which

$$\alpha - \eta = M_2, \qquad (17)$$

$$I_0\alpha/(\alpha - \eta) = \exp N_2. \qquad (18)$$

Thus, from Eqs. (15) and (18),

$$\alpha/\eta = (M_1 - 1) \exp N_2/N_1, \qquad (19)$$

so that knowing $\alpha - \eta$ and α/η, the individual coefficients are determined:

$$\alpha = (\alpha - \eta)/(1 - \eta/\alpha), \qquad (20)$$

$$\eta = (\alpha - \eta)/(\alpha/\eta - 1). \qquad (21)$$

This method is capable of yielding very accurate values of the <u>net</u> ionization coefficient, even in more complex situations where electron detachment is occurring (Davies, 1976). However, the procedure for extracting the individual coefficients is rather indirect, involves determinations of intercepts, and is subject to large uncertainties.

Pulsed Townsend Method

This method, first introduced by Hornbeck (1948), uses essentially the same apparatus as the steady-state method, with the exception that a pulsed source of electrons is employed instead of a continuous source, as illustrated in Fig. 2. This allows the separation of the electron component from the ion components by time resolving the measured signals and

taking advantage of the large difference in the drift velocities of electrons and ions. This method has been developed by Raether and co-workers
(1964) for the measurement of attachment coefficients in the presence of
ionization. In their arrangement, a resistor in the anode circuit is
used to measure the flow of charge between cathode and anode following
the injection of an electron pulse from the cathode. In this case, the
relevant analytical expressions are Eqs. (8), (11), and (12). Thus, the
three components of current measured in the external circuit
when the time constant of the measuring circuit RC $\ll \tau_e$ are:

$$I_e(t) = (en_o \; \delta t \; w_e/\tau_e) \; \exp \; [(\alpha - \eta)w_e t], \tag{22}$$

$$I_+(t) = (en_o \; \delta t \; w_e/\tau_+)[\alpha/(\alpha - \eta)]\{\exp \; [(\alpha - \eta)d] - \exp \; [(\alpha - \eta)w_+ t]\}, \tag{23}$$

$$I_-(t) = (en_o \; \delta t \; w_e/\tau_-)[\eta/(\alpha - \eta)]\{\exp \; [(\alpha - \eta)(d - w_- t)] - 1\}, \tag{24}$$

where τ_e, τ_+, and τ_- are the transit times of electrons, positive ions,
and negative ions, respectively. At time $t = \tau_e$, the following relations
are obtained from Eqs. (22)-(24):

$$\alpha/(\alpha - \eta) = (\tau_+/\tau_e)[I_+(\tau_e)/I_e(\tau_e)], \tag{25}$$

$$\alpha/\eta = (\tau_+/\tau_-)[I_+(\tau_e)/I_-(\tau_e)]. \tag{26}$$

Thus, in principle, the individual coefficients may be derived from the
total waveform. However, in practice, it is difficult to distinguish
between the positive- and negative-ion components at time $t = \tau_e$, particularly if there are more than a single ion species of either polarity
present simultaneously; moreover, it is also difficult to determine the
smaller of the negative- and positive-ion transit times in the usual
event that they are unequal. Consequently, this approach has not been
used extensively.

On the other hand, a variant of this technique has been developed,
primarily by Grunberg (1969), which offers more precision. In contrast
to the measurements of Raether et al., the time constant of the measuring
circuit is deliberately made much larger than the ion transit time, i.e.,

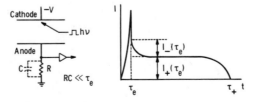

Fig. 2. Schematic diagram of the fast-response, pulsed Townsend
approach and predicted waveforms for the case where $\alpha > \eta$.

182

$RC \gg \tau_{\pm}$, leading to predicted waveforms as shown in Fig. 3. The relevant expressions for the contributions to the voltage developed in the external circuit due to the drift of electrons, positive ions, and negative ions are obtained by time–integrating the expressions given in Eqs. (22)–(24) over the appropriate species transit time:

$$V_e(\tau_e) = \frac{e \, n_o \, \delta t \, w_e}{C \, (\alpha - \eta) \, d} \, \{ \exp \, [(\alpha - \eta)d] - 1 \}, \tag{27}$$

$$V_+(\tau_+) = \frac{e \, n_o \, \delta t \, w_e \, \alpha}{C \, (\alpha - \eta) \, d} \, \left\{ d \, \exp \, [(\alpha - \eta)d] - \frac{\exp \, [(\alpha - \eta)d] - 1}{(\alpha - \eta)} \right\}, \tag{28}$$

$$V_-(\tau_-) = \frac{e \, n_o \, \delta t \, w_e \, \eta}{C \, (\alpha - \eta) \, d} \, \left\{ \frac{\exp \, [(\alpha - \eta)d] - 1}{(\alpha - \eta)} - d \right\}, \tag{29}$$

and a total voltage $V_T = V_e + V_+ + V_-$ given by:

$$V_T = \frac{e \, n_o \, \delta t \, w_e}{C} \, \left\{ \frac{\alpha}{(\alpha - \eta)} \exp \, [(\alpha - \eta)d] - \frac{\eta}{(\alpha - \eta)} \right\}. \tag{30}$$

Evidently, this expression is identical to that derived for the steady-state current growth method [Eq. (6)]. Thus, accurate values of the net ionization coefficient may be derived using exactly the same procedure as described for the steady–state method. However, despite the temporal resolution of the electron and total ion components of the total signal, determination of the individual ionization and attachment coefficients must still be made using either the indirect procedure described for the steady–state method or a curve–fitting routine, both of which are subject to large uncertainties.

Shielded–Collector Pulsed Drift Tube Method

The principle of this method is illustrated in Fig. 4 and utilizes a pulsed drift tube in which both cathode and anode collectors are shielded from the main body of the drift region by highly–transparent conducting

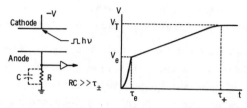

Fig. 3. Schematic diagram of the time–integrated pulsed Townsend approach and predicted waveforms.

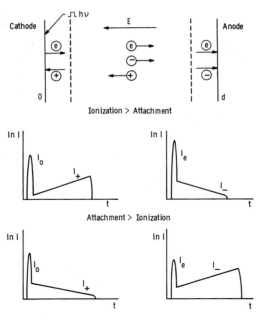

Fig. 4.   Schematic diagram of the principle of the shielded-
          collector drift tube and predicted collector currents at
          the cathode and anode.

screens located in close proximity to either collector. Thus, the collectors are very effectively shielded from induced currents due to the
motion of charge between the two screens. In this way, all three components (i.e., the electron, positive ion, and negative ion components)
of the total current may be directly resolved. Following the release of
a pulse of electrons from the cathode, the signal collected at the
cathode comprises that due to the initial electron pulse before it drifts
through the cathode screen followed at later time by the positive-ion
pulse after it has drifted through the cathode screen. At the anode, the
collected signal comprises the amplified or attenuated initial electron
pulse (depending on the value of E/N) followed at later time by the
negative-ion pulse, both signals being recorded only after the drifting
electrons and negative ions have passed through the anode screen. Predicted cathode and anode current waveforms are shown in Fig. 4 for the
cases where ionization is larger than and less than attachment.

Examples of actual measured cathode and anode waveforms taken in
CCl_4 using a shielded-collector drift tube (Davies, 1990) are shown in
Fig. 5. Each photograph is a composite of superimposed cathode (upper)
and anode (lower) waveforms, with (a), (b) recorded at a value of E/N
= 850 Td where $\alpha < \eta$, and (c), (d) recorded at a value of E/N = 950 Td
where $\alpha > \eta$. The upper two photographs (a) and (c) clearly show the

184

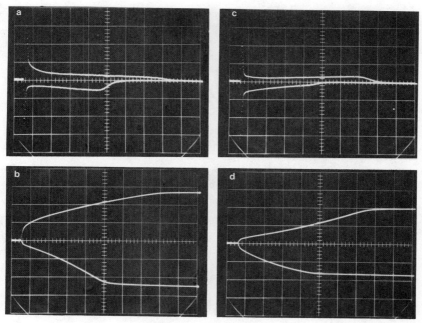

Fig. 5. Examples of cathode and anode waveforms obtained for a drift distance of 5.2 cm in CCl_4 under conditions where (a), (b): $\alpha < \eta$ (E/N = 850 Td, and N = 1.092×10^{16} cm^{-3}); (c), (d): $\alpha > \eta$ (E/N = 950 Td, and N = 9.45×10^{15} cm^{-3}). The photographs (b) and (d) are the integrated current (or charge) waveforms corresponding to the current waveforms (a) and (c). The time scale is 100 ns/point (~ 6 μs/ division).

resolution of the positive- and negative-ion current components and their widely different transit times. However, in general, both the positive- and negative-ion components can be complex in that each may be composed of more than one ion species which alternatively can be primary ions formed directly as a result of electron-molecule collisions or product ions formed as a result of ion-molecule reactions during the drift of the primary ions. As a result of this probable ion waveform complexity, it is convenient to determine the ionization and attachment coefficients from the time-integrated or charge waveforms derived from the current waveforms as shown in the lower two photographs (b) and (d) of Fig. 5.

The procedure for the determination of the individual ionization and attachment coefficients is outlined in Fig. 6. Current waveforms are measured both at the cathode and anode in separate experiments but at constant E/N, from which the charge waveforms are derived by integration. The expressions describing each of the charge components are derived from Eqs. (8), (11), and (12). At the cathode, these are (Davies, 1990):

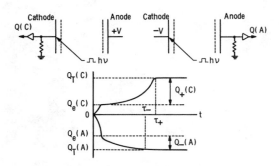

Fig. 6. Schematic diagram of the shielded-collector drift tube
procedure for the determination of the ratio α/η.

$$Q_e(C) = Q_o, \tag{31}$$

$$Q_+(C) = \frac{f_e\, f_i\, Q_o\, \alpha}{(\alpha - \eta)}\, \{\exp\,[(\alpha - \eta)d] - 1\}, \tag{32}$$

$$Q_T(C) = Q_o\, \left\{ \frac{f_e\, f_i\, \alpha}{(\alpha - \eta)}\, \exp\,[(\alpha - \eta)d] - \frac{f_e\, f_i\, \eta}{(\alpha - \eta)} \right\}, \tag{33}$$

where we have written $Q_o = e\, n_o\, \delta t\, w_e$ for the initial electron charge
pulse released from the cathode and f_e and f_i are the fractions of
electrons and ions, respectively, transmitted by each screen. The
corresponding expressions for the charge waveforms at the anode are
(Davies, 1990):

$$Q_e(A) = f_e^2\, Q_o\, \exp\,[(\alpha - \eta)d], \tag{34}$$

$$Q_-(A) = \frac{f_e\, f_i\, Q_o\, \eta}{(\alpha - \eta)}\, \{\exp\,[(\alpha - \eta)d] - 1\}, \tag{35}$$

$$Q_T(A) = Q_o\, \left\{ \left[\frac{f_e\, f_i\, \eta}{(\alpha - \eta)} + f_e^2 \right]\, \exp\,[(\alpha - \eta)d] - \frac{f_e\, f_i\, \eta}{(\alpha - \eta)} \right\}. \tag{36}$$

The expressions given by Eqs. (33) and (36) for the total cathode and
anode charge, respectively, have the same form as Eq. (6) derived for the
steady-state method; in fact, if we set $f_e = f_i = 1$, then Eqs. (33) and
(36) reduce identically to Eq. (6), as expected.

Thus, the same procedure outlined for the steady-state method can be
used to determine accurate values for the net ionization coefficient from
either the total cathode or anode charge waveforms. However, the separation of positive-ion and negative-ion charge components achieved in the
shielded-collector pulsed drift tube enables the ratio α/η to be determined directly [by arranging (Davies, 1990) for proper normalization
of the initial charge Q_o] from:

$$\alpha/\eta = Q_+(C)/Q_-(A). \tag{37}$$

Since both $(\alpha - \eta)$ and α/η are determined, the individual coefficients α and η are also determined from Eqs. (20) and (21). Thus, the analysis of the experimental data is direct and simple and the random uncertainties associated with the derived coefficients are well defined. The major source of systematic error in the method arises from the assumption that the screen transmission is the same for positive and negative ions. While this is expected to be true to first order, the assumption may be tested experimentally at the limiting value of E/N where $\alpha = \eta$. At this value of E/N (which may be determined from the independent measurements of the net ionization coefficient) the ratio of $Q_+(C)/Q_-(A)$ should be equal to unity if the transmission of the screens is the same for positive and negative ions [cf. Eqs. (32) and (35)]. This result has been observed to be true for all the attaching gases we have studied so far.

MEASUREMENTS OF ATTACHMENT COEFFICIENTS USING THE SHIELDED-COLLECTOR PULSED DRIFT TUBE

The shielded-collector pulsed drift tube (Davies, 1990) has been used for measurements of attachment coefficients in a variety of gases and gas mixtures including dry and humid air and gases used in laser and etching/deposition discharges. A summary of the data obtained in some of these gases is now described to illustrate the capabilities of this technique.

Dry and Humid Air

Figure 7 shows the measurements of two-body attachment coefficient determined over the range from 60 to 200 Td both in dry air and in dry air-2% water vapor mixtures. In contrast to previous work (Raja Rao and Govinda Raju, 1971; Moruzzi and Price, 1974), negative ions are detected at values of E/N > 100 Td, i.e., well above the onset of ionization. Over the range of values of E/N covered by the present measurements, the primary negative ion in dry air is expected (Moruzzi and Phelps, 1966) to be O^- formed in the dissociative reaction

$$e + O_2 \longrightarrow O^- + O,$$

whereas for the case of humid air, an additional primary ion H^- is produced (Moruzzi and Phelps, 1966) by dissociative attachment of H_2O according to the reaction

$$e + H_2O \longrightarrow H^- + OH.$$

Further selected reactions involving the O^- ions, H^- ions, and their progeny resulting in the formation of more complex ions are (Parkes,

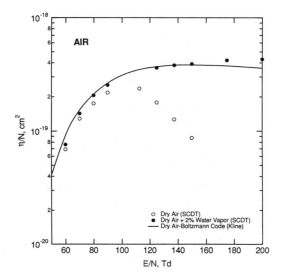

Fig. 7. Attachment coefficient in dry and humid air as a function of E/N.

1971; Fehsenfeld and Ferguson, 1974; Fehsenfeld et al., 1969; Dotan et al., 1977; Dunkin et al., 1970; Betowski et al., 1975)

$$O^- + O_2 + M \rightarrow O_3^- + M$$

$$O^- + H_2O + M \rightarrow O^- \cdot H_2O + M$$

$$O^- + CO_2 + M \rightarrow CO_3^- + M$$

$$O_3^- + H_2O + M \rightarrow O_3^- \cdot H_2O + M$$

$$O_3^- + CO_2 \rightarrow CO_3^- + O_2$$

$$O^- + O_2 \rightarrow O_2^- + O$$

$$H^- + H_2O \rightarrow OH^- + H_2$$

$$OH^- + H_2O + M \rightarrow OH^- \cdot H_2O + M$$

$$OH^- + CO_2 + M \rightarrow OH^- \cdot CO_2 + M.$$

On the other hand, the primary ions O^- and H^- can also undergo either associative detachment or collisional detachment reactions according to (Comer and Schulz, 1974; Lindinger et al., 1975; Rayment and Moruzzi, 1978; Doussot et al., 1982; Frommhold, 1964; Ryzko and Astrom, 1967; Eccles et al., 1970; O'Neill and Craggs, 1973; Berger, 1981)

$$O^- + N_2 \rightarrow N_2O + e$$

$$O^- + M \rightarrow O + M + e$$

$$H^- + O_2 \rightarrow HO_2 + e.$$

188

Based on the known rates for these reactions (Parkes, 1971; Fehsenfeld and Ferguson, 1974; Fehsenfeld et al., 1969; Dotan et al., 1977; Dunkin et al., 1970; Betowski et al., 1975; Comer and Schulz, 1974; Lindinger et al., 1975; Rayment and Moruzzi, 1978; Doussot et al., 1982; Frommhold, 1964; Ryzko and Astrom, 1967; Eccles et al., 1970; O'Neill and Craggs, 1973; Berger, 1981) (albeit at thermal energies), the observed ions corresponding to the measurements in CO_2 - free dry air are expected to be O_3^-. However, the attachment coefficient data show a very sharp decrease as a function of E/N for values of E/N > 100 Td which is contrary to the behavior predicted from solutions to the Boltzmann equation (Kline, private communication). On the other hand, the data at the lower values of E/N are in very good agreement with Boltzmann code predictions. The decreasing attachment coefficient with increasing E/N observed experimentally is interpreted as due to the presence of detachment, and that the observed values are to be regarded as effective values. For arbitrary density N, this effective attachment coefficient is a function of the true attachment coefficient (for O^- formation), the charge transfer coefficient, and the detachment coefficient (Davies, 1976). The onset for associative detachment occurs at a value of E/N ~ 5 Td whereas that for collisional detachment occurs at E/N ~ 100 Td. Unfortunately, reported measured detachment rates vary by more than an order of magnitude for associative detachment and by more than three orders of magnitude for collisional detachment. In view of the agreement of the present measured effective attachment coefficients with predicted attachment coefficients for E/N values up to 100 Td, it is concluded that the detachment process is collisional rather than associative. Since the ion conversion channel for the O^- is a three-body process compared with the two-body collisional detachment reaction, the effective attachment coefficient is expected to be pressure dependent, reflecting the competing reaction channels for the O^- ions.

In the case of CO_2 - free humid air, the presence of H_2O in the humid air mixtures is observed to have a large effect on the measured attachment coefficient for values of E/N > 100 Td, but for E/N < 100 Td the attachment coefficient is only slightly increased. Based on reaction rate data (again at thermal energies), the dominant reaction path of ions derived from O^- is expected to lead to the formation of $O^- \cdot H_2O$ and higher order clusters. On the other hand, for the ratio of H_2O/O_2 ~ 0.1 pertaining to the present mixtures, the H^- ions are most likely to suffer detachment reactions. Thus, even though the attachment rate coefficient for H^- formation is approximately a factor of two larger than that for O^- formation at comparable electron mean energies, the contribution of H^- derived ions to the total negative ion spectrum is expected to be small.

However, the presence of H_2O provides an important additional ion
conversion channel for the O^- ions which is not present in dry air. It
is proposed that the formation of cluster ions reduces the probability of
collisional detachment as a result of (1) the decreased mean energy of
the more massive cluster ions and (2) the additional degrees of freedom
for energy conservation in ion-molecule collisions provided by successive
detachment of H_2O molecules. Although these measurements in humid air
are in good agreement with Boltzmann code predictions for O^- attachment
(i.e., neglecting detachment), they must still be regarded as effective
coefficients.

Further measurements in humid air mixtures having different water
vapor concentrations over the range from 0 to 2% would provide further
information on the branching ratio of O^- ion conversion to detachment as
a function of E/N. The analysis of such data is particularly straightforward (and less ambiguous) if the production rate of O^- is independent
of water vapor concentration at constant E/N. For this to be true, the
contribution of the electron collision cross sections in water vapor in
determining the electron energy distribution in humid air over the range
of E/N > 60 Td must be small. That such is the case is suggested by
comparison of our measurements of the electron drift velocity in this
range of E/N for dry air and humid air with 2% H_2O.

$\underline{CO_2}$

Figure 8 shows our measurements of attachment coefficients in CO_2
compared with previous measurements (Moruzzi and Price, 1974; Bhalla and
Craggs, 1960; Alger and Rees, 1976; Davies, 1978). Previous data are
broadly clustered in two sets and the present results are consistent with
the lower of these sets and also with predictions (Davies, 1978) of the
two-term Boltzmann equation using an accepted cross section set (Lowke et
al., 1973). The primary negative ion formed in CO_2 is expected (Moruzzi
and Phelps, 1966) to be O^- produced in the dissociative attachment reaction

$$e + CO_2 \rightarrow O^- + CO.$$

However, these O^- ions are rapidly converted to CO_3^- ions in three-body
reactions according to the reaction

$$O^- + CO_2 + CO_2 \rightarrow CO_3^- + CO_2.$$

Measurements of the negative-ion mobility determined in the present work
agrees within combined uncertainties with the mass-identified CO_3^-
measurements of Ellis et al. (1976) and of Moseley et al. (1976).

$\underline{HCl}$

Measurements of attachment coefficients in pure HCl as a function of
E/N are shown in Fig. 9 where they are compared with previous data

190

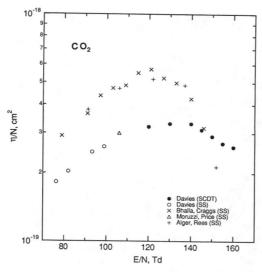

Fig. 8.   Attachment coefficient in CO_2 as a function of E/N.

(Bailey and Duncanson, 1930).   For values of E/N < 100 Td no positive
ions are detected and the data in this region correspond to values of E/N
below the onset of ionization.   The experimental data are also compared
with predictions of the Boltzmann equation using the two-term approximation and a cross section set derived from our additional swarm
measurements on electron mobility and ionization coefficient (not
described here).   Three predicted curves are shown corresponding to H^-,
Cl^-, and the sum of the two.   The primary ions H^- and Cl^- are formed in
the dissociative attachment reactions

$$e + HCl \longrightarrow Cl^- + H$$

$$e + HCl \longrightarrow H^- + Cl$$

with onset energies of 0.67 eV and 5.6 eV, respectively.

The predicted curves shown in Fig. 9 have been obtained using the
cross section shapes reported by Azria et al. (1974).   In the case of the
Cl^- cross section there is excellent agreement among four independent
measurements (Azria et al., 1974; Ziesel et al., 1975; Abouaf and
Teillet-Billy, 1977; Allan and Wong, 1981).   However, the magnitudes of
the cross sections have been increased uniformly by approximately a
factor or two from those reported by Azria et al.   The resulting
magnitude at the peak of the Cl^- cross section is 2.3×10^{-17} cm^2.
Attempts to reconcile the measured attachment coefficients with those
predicted from the Boltzmann code for the total ($H^- + Cl^-$) attachment
coefficient by adjusting the magnitudes of the cross sections have not

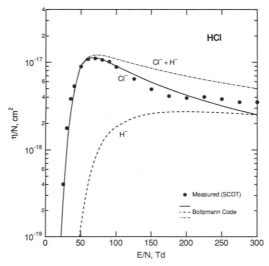

Fig. 9. Attachment coefficient in HCl as a function of E/N.

been successful. Thus, in view of the very good agreement between our
measured values and those predicted for Cl^- production, our interpretation of the present experimental measurements is that they pertain
to Cl^- only. In fact, the agreement is within 20% for values of E/N in
the range from 40 to 250 Td and within 2% at the peak. Moreover, the
position of the peak is in exact agreement with experiment. The
increasing discrepancy between predicted and measured values at values of
E/N > 250 Td may be indicative of the presence of an additional process.
The only likely process in this range of E/N is the pair production
process

$$e + HCl \longrightarrow H^- + Cl^-,$$

and further work is required to determine the role of this process.

The fate of the H^- ions under the present swarm conditions is not
known at present but follows a trend observed in other gases (Davies et
al., 1989) in which this ion is known to be formed under single collision
conditions using monoenergetic electron beams at low gas density. Thus,
we propose that the H^- ions suffer collisional detachment in a swarm
environment according to

$$H^- + HCl \longrightarrow H + HCl + e.$$

From the work of Azria et al. (1974) it is known that the H^- ions are
formed with appreciable kinetic energy (>3 eV) leading to a high

probability for collisional detachment. Further, the production of H⁻ occurs at sufficiently high values of E/N for the ions to gain appreciable energy from the applied field. Evidence in support of this interpretation is our observation of only a single ion specie in the negative-ion waveforms. The low-field mobility of this ion is in very good agreement with that obtained by McDaniel and McDowell (1959).

CCl_4

Figure 10 shows the measurements of attachment coefficient obtained in CCl_4 as a function of E/N. From the negative-ion waveforms, only one ion specie is observed. There is general agreement (Reese et al., 1956; Fox and Curran, 1961; Dorman, 1966; Scheunemann et al., 1980) that the dominant ion specie produced by electron impact on CCl_4 under single collision conditions is Cl^- formed by dissociative attachment. Although there is little doubt that the primary ion formed in the present work is Cl^-, the identity of the final species recorded is undetermined. No positive ions are detected for values of E/N ≤ 650 Td, and the data in this region correspond to values of E/N below the onset of ionization.

The present data are compared with the values determined by Geballe and Harrison (1955) from spatial current growth measurements in Fig. 10. Their data exhibit anomalous behavior in the vicinity of the limiting value of E/N (i.e., 880 Td). Such behavior is difficult to understand

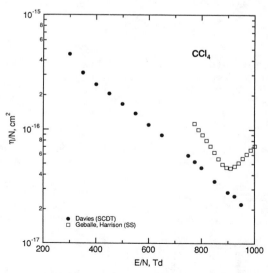

Fig. 10. Attachment coefficient in CCl_4 as a function of E/N.

and is attributed to uncertainties arising from the analysis of the data
particularly since the measurements are concentrated in the vicinity of
the limiting value of E/N where such uncertainties are largest.

SUMMARY

The three methods available for the determination of attachment
coefficients in the range of E/N where ionization is occurring have been
compared with emphasis on their respective ability to yield unambiguous
values of well-defined uncertainty. All three methods are capable of
giving high-precision measurements of the net ionization coefficient.
However, it has been shown that for the determination of the individual
coefficients, resolution of the three components of charge - electron,
positive ion, and negative ion - is highly desirable for unambiguous
analysis of the raw data.

The steady-state method provides no resolution of charge carriers
and is restricted to the range of E/N above the onset of ionization. The
pulsed drift tube removes this latter restriction. However, in its usual
form (i.e., the pulsed Townsend method), only the electron and total ion
charge is resolved and analysis of the raw data suffers from the same
limitations as the steady-state method for the determination of the
individual attachment and ionization coefficients. On the other hand,
incorporation of shielded collectors within the pulsed drift tube enables
all three charge types to be resolved. This not only provides an
unambiguous method of detecting the occurrence of attachment in the
presence of ionization but also leads to a very direct and simple
analysis of the raw data to yield the individual coefficients.

REFERENCES

Abouaf R., and D. Teillet-Billy, 1977, J. Phys. B $\underline{10}$, 2261.
Alger S. R., and J. A. Rees, 1976, J. Phys. D $\underline{9}$, 2359.
Allan, M., and S. F. Wong, 1981, J. Chem. Phys. $\underline{74}$, 1687.
Azria, R., L. Roussier, R. Paineau, and M. Tronc, 1974, Rev. Phys. App.
    $\underline{9}$, 469; M. Tronc, R. Azria, Y. LeCoat, and D. Simon, 1979, J. Phys.
    B $\underline{12}$, L467; R. Azria, Y. LeCoat, D. Simon, and M. Tronc, 1980, J.
    Phys. B $\underline{13}$, 1909.
Bailey, V. A., and W. E. Duncanson, 1930, Phil. Mag. $\underline{10}$, 145.
Berger, G., 1981, "Proc. XV Int. Conf. Phen. Ionized Gases," Minsk, p.
    571.
Betowski, D., J. D. Payzant, G. I. Mackay, and D. K. Bohme, 1975, Chem.
    Phys. Lett. $\underline{31}$, 321.
Bhalla, M. S., and J. D. Craggs, 1960, Proc. Phys. Soc. London $\underline{76}$, 369.
Blevin, H. A., K. J. Nygaard, and K. R. Spriggs, 1981, J. Phys. D $\underline{14}$,
    841.
Comer, J., and G. J. Schulz, 1974, Phys. Rev. A $\underline{10}$, 2100.
Davies, D. K., 1976, J. Appl. Phys. $\underline{47}$, 1916.
Davies, D. K., 1978, J. Appl. Phys. $\underline{49}$, 127.
Davies, D. K., 1990, submitted to Rev. Sci. Inst.

Davies, D. K., L. E. Kline, and W. E. Bies, 1989, J. Appl. Phys. 65, 3311.

Dorman, F. H., 1966, J. Chem. Phys. 34, 3856.

Dotan, I., J. A. Davidson, G. E. Streit, D. L. Albritton, and F. C. Fehsenfeld, 1977, J. Chem. Phys. 67, 2874.

Doussot, C., F. Bastien, E. Marode, and J. L. Moruzzi, 1982, J. Phys. D 16, 2451.

Dunkin, D. B., F. C. Fehsenfeld, and E. E. Ferguson, 1970, J. Chem. Phys. 53, 987.

Eccles, M. J., B. C. O'Neill, and J. D. Craggs, 1970, J. Phys. B 3, 1724.

Ellis, H. W., R. Y. Pai, I. R. Gatland, E. W. McDaniel, R. Wernlund, and M. J. Cohen, 1976, J. Chem. Phys. 64, 3935.

Fehsenfeld, F. C., and E. E. Ferguson, 1974, J. Chem. Phys. 61, 3181.

Fehsenfeld, F. C., E. E. Ferguson, and D. K. Bohme, 1969, Planet. Space Sci. 17, 1759.

Fox, R. E., and R. K. Curran, 1961, J. Chem. Phys. 34, 1595.

Frommhold, L., 1964, Fortschr. Phys. 12, 597.

Geballe, R., and M. A. Harrison, 1955, L. B. Loeb, "Basic Processes of Gaseous Electronics," Univ. California Press, p. 414.

Grunberg, R., 1969, Z. Naturforsch. 24a, 1039.

Harrison, M. A., and R. Geballe, 1953, Phys. Rev. 91, 1.

Hornbeck, J. A., 1948, Phys. Rev. 73, 570.

Huxley, L. G. H., and R. W. Crompton, 1974, "The Diffusion and Drift of Electrons in Gases," John Wiley and Sons, New York.

Kline, L. E., private communication.

Lindinger, W., D. L. Albritton, F. C. Fehsenfeld, and E. E. Ferguson, 1975, J. Chem. Phys. 63, 3238.

Llewellyn-Jones, F., 1967, "Ionization Avalanches and Breakdown," Methuen and Co., London.

Lowke, J. J., A. V. Phelps, and B. W. Irwin, 1973, J. Appl. Phys. 44, 4664.

McDaniel, E. W., and M. R. C. McDowell, 1959, Phys. Rev. 114, 1028.

Moruzzi, J. L., and A. V. Phelps, 1966, J. Chem. Phys. 45, 4617.

Moruzzi, J. L., and D. A. Price, 1974, J. Phys. D 7, 1434.

Moseley, J. T., P. C. Cosby, and J. R. Peterson, 1976, J. Chem. Phys. 64, 4228.

O'Neill, B. C., and J. D. Craggs, 1973, J. Phys. B 6, 2625.

Parkes, D. A., 1971, Trans. Faraday Soc. 67, 711.

Purdie, P. H., and J. Fletcher, 1989, J. Phys. D 22, 759.

Raether, H., 1964, "Electron Avalanches and Breakdown in Gases," Butterworth and Co., London.

Raja Rao, C., and G. R. Govinda Raju, 1971, J. Phys. D 4, 494.

Rayment, S. W., and J. L. Moruzzi, 1978, Int. J. Mass Spec. Ion Phys. 26, 321.

Reese, R., V. Dibeler, and F. Mohler, 1956, J. Res. Natl. Bur. Stand. 57, 367.

Ryzko, H., and E. Astrom, 1967, J. Appl. Phys. 38, 328.

Scheunemann, H. U., E. Illenberger, and H. Baumgartel, 1980, Ber. Bunsenges. Phys. Chem. 84, 580.

Townsend, J. S., 1925, "Motion of Electrons in Gases," Clarendon Press, Oxford.

Ziesel, J. P., I. Nenner, and G. J. Schulz, 1975, J. Chem. Phys. 63, 1943.

ION TRANSPORT AND ION-MOLECULE REACTIONS OF NEGATIVE IONS IN SF_6

Yoshiharu Nakamura

Faculty of Science and Technology
Keio University
3-14-1 Hiyoshi, Yokohama 223, Japan

A double shutter drift tube which is practically the same as used in measurements of electron swarm parameters in gases is used to investigate negative ion transports in SF_6. An explanation is given to the signal obtained by the drift tube in SF_6, and reduced mobility and longitudinal diffusion coefficient of negative ions (F^-, SF_5^-, and SF_6^-) and electron drift velocity are determined. Results of the measurements are given for an E/N range from 20 to 600 Td for negative ions and from 280 to 700 Td for electrons (1 Td = 10^{-17} V-cm^2.)

An analysis of time-of-flight (TOF) spectra of negative ions in SF_6 is given, and the results of negative ion-neutral reaction coefficients are presented for E/N from 180 to 450 Td. The present results are compared with other results.

INTRODUCTION

Sulfur hexafluoride (SF_6) among other gases has long attracted concerns of many scientists and engineers because of its wide applications in the fields of high voltage engineering and excimer lasers (Teich, 1981; Maller and Naidu, 1981; Gallagher et al., 1983). Numerous efforts have been made to determine electron swarm data in the gas. Also there have been several attempts (Kline et al., 1979; Yoshizawa et al., 1979; Novak and Frechette, 1984) to obtain a set of cross sections for the SF_6 molecule which is consistent with electron transport coefficients in the gas. The evaluation, however, has been hindered by the fact that there have been only a limited number of measurements of electron drift velocity in SF_6 (Harris and Jones, 1971; Naidu and Prasad, 1972; Teich and Sangi, 1972; Aschwanden, 1985) and that the consistency among them was not necessarily good.

Insulating property of SF_6 stems from its strong electron affinity and, therefore, negative ions in SF_6 (SF_6^-, other ionic dissociation products such as SF_5^- and F^-, and cluster ions) also play important roles. Mobilities of negative ions, with masses identified (Patterson, 1970) and with masses unidentified (Fleming and Rees, 1969; Naidu and Prasad, 1970), in the parent SF_6 gas have been studied. Those measurements, however, were limited in low E/N.

Another recent use of SF_6 stems from the chemically active property of an SF_6 plasma. It is believed that the atomic fluorine formed in SF_6 plasma acts as the main etchant for silicon. Plasma diagnostics of a SF_6 rf discharge (Picard et al., 1986) showed that there were three kinds of e-SF_6 collisions contributing to the production of the atomic fluorine, i.e., the dissociative ionization, the dissociative attachment and electron impact dissociation.

Ion conversion reactions may also contribute to the formation of atomic fluorine.

We have tried to apply a double shutter drift tube to measurements of the transport properties of electrons and negative ions in SF_6 (Nakamura, 1988). The drift tube was proved to be capable of forming three species of negative ions selectively and, therefore, of determining transport parameters of respective negative ions as well as electrons in the gas.

In the following, an explanation is given to the signal obtained by using the drift tube in SF_6. And the reduced mobility and the longitudinal diffusion coefficient of negative ions and the electron drift velocity in SF_6 are given over the E/N from 20 to 600 Td on negative ions and from 280 to 700 Td on electrons. Three negative ions are identified to be F^-, SF_5^-, and SF_6^-, according to their reduced mobilities at low electric field.

The negative ion signal often showed clear evidence of ion-neutral reactions. An analysis of time-of-flight (TOF) spectra of negative ions in SF_6 is also given to reduce reaction coefficients for the following processes (Nakamura and Kizu, 1987);

$$SF_6^- + SF_6 \longrightarrow SF_5^- + F + SF_6 \text{ (coefficient } \mu_1),$$
$$SF_5^- + SF_6 \longrightarrow F^- + SF_4 + SF_6 \text{ (coefficient } \mu_2),$$

and

$$F^- + SF_6 \longrightarrow SF_6^- + F \text{ (coefficient } \mu_3).$$

Results of the measurement over the E/N from 180 to 450 Td is presented. The present results are compared with other results.

Double Shutter Drift Tube

The cross section of the electrode system of the double shutter drift tube is shown in Fig. 1.

The system consists of a gold film photocathode (PK) deposited on a quartz window, a pair of electrical shutters (S1 and S2), a collector (C), and five guard rings. The inner diameter of the guard rings is 5 cm. Each electrical shutter consists of two parallel wire grids (Tyndall et al., 1928, 1931). The grid spacing is 1 mm and the wire spacing of the grid is 0.5 mm. The second shutter (S2) and the collector are mounted on a linear motion feedthrough which varies the distance between the two shutters (the drift distance) from 1 to 5 cm.

The S1- and S2-pulse are rectangular and have the same period. Typical pulse width was 1 to 4 μs for ion measurements and 100 ns or less for electron measurements. The delay time of the S2-pulse relative to the S1-pulse is slowly varied, and the current to the collector (C), which is roughly proportional to the number of the charge at the S2 when it is open, is observed (the time-of-flight, TOF, spectrum).

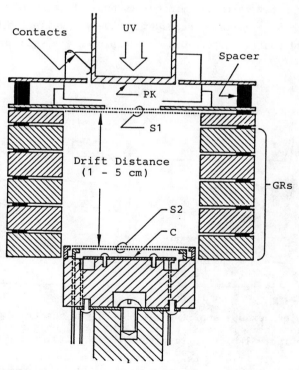

Fig. 1. The cross section of the double shutter drift tube.

Time-of-Flight Spectra

Typical examples of the collector current signals as a function of the delay time (the TOF spectra) are shown in Figs. 2 and 3. As is shown in those figures each trace, which is vertically displaced so as not to overlap each other, had a sharp minimum and several peaks.

The relative magnitude of minimum and the peak sensitively depended on the combination of the blocking bias to S1 (SB1) and the amplitude of the S1-pulse (PH1). The bias to S2 and the amplitude of the S2-pulse, on the other hand, had little effect on the shape of the trace. Figure 2(a) shows the effect of PH1 on the TOF spectrum at relatively low bias voltage (SB1=0.8 volt). Each trace started at the time slightly earlier than the S1-pulse was applied, and it reached a minimum when the two shutters were simultaneously open (i.e., when the S1- and S2-pulses were applied exactly at the same time). While the minimum became deeper, the first peak gradually gained dominancy with increasing PH1. The second peak in Fig. 2(a) in fact consisted of two neighboring peaks and that was clearly seen in the spectra at higher pressure and with longer drift distance as shown in Fig. 2(b), where the minimum of the trace was not recorded deliberately. Figure 3 shows the effect of the S1-bias voltage (SB1) on the TOF spectrum. The peaks and the minimum initially decreased their magnitudes with increasing SB1. When the bias voltage was increased over a few volts the minimum turned into a sharp peak and further increase in the bias voltage increased its height while the height of slower peaks decreased continuously. Only a sharp peak was seen when the bias

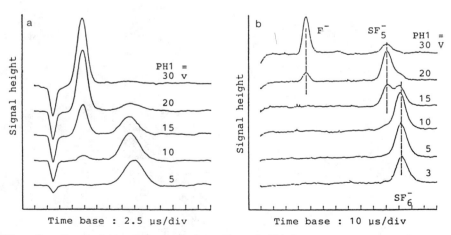

Fig. 2. Typical examples of the time-of-flight spectra. Examples show the effect of the S1-pulse height (PH1) on the TOF spectra at constant S1-bias voltage (SB1 = 0.8 volt). (a) E/N = 350 Td, N = 1.110×10^{16} cm^{-3}, Shutter pulse width = 1 μs, Drift distance = 1 cm, (b) E/N = 200 Td, N = 3.54×10^{16} cm^{-3}, Shutter pulse width = 4 μs, Drift distance = 4 cm.

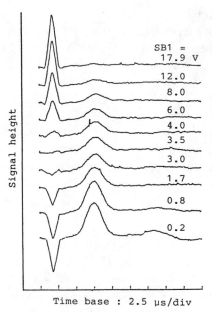

SB1 =
17.9 V

12.0

8.0

6.0

4.0

3.5

3.0

1.7

0.8

0.2

Time base : 2.5 μs/div

Fig. 3. Typical examples of the time-of-flight spectra. Examples show
the effect of the S1-blocking bias voltage (SB1) on
the TOF spectra at the constant S1-pulse height (PH1 = 20
volt). E/N = 350 Td, N = 8.8 x 10^{15} c, $^{-3}$, Shutter pulse
width = 1 μs, Drift distance = 1 cm, Pulse height to S1 =
20 volt, Pulse height to S2 = 8 volt.

voltage was about 18 volts. The FWHM values of the minimum and the sharp peak
were the same, and they coincided exactly with the width of the shutter pulse.

The time-of-flight spectra observed in the present apparatus can be
understood in the following manner.

At higher shutter bias the S1-shutter blocks all of the charged particles and only electrons can be admitted into the drift space by applying
the S1-pulse, and an isolated electron swarm will drift and diffuse towards
the second shutter. At lower shutter bias electrons can pass through the
shutter S1 freely. When the S1-pulse is on, the pulse accelerates electrons
between the grids of the S1 and raises energy of electrons depending upon
the pulse height. Electron attaching processes are highly dependent on the
energy of electron (McGeehan et al., 1975), and thus electrons attach to SF_6
to form SF_6^-, SF_5^-, or F^- depending on the S1-pulse height. The isolated
negative ions swarm thus produced within the S1 shutter drifts and diffuses
towards the second shutter. When electrons attach to SF_6 molecules to form
negative ions, an instantaneous reduction in electron density will also take
place for the duration of the S1-pulse. Since this must travel at the electron drift velocity (order of 10^7 cm/s), its transit time between the two
shutters must be much shorter than the typical width of the shutter pulse

(1 to 4 µs). This reduction of electron density, therefore, arrives at the S2 without any appreciable diffusion effect. Since the opening time of the second shutter is the same as the first shutter, the resulting time-of-flight spectrum of electrons must be an isosceles triangle. This is what we have observed in our measurement.

Transport Coefficients of Negative Ions

Figure 4 shows negative ion spectra taken at four different drift distances at the same E/N. The linear plot of arrival time of a peak against the drift distance gives the drift velocity of the corresponding ion. The linear plot of the square of the characteristic width of a peak against the drift distance gives the longitudinal diffusion coefficient of the ion (Nakamura, 1987).

The actual measurements were carried out at the S1-pulse height when there was only one dominant peak in each trace. Measurements were done at the gas number density N ranging from 8.0×10^{15} to 5.3×10^{16} cm^{-3} and over the E/N ranging from 20 to 600 Td. The results were converted into the traditional reduced mobility (reduced to the STP) and shown in Fig. 5 for two slow ions and in Fig. 6 for the fastest. Each ion showed a clear peak in its mobility at the E/N about 300 Td and, to the author's knowledge, there was no measurement in the past showing mobility peaks of negative ions in SF_6.

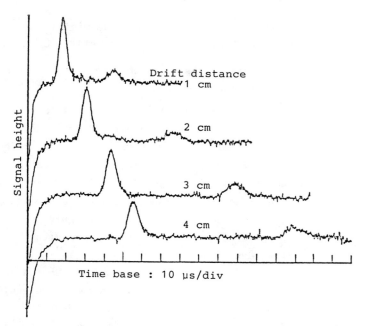

Fig. 4. Typical examples of the time-of-flight spectra taken at four different drift distances. E/N = 170 Td, N = 3.18 x 10^{16} cm^{-3}, Pulse height to S1 = 30 volt.

202

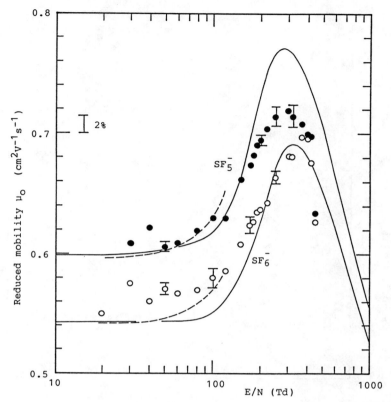

Fig. 5. Measured reduced mobilities of SF_6^- (open circle) and SF_5^- (closed circle) in SF_6 gas. Short bars indicate the standard deviation of the mobility measured at several different pressures.

The present mobilities of two slow ions in low E/N were compared with those of mass analyzed measurement done by Patterson [broken line, (Patterson, 1970)], and they were identified as SF_6^- and SF_5^-. These were also compared with the results of calculation done by Brand and coworkers [solid line, (Brand et al., 1983)] assuming the (9,6,4)-potential between negative ions and neutrals. A rather good agreement was observed for SF_6^-, but the calculation gave a higher mobility peak for SF_5^- than the present measurement. In E/N higher than 600 Td the mobility measurement for the two slow ions became practically impossible.

A simple estimation of the reduced mass using the Langevin equation and the fact that its appearance needed a high S1-pulse strongly suggested that the fastest ion must be F^-. There were no measurements of such high mobility of negative ion in SF_6 in the past.

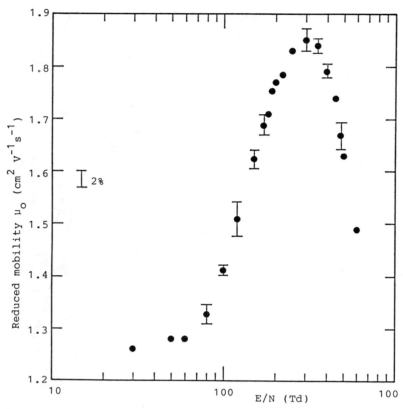

Fig. 6. Measured reduced mobility of the fastest ion. The ion was
considered to be F⁻ as described in the text.

It was also possible, for the first time, to determine the longitudinal
diffusion coefficient times the gas number density for three negative ions
(ND_L) from the TOF spectra shown in Fig. 4 (Nakamura, 1988), although the
results are not included in the text.

Electron Drift Velocity

When the width of the S1- and S2-pulse was reduced to about 100 ns or
less, the TOF spectra clearly showed that the mobility of the V-shaped dip or
the reversed peak must be electronic, as shown in Fig. 7. It was confirmed
that both isosceles peak and dip gave the same velocity. The present result
of electron drift velocity is shown in Fig. 8 and is compared with previous
measurements available so far. It shows a good agreement with that obtained
by Teich and coworkers [broken line, (Teich and Sangi, 1972)] and by
Aschwanden [fine dotted line, (Aschwanden, 1985)] from the pulsed Townsend
measurements. Naidu and coworkers [cross, (Naidu and Prasad, 1972)] also
measured electron drift velocity by a TOF method, but their result shows clear
deviation at higher E/N from the present result, although their principle of
the measurement was almost the same as the present. The reason for this
disagreement was not clear to the author.

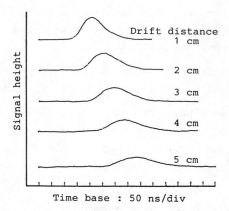

Time base : 50 ns/div

Fig. 7. Examples of the time-of-flight spectra of electrons in SF_6. TOF spectra showing the drift of electron swarm, E/N = 370 Td, N = 7.8 x 10^{15} cm^{-3}, Pulse width = 80 ns, Shutter bias to S1 = 15 volt, Pulse height to S1 = 15 volt.

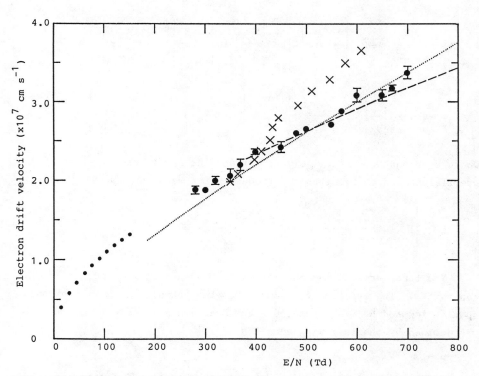

Fig. 8. Measured drift velocity of electrons in SF_6. The drift velocity is shown by the mean value (closed circle) and the standard deviation (short bars) of the data.

Reaction Coefficients of Negative Ions

During the present ion swarm measurements, the TOF spectra which showed clear sign of ion conversion reactions were also observed and we have tried to determine those reaction coefficients.

Reaction coefficients of negative ions were determined by fitting the analytical expressions to the experimental TOF spectra.

When two species of ions (ion1 and ion2) exist and ion1 reacts with the parent gas to form ion2 by charge transfer in the drift space, the respective components of the current flowing into the collector, $I_1(t)$ and $I_2(t)$, can be expressed as follows (Beaty and Patterson, 1965).

$$I_1(t) = \frac{1}{\sqrt{4\pi D_{L1}t}} e^{-(\nu+\gamma D_{T1})t} \int_{-\infty}^{\infty} f_1(x) \exp\left(-\frac{(x-W_1 t)^2}{4D_{L1}t}\right) \frac{x+W_1 t}{2t} dx$$

$$I_2(t) = \frac{1}{\sqrt{4\pi D_{L2}t}} e^{-\gamma D_{T2}t} \int_{-\infty}^{\infty} \left\{ f_2(x) \exp\left(-\frac{(x-W_2 t)^2}{4D_{L2}t}\right) \frac{x+W_2 t}{2t} dx \right.$$

$$+ \frac{\nu W_2}{2(W_1-W_2)} \exp\left[\frac{\nu W_2 t}{W_1 - W_2}\right] \int_{-\infty}^{\infty} \left\{ \frac{1+s}{2s} \exp\left(-\frac{sx}{b}\right) \left(\mathrm{erf}\left[\frac{x-W_2 st}{4D_{L2}t}\right] - \mathrm{erf}\left[\frac{x-W_1 st}{4D_{L1}t}\right] \right) \right.$$

$$\left. + \frac{s-1}{2s} \exp\left[\frac{sx}{b}\right] \left(\mathrm{erf}\left[\frac{x+W_2 st}{4D_{L2}t}\right] - \mathrm{erf}\left[\frac{x+W_1 st}{4D_{L1}t}\right] \right) \right\} f_1(x) \exp\left[\frac{x}{b}\right] dx \right)$$

where W, D and ν, respectively, represent the drift velocity, diffusion coefficient and charge transfer frequency for each ion. And

$$s = \left(1 + \gamma b^2 + \frac{2\nu b}{W_1 - W_2}\right)^{\frac{1}{2}} \quad \text{and} \quad b = \frac{2D_{L1}}{W_1} = \frac{2D_{L2}}{W_2}.$$

The transverse diffusion loss of ions was taken into account by a constant γ. The function f(x) describes the boundary condition of each species imposed by the first shutter. Reaction coefficient, μ, of each ion was defined by the following equation,

$$\mu = \nu / W1 .$$

As described in the foregoing sections, the relative magnitudes of peaks of TOF spectra depended sensitively on both of the blocking bias and the height of the gate pulse to the first shutter, S1. And simply by changing the magnitude of the S1-pulse we were able to choose one or two target species of negative ions injected into the drift region.

Typical examples of the TOF spectra observed in the present measurement are shown in Figs. 9 and 10.

In Fig. 9, an example of the experimental TOF spectra taken at high S1-pulse (about 20 V) is shown by dots. As can be seen in the figure, the spectrum consists of two dominant peaks and broad current component between them. The two peaks were identified to be F^- and SF_5^- in the preceding measurement. This spectrum clearly shows that ions emitted from S1 reacted with neutrals and formed different ion species in the drift space. In Fig. 10, another example taken at low S1-pulse (about 10 V) is shown again by dots, where the two slower ions (SF_5^- and SF_6^-) were involved.

Dominant reactions in SF6 involving those three species of negative ions were shown by the following schemes (O'Neill and Craggs, 1973):

dissociative charge transfer (μ_1):
$$SF_6^- + SF_6 \longrightarrow SF_5^- + F + SF_6,$$
dissociative charge transfer (μ_2):
$$SF_5^- + SF_6 \longrightarrow F^- + SF_4 + SF_6,$$

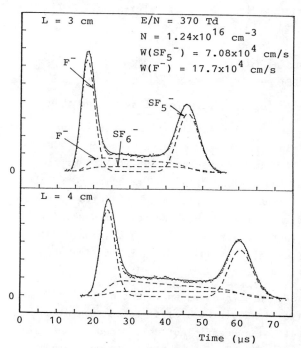

Fig. 9. TOF spectra showing F^-, SF_5^- and their reaction products at two different drift distances. E/N = 370 Td and N = 1.24 x 10^{16} cm^{-3}.

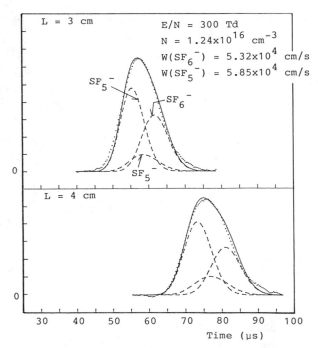

Fig. 10. TOF spectra showing SF_5^-, SF_6^- and their reaction product at two different drift distances. E/N = 300 Td and N = 1.24×10^{16} cm^{-3}.

and charge transfer (μ_3):

$$F^- + SF_6 \longrightarrow SF_6^- + F,$$

where the notation of coefficients follows after them.

   We have tried to fit the analytical expressions to the experimental TOF spectra by considering the above three reactions. The latter two reactions should be involved in the spectra shown in Fig. 9. And it should be noted that the unique separation of the two reaction coefficients was realized by simultaneous fitting to two spectra taken at two different drift distances. The results of the fit were shown by solid lines and the four components corresponding to two peaks (F^- and SF_5^-) originated at the first shutter and two reaction products (F^- and SF_6^-) in the drift space were shown by broken lines.

   SF_5^- and SF_6^- are involved as injected species from the S1 in Fig. 10. Although the two peaks were too close to be separated, the mobility and the longitudinal diffusion coefficient determined in the previous measurement made the fitting procedure fairly easy. As was seen in the figure, a slight disagreement was usually observed at the later part of the spectrum. This disagreement has not been fully analyzed yet, but it could be explained by cluster formation in the drift space.

The present results of three reaction coefficients (μ_1: circle, μ_2: square, μ_3: triangle) are shown in Fig. 11. The experimental point with error bar represents the mean value of the results at several different gas pressures and the error bar shows their maximum scattering. The solid lines are to guide the reader's eye. The present results were also compared with those published previously. Fair agreement for μ_1 was observed with the results obtained by McGeehan et al. (1975) and de Urquijo-Carmona et al. (1986), especially in their higher E/N. Only O'Neill and coworkers [broken lines, (O'Neill and Craggs, 1973)] had measured the three reaction coefficients at E/N > 340 Td.

Concluding Remarks

A novel application of a double shutter (four-gauge) drift tube was demonstrated to study electron and negative ion transport in SF_6.

The present results are summarized as follows:

(1) Mobilities and longitudinal diffusion coefficients of SF_6^-, SF_5^-, and F^- in the parent SF_6 gas were determined over wide ranges of E/N. The ion species in the drift space was easily selected by simply varying the pulse height and the bias applied to the first shutter. Swarm parameters of F^- ions in SF_6 were measured for the first time.

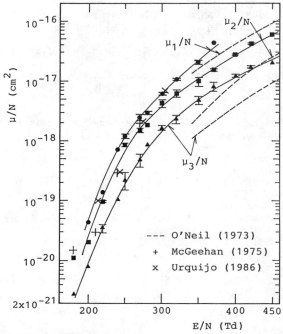

Fig. 11. Negative ion-molecule reaction coefficients in SF_6. The present results are shown by circle (μ_1), square (μ_2) and triangle (μ_3).

209

(2) Reduced mobility peak for each negative ion was confirmed for the first time in SF_6.

(3) The electron drift velocity in SF_6 was measured over the E/N from 280 to 700 Td, and the present result showed good agreement with those measured by Teich and coworkers and by Aschwanden.

(4) Time-of-Flight spectra of negative ions were analyzed, and reaction coefficients of charge transfer of SF_6^-, SF_5^- and F^- in the parent SF_6 gas were successfully determined over the E/N range from 180 to 450 Td.

Applications of the present apparatus may not be restricted to SF_6, and we are planning to study several halogenated gases and oxygen in the near future.

REFERENCES

Aschwanden, Th., 1985, PhD. Thesis.
Beaty, E. C., and P. L. Patterson, 1965, Phys. Rev. 137, 346.
Brand, K. P., and H. Jungblut, 1983, J. Chem. Phys. 78, 1999.
deUrquijo-Carmona, J., I. Alvarez, and C. Cisneros, 1986, J. Phys. D, 19, 207.
Fleming, I. A., and J. A. Rees, 1969, J. Phys. B 2, 777.
Gallagher, J. W., E. C. Beaty, J. Dutton, and L. C. Pitchford, 1982, J. Phys. Chem. Ref. Dat 12, 109.
Harris, F. M., and G. J. Jones, 1971, J. Phys. B 4, 1536.
Kline, L. E., D. K. Davies, C. L. Chen, and P. J. Chantry, 1979, J. Appl. Phys. 50, 6789.
Maller, V. N., and M. S. Naidu, 1981, "Advances in High Voltage Insulation and Arc Interruption in SF_6 and Vacuum" (Pergamon Press, Oxford).
McGeehan, J. P., B. C. O'Neill, A. N. Prasad, and J. D. Craggs, 1975, J. Phys. D 8, 153.
Naidu, M. S., and A. N. Prasad, 1970, J. Phys. D 3, 951.
Naidu, M. S., and A. N. Prasad, 1972, J. Phys. D 5, 1090.
Nakamura, Y., 1987, J. Phys. D 20, 933.
Nakamura, Y., 1988, J. Appl. Phys. D 21, 67.
Nakamura, Y., and T. Kizu, 1987, Proceedings of 5th International Swarm Seminar, Birmingham, 126.
Novak, J. P., and M. F. Frechette, 1984, J. Appl. Phys. 55, 107.
O'Neill, B. C., and J. D. Craggs, 1973, J. Phys. B 6, 2634.
Patterson, P. L., 1970, J. Chem. Phys. 53, 696.
Picard, A., G. Turban, and B. Grolleau, 1986, J. Phys. D 19, 991.
Teich, T. H., 1981, "Electron and Ion Swarms," Edited by Christophorou, L.G., (Pergamon Press, New York), 241.
Teich, T. H., and B. Sangi, 1972, Proceedings of the International Symposium of High Voltage Technology, 391.
Tyndall, A. M., and C. F. Powell, 1931, Proc. Roy. Soc. A 134, 125.
Tyndall, A. M., L. H. Starr, and C. F. Powell, 1928, Proc. Roy. Soc. A 121, 172.
Yoshizawa, T., Y. Sakai, H. Tagashira, and S. Sakamoto, 1979, J. Phys. D 12, 1839.

A SURVEY OF RECENT RESEARCH ON ION TRANSPORT IN SF_6

J. de Urquijo, I. Alvarez, C. Cisneros and H. Martínez

Instituto de Física
Universidad Nacional Autónoma de México
P.O. Box 139-B
62191 Cuernavaca, Mor. México

ABSTRACT

A compilation of recent experimental work on the mobility and longitudinal diffusion of SF_6 ions in their parent gas is presented and critically evaluated. Some progress in the extension of the E/N and pressure ranges over earlier work is observed. With the exception of very recent, mass-identified data for two positive ions, all other data have been derived from experiments lacking mass-analysis. The need for mass-analyzed data is stressed.

INTRODUCTION

The wide range of applications of sulfur hexafluoride under swarm or discharge conditions has made this gas the subject of research for several decades. Considerable progress in the determination of electron swarm data (e.g. ionization, and attachment coefficients, electron mobilities and diffusion coefficients) can be appreciated from recent reviews and papers relative to this subject (Christophorou and Hunter, 1984; Teich, 1981; Gallagher et al., 1983; Phelps and van Brunt, 1988; Itoh et al., 1988). Of similar importance for both basic and applied science is the determination of accurate sets of ion transport properties, such as the mobility and diffusion coefficients of the various ion fragments and clusters formed in parent SF_6. The mobility K, related to electric field intensity E is $K = v_d/E$, where v_d is the drift velocity of the ion swarm in the direction of the field, and is a function of the parameter E/N, where N is gas number density, and the unit of E/N is the Townsend (1 Td=10^{-17} V cm^2). Mobility and diffusion data are, on the one hand, used to derive the ion-molecule interaction potential through their dependence on E/N and gas temperature

(Gatland, 1984) and for the calculation of ion-ion recombination coefficients, ion-molecule reaction rates and cross sections (McDaniel and Mason, 1973). On the other hand, these parameters are also needed in the modeling and quantitative understanding of phenomena in electrical discharges in gases (Thompson et al., 1986; Picard et al., 1986).

In contrast to the abundant set of electron swarm data, there is still a scarcity of ion transport data in SF_6. This situation is due to the fact that techniques and apparatus for producing and detecting electrons in a swarm experiment are generally less sophisticated than those for ions, where ion sources and mass spectrometers are needed for an unambiguous determination of ion transport data. This is also the case for transport theory, where the small mass of the electron as compared with that of the neutrals permits some simplifications for electron transport theory that are not valid for the case of ion transport where, in addition to this, rotational and vibrational degrees of freedom involved in molecular systems make theory even more complex.

A comprehensive review of experimental techniques and theories available up to 1972 has been provided by McDaniel and Mason, (1973); more recent work can be found in the compilation of Lindinger et al., (1984).

The aim of this paper is to present a summary and critical evaluation of the experimental methods and results of recent research on the mobility and diffusion of SF_6 ions in their parent gas from 1983 to date. A previous compilation of mobility data up to 1982 has been given by Brand and Jungblut (1983). Research on transverse diffusion can be found in the review of Rees (1974).

SUMMARY OF RECENT RESEARCH WORK

As a guide for the present discussion, Fig. 1 shows a simplified reaction scheme for low and moderate energy electrons in SF_6, developed by Teich (1981) and based mostly on the work of O'Neill and Craggs (1973); McGeehan et al. (1975); Fehsenfeld (1970); and Patterson (1970). Aschwanden (1985) recently incorporated the attachment reaction leading to formation of F^- (Curran, 1961; Hayashi and Nimura, 1984).

Let us focus our attention on the rather complex ion-molecule reaction scheme for the ions. With the exception of F^-, the mobilities of SF_5^-, SF_6^-, $SF_6^- SF_6$ and $SF_6^- (SF_6)_2$ were measured by Patterson with a drift tube-mass spectrometer with fixed drift distance up to E/N of about 140 Td. To date, these are the only mass-analyzed negative ion mobilities reported. Further work in this direction has relied on a comparison of the data with those of Patterson in order to infer the identity of the ions (de Urquijo-Carmona, 1983; de Urquijo-Carmona et al., 1985b; Cornell and Littlewood, 1986; de

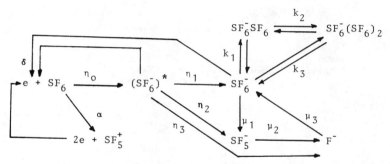

Fig. 1. Simplified reaction scheme for SF_6. α, η, δ, μ, ionization,
         attachment, detachment and charge transfer coefficients;
         k_1, k_2, k_3, reaction rates.

Urquijo-Carmona et al., 1986; Nakamura, 1988). On the other hand, Patterson's data were the base upon which Brand and Jungblut (1983) identified the
available mobility data from which they calculated the interaction potentials between these ions and SF_6 neutrals. No mass-analyzed data were yet
available for the positive ions, thus leading Brand and Jungblut to assume a
mobility equal to that for SF_5^-.

     A summary of recent research on ion transport in SF_6 is shown in Table 1.
An inspection of this table indicates that over the present period of survey:
(a) The data cover wide ranges of E/N and gas pressure. (b) Longitudinal
diffusion coefficients for several ion species are reported for the first
time. (c) All mobility data, with the exception of those for SF_5^+ and SF_3^+ (de
Urquijo-Carmona et al., 1990) have been derived from experiments lacking
mass-analysis. In all other cases the authors have resorted to a comparison
with either Patterson's data or Brand and Jungblut's calculations to identify
their ions. (d) Drift tube techniques were used in two cases, while the
remaining set of data were obtained from pulsed Townsend
experiments.

EXPERIMENTAL TECHNIQUES

     The techniques used for the recent measurements of the mobility and
longitudinal diffusion of SF_6 ions in their parent gas were drift tubes of
the (a) double-shutter (DSDT) and (b) mass spectrometer types, and (c) the
pulsed Townsend method (PTM).

The Drift Tube Technique

     Drift tubes for the study of ion transport in gases have been used
extensively for several decades (McDaniel and Mason, 1973; Lindinger et al.,
1984). This technique underwent substantial improvements in the 1960's by
the incorporation of mass spectrometers to sample the ions arriving at the

Table 1. Summary of Recent Research on Ion Transport in SF_6[a]

Apparatus	Parameters		Ion species		E/N range	Pressure range	Accuracy	Reference
	from	Measured	Ident.	Infer.	(Td)	(kPa)	(%)	
PT	CFIT, TT	K		$SF_6^-SF_6$	7.6-243	9.3-25.7	5	b
PT	CFIT	K		Pos.	212-910	0.13-2.7	1-3	c
PT	CFIT	K, D_L		SF_5^+	909-5450	9.3-91.8 Pa	2-5[a]	d
PT	CFIT	K		SF_5^- SF_6^-	212-303	0.53-1.1	4-6	e
PT	TT	v_d (K)		$SF_6^-SF_6$	3-185	13.3-80	3	f
DSDT	AATS	K		SF_5^- SF_6^- F^-	20-600	N= 8×10^{15}- 5.3×10^{16} cm^{-3}	2	g
		D_L					20	
DTMS	AATS	K	SF_3^+		20-510	3.3-13.3	2	h
		D_L	SF_5^+		30-360	Pa	10-25	

[a] 25% accuracy for D_L. PT, Pulsed Townsend apparatus; DSDT, double-shutter drift tube; CFIT, curve fitting to ion transients; TT from transit time; K, mobility; D_L longitudinal diffusion coefficient; v_d, drift velocity. Temperature range: 295-300 K AATS, analysis of arrival-time spectra.

[b] de Urquijo-Carmona, 1983.

[c] Aschwanden, 1985.

[d] de Urquijo-Carmona et al., 1985b.

[e] de Urquijo-Carmona et al., 1986.

[f] Cornell and Littlewood, 1986.

[g] Nakamura, 1988.

[h] de Urquijo-Carmona, et al., 1990.

end of the drift space, and by the use of movable ion sources in order to eliminate end effects and other additional parameters influencing the measured ion transit time, from which the ion drift velocity is evaluated (McDaniel and Mason, 1973). More recent improvements include ion-selected sources, but they have not been applied to the present case (Lindinger et al., 1984). A sketch of the DSDT and the DTMS is shown in Figs. 2a and 2b, respectively. Both types of drift tube bear several parts in common, such as an ion source to produce the ions and a set of guard rings to produce a highly uniform electric field in the longitudinal direction (z). Also, in both cases the information derived from the experiment is an arrival-time spectrum (ATS) of the ions, such as that shown in Fig. 3. However, in the presence of several ion species in the drift space, the DSDT will produce a

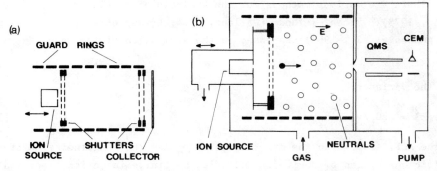

Fig. 2. Sketches of (a) a double-shutter drift tube and (b) a drift
tube-mass spectrometer. QMS: quadrupole mass spectrometer;
CEM: channeltron multiplier.

set of arrival time spectra corresponding to each ion species whereas, in
the case of the DTMS, only the spectrum of the selected ion will be observed. In fact, this is one of the major advantages of the DTMS over the
DSDT, since unambiguous, mass-analyzed mobility and diffusion data can be
obtained.

The experiment is performed in both cases by admitting short bursts of
ions into the drift space at regular intervals and letting them drift and
diffuse under the action of the electric field and collide with the neutrals
until they arrive at the end of the drift space. Then, for the DSDT case, a
short, delayed pulse applied to the second shutter grids, allows the ions in
their neighborhood to be detected by the collector plate; for the DTMS case,
a sample of the ions passes through a small orifice into
a mass spectrometer followed by a particle detector.

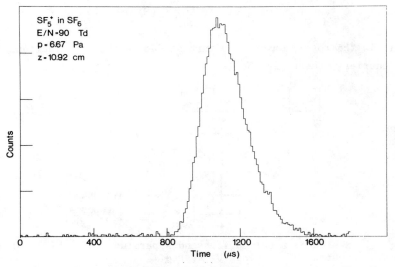

Fig. 3. Typical arrival-time spectrum from a DTMS.

There are various methods for the analysis of an ATS (McDaniel and Mason, 1973). Recent methods (Gatland, 1984; Løvaas et al., 1987; Stefánsson et al., 1988) are based on the evaluation of the mean

$$\langle t \rangle = z/v_d + b_1 \qquad (1)$$

and the variance

$$\sigma^2 = 2zD_L/v_d^3 + b_2 \qquad (2)$$

of the ATS, from which the drift velocity v_d and the longitudinal diffusion coefficient D_L can be derived. End effects, finite injection times and mass spectrometer effects can be eliminated by a differencing technique (McDaniel and Mason, 1973; Løvaas et al., 1987) consisting of measuring ATS at several source positions. These effects are contained in the b_1 and b_2 terms of Eqs. (1) and (2), respectively.

The Pulsed Townsend Method (PTM)

Basically, the PTM is based on the observation of the total displacement current due to the individual motion of all charge carriers drifting across a parallel-plate discharge gap filled with gas of uniform density, to which a uniform electric field is applied between the plates. A sketch of a typical PT apparatus is shown in Fig. 4. The initiation of the carrier swarm current is accomplished by a short pulse (3-20 ns) of UV light which releases a large number of photoelectrons from the cathode. These electrons and their reaction products with the neutrals drift to their corresponding electrodes under the action of the electric field. These displacement currents will thus produce a total voltage drop $U_R(t)$ across measuring resistor R, to which a parasitic capacitance is effectively connected. This voltage is (Raether, 1964)

$$U_R(t) = C^{-1}\exp(-t/RC) \int_0^t \exp(\theta/RC)\, I_t(\theta)\, d\theta \qquad (3)$$

where I_t is the total, observable current due to the contribution of all charge carriers present in the gap.

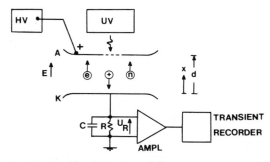

Fig. 4. Sketch of a pulsed Townsend apparatus. HV: high voltage source; UV: pulsed UV source.

Because of the substantial difference between electron and ion drift velocities, the total current pulse shape can be decomposed into a fast, electron component and a slow ion component. According to the value of the RC time constant in comparison with charge carrier growth/decay, the experiment can be performed under two different conditions. These are the differentiated pulse condition, for which Eq. (3) becomes

$$U_R(t) \simeq RI_t(t), \tag{4}$$

if the RC time constant is small (a few ns or less), and the integrated pulse condition, when the product RC can be as large as several seconds.

There are at present a number of well accounted reviews on the PTM, and the interested reader is referred to them (Christophorou and Hunter, 1984; Aschwanden, 1985; Raether, 1964; Huxley and Crompton, 1974; Teich and Zaengl, 1979; de Urquijo-Carmona, 1980). The above limiting conditions have been used widely for the determination of electron and ion swarm parameters (Aschwanden, 1985; Raether, 1964; Teich and Zaengl, 1979; de Urquijo-Carmona, 1980; Grünberg, 1969; Teich and Branston, 1973; Blevin et al., 1981; Hunter et al., 1986).

As regards our present discussion, we shall briefly consider the cases relevant to the derivation of ion transport coefficients from the analysis of ionic transients.

A recent model (de Urquijo-Carmona et al. 1985a) considering the occurrence of ionization, attachment, ion drift and longitudinal diffusion in a homogeneous-field discharge gap was applied to the analysis of observed ion transients in SF_6 at high E/N (909-5450 Td) by de Urquijo-Carmona et al. (1985b). Attachment processes at such high E/N are either negligible or virtually absent and, therefore, were not considered in their analysis. The ion transients were fitted according to a formula given in de Urquijo-Carmona et al. (1985a) and an example of the fitting procedure is shown in Fig. 5. Values of the ionization coefficient α /N, v_d and ND_L were obtained with the overall uncertainties of 9%, 3% and 25%, respectively (see also Figs. 12 and 13 for data). The α /N values are in good agreement with those recently calculated by Phelps and van Brunt (1988) up to 3000 Td. The reduced mobility values, derived from v_d by

$$K_o = (v_d/N_o)/(E/N), \tag{5}$$

and the density-normalized longitudinal diffusion coefficients ND_L will be discussed in the next section. Here, $N_o = 2.69 \times 10^{19}$ cm^{-3}.

In principle, the PTM lacking mass analysis suffers from a major disadvantage over the DTMS as far as ion transport research is concerned. However, the PTM allows experiments to be carried out over wide gas pressure ranges

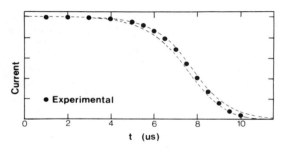

Fig. 5.   Positive ion transient in SF_6 at E/N = 1820 Td, p = 40 Pa,
from de Urquijo-Carmona et al., 1985b.   The broken curves
enclose the region of maximum combined uncertainties of the
fitting process.

(say $10-10^5$ Pa), significantly above the upper limits imposed on drift tubes
(about 1 kPa).   The experiments to be described below were carried out at
pressures near and well above the latter limit for drift tubes.

An inspection of the plot of all available negative ion mobility data
derived from the PTM (see Brand and Jungblut's plots in their paper) prior to
1983 reveals that the use of elementary models based on the drift of a single,
stable negative ion species leads to mobility values with an E/N dependence
for which no explanation can be offered in terms of existing mobility theories.   In fact, Aschwanden (1985) found that by using such elementary methods
to fit ion transients, the negative ion mobilities showed a strong pressure
dependence.   However, it is well known that SF_6 negative ions react with the
neutrals according to the reaction scheme of Fig. 1.   Then, a comprehensive
model accounting for the relevant processes occurring in the drift gap would
have to be developed in order to fit the ion transients to model and use the
full reaction scheme with success.   However a limited number of reactions can
be used over particular ranges of E/N and pressure to derive useful, quantitative information, as in the case of the fitting to ion transients according
to a portion of the reaction scheme of McGeehan et al. (1975).   This was used
by de Urquijo et al. (1986) in an attempt to explain observed ion transients
in SF_6 in the range 212-303 Td at pressures of 0.53 and 1.1 kPa.

Figure 6 shows plots of relative contribution of F^-, SF_5^- and SF_6^- to the
total ion current measured in a double-gap, steady state Townsend apparatus
with mass spectrometer of McGeehan et al, from which they derived their
reaction scheme.   Inspection of these curves shows that over the range 200-300
Td, the contribution of F^- to the total ion current is small (<1%), and that
SF_5^- and SF_6^- are the predominant ions.   This finding was used by de Urquijo et
al. (1986) to fit ion transients from a time-resolved experiment by assuming a
model considering the occurrence of the reaction

$$SF_6^- + SF_6 \xrightarrow{\mu_1} SF_5^- + F + SF_6$$

218

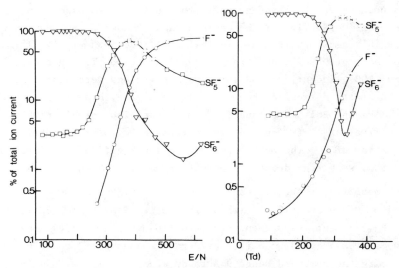

Fig. 6. Negative ion concentrations as a function of E/N from the double Townsend ionization chamber experiment of McGeehan et al., (1975) Left: p=168 Pa, Right: 1.9 kPa.

was developed, where μ_1 is the dissociative charge transfer coefficient. Additionally, formation of SF_5^- and SF_6^- by dissociative and resonant electron attachment was considered, as well as the contribution of the current due to positive ions.

A sample transient and its fitting curve is shown in Fig. 7, where the contributions of the three ions to the total current are shown by dotted lines.

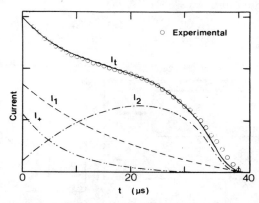

Fig. 7. Ion transient in SF_6 at E/N=303 Td, p=526 Pa. Solid line: total, calculated current. I_1, I_2, currents due to slower and faster negative ions, assumed to be SF_6^- and SF_5^-, respectively. I_+, positive ions. [From (de Urquijo-Carmona et al., 1986)].

219

As pointed out by these authors, the most relevant contributions of their work were the derivation of the charge transfer coefficient, which was found to be in fair agreement between the values of McGeehan et al. (1975) up to 240 Td, and those of O'Neill and Craggs (1973) above 300 Td. Moreover, Nakamura and Kizu (1987) recently found that their μ_1 values were also in fair agreement. The total attachment coefficients have been found to be in good agreement by a recent Boltzmann equation analysis of Itoh et al. (1988). Because of the neglect of diffusion and, in retrospective, of the reaction leading to the formation of F^-, the mobilities of the two negative ions, assumed to be SF_5^- and SF_6^-, were regarded as 'apparent' (see Figs. 9 and 10).

The mobility of two ion clusters, namely $SF_6^-SF_6$ and $SF_6^-(SF_6)_2$, was first measured by Patterson (1970) with a fixed-drift length DTMS up to about 140 Td and 133 Pa. Recent measurements (de Urquijo-Carmona, 1983; Cornell and Littlewood, 1986) of the mobility of negative ions using the PTM at high pressures in the combined range 9.3 to 80 kPa provide a strong indication that their unidentified mobilities correspond to $SF_6^-SF_6^-$, in view of the good agreement with that measured by Patterson for this ion. Measurements of de Urquijo-Carmona (1983) were carried out under the differentiated pulse condition (Figs. 8a, b), and those of Cornell and Littlewood (1986) under the integrated pulse condition (Fig. 8c).

DISCUSSION OF THE DATA

This section presents a discussion of the recent mobility and longitudinal diffusion coefficients of SF_6 ions in their parent gas. Wherever applicable, data reported prior to 1983 are included in order to enhance the discussion. Mobility data are presented in the usual form of reduced mobilities given by Eq. (5), and longitudinal diffusion coefficients in terms of the product ND_L.

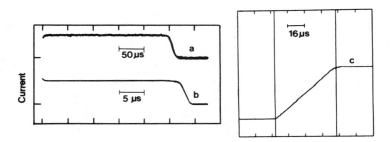

Fig. 8. Negative ion transients in SF_6. (a),(b), from the differentiated pulse condition (de Urquijo-Carmona, 1983), and (c) from the integrated pulse condition (Cornell and Littlewood, 1986). (a) 45.5 Td, 25.7 kPa; (b) 242 Td, 9.3 kPa, (c) 135 Td, 26.7 kPa.

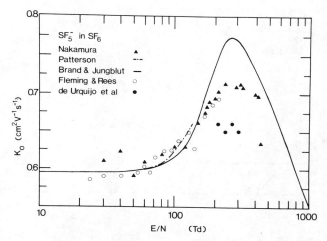

Fig. 9.   Reduced mobility data of SF_5^- in SF_6 as a function of E/N.
Recent data are due to Nakamura (1988) and de Urquijo-Carmona
et al., (1986).

SF_5^- in SF_6

Data for the reduced mobility of SF_5^- in SF_6 have been plotted in Fig. 9
as a function of E/N. The only mass-analyzed data are still those of
Patterson (1970) from which recent data of Nakamura (1988) in the overlap
range were inferred to correspond to this ion. Data of this author show a
clear maximum, well below that calculated by Brand and Jungblut for a (9, 6,
4) interaction potential. In connection with this, it may have been that
this higher maximum was obtained from a fitting to the unresolved trend of
negative ion mobilities derived from pulsed Townsend experiments using the
most elementary model of a single, stable ion species drifting across the
gap, without resorting to any charge-transfer reactions. In fact,
Nakamura's data are very recent (1988). The few K_o values of de Urquijo-
Carmona et al. (1986) are about 8% lower than those of Nakamura, and show no
clear trend. As stated above, these authors regarded these mobilities as
'apparent' because of the neglect of diffusion, and also of the formation of
F^- both from charge transfer and dissociative attachment (see Fig. 1). For
comparison with these recent data, the mobilities of Fleming and Rees (1969)
are also shown, and seem to be in good agreement with those of Nakamura in
the overlap range.

SF_6^- in S_6

Again, the only mass-analyzed reduced mobilities for this ion are those of
Patterson (1970). Recent values of Nakamura (1988) also show a maximum at E/N
between 300-400 Td. However, his data seem to be higher than those of Patterson
in the overlap region, and from which this author inferred the identity of the

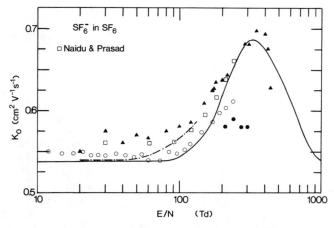

Fig. 10. Reduced mobility of SF_6^- in SF_6 as a function of E/N. For other symbols, refer to Fig. 9.

ion. In contrast to the above ion, the few data of Nakamura around the maximum agree fairly with the calculated curve of Brand and Jungblut (1983), although the suggested trend for E/N above 400 Td is clearly different. Data of de Urquijo-Carmona et al. (1986), again lower by about 8% than those of Nakamura, show no clear trend and have been regarded as 'apparent'. The inclusion of previous values of Fleming and Rees (1969) and Naidu and Prasad (1970) shows a large scatter between all the mobility data for this ion.

$SF_6^-SF_6$ in SF_6

Reduced mobility data for this cluster ion are plotted as a function of E/N in Fig 11. Recent values are those of de Urquijo-Carmona (1983) and of Cornell and Littlewood (1986), for which a discussion on their derivation

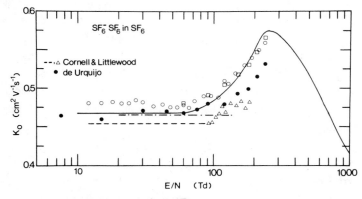

Fig. 11. Reduced mobilities of $SF_6^-SF_6$ in SF_6 as a function of E/N.
Recent data are due to de Urquijo-Carmona (1983) and Cornell and Littlewood (1986). For other symbols, refer to Fig. 9.

has been provided already. In both cases, these authors identified the ion by comparison with the mass-analyzed mobilities of Patterson (1970), and found good agreement up to E/N about 70 Td for de Urquijo-Carmona's values and 100 Td for those Cornell and Littlewood. In either case, a departure from a constant value is apparent for E/N about 100 Td. Brand and Jungblut (1983) have predicted a well defined maximum, but none of the experimental data shown in this plot give an indication of it.

A comparison of the recent data, derived for the combined pressure range 9.3–80 kPa (de Urquijo-Carmona, 1983; Cornell and Littlewood, 1986) and previous work of Naidu and Prasad (1970) at pressures up to 1.3 kPa (Fleming and Rees give no details, but their experiment was essentially the same) indicates that the mobility curves of the latter authors which agree fairly well in the overlap range, are systematically higher than those of the recent work carried out at high pressures. We speculate about this strong difference by suggesting that the higher mobility values could be the unresolved mean of the drift of this ion partly as SF_6^-, since these cluster ions were found to be easily destroyable by Patterson, who suggested that the reaction leading to $SF_6^-SF_6$ was reversible (Patterson, 1970).

Positive Ions

Mass-analyzed mobility data for SF_5^+ and SF_3^+ in SF_6 have become available from a recent DTMS experiment (de Urquijo-Carmona et al., 1990). Other recent data derived from the PTM are those of Aschwanden (1985) and de Urquijo-Carmona et al. (1985b) which have been plotted in Fig. 12, together with the earlier data of Fleming and Rees (1969). The latter data are found in good agreement with the mass-analyzed data of de Urquijo-Carmona et al. (1990) up to E/N about 200 Td. Moreover, an extrapolation to zero field of the SF_5^+ data gives a zero-field mobility $K_{oo} = 0.591 \pm 0.006$ $cm^2V^{-1}s^{-1}$, which agrees well with the previously quoted value of Fleming and Rees of 0.59(0) $cm^2V^{-1}s^{-1}$. Aschwanden's data (Aschwanden, 1985) disagree strongly in their trend for E/N < 300 Td with respect to the SF_5^+ data. This disagreement is believed to be due to the fact that below E/N ~ 360 Td, the critical field strength for which $\alpha \simeq \eta$, the contribution of the positive ion current to the total, observable current is a decaying exponential (Aschwanden, 1985; Raether, 1964; Teich and Zaengl, 1979; de Urquijo-Carmona, 1980) (see also Fig. 7), from which the positive ion drift velocity v_+ is derived indirectly from time constant $1/(\alpha - \eta)v_+$, where the value of the effective ionization coefficient $(\alpha - \eta)$ enters sensitively in the result. On the other hand, above E/N = 360 Td, $\alpha > \eta$, and the role of contribution of negative and positive ion currents reverses; the positive ion transit time and hence the drift velocity can be determined with more accuracy from the fall of the total current pulse. Indeed, above 400 Td, Aschwanden's mobilities for the positive ions are in good agreement with those of SF_5^+.

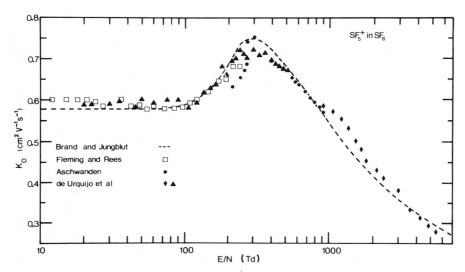

Fig. 12. Reduced mobilities of SF_5^+ in SF_6 as a function of E/N. Diamonds, mass-analyzed data of de Urquijo-Carmona et al. (1990). Triangles, data from de Urquijo-Carmona et al. (1985b).

Another recent set of data for positive ions in SF_6 was derived from the PTM by de Urquijo-Carmona et al. (1985b) for E/N in the range 909-5450 Td. These data are also shown in Fig. 12, where they depart from the smooth curve of Brand and Jungblut (1983) up to 1600 Td, thereafter the agreement improving.

As a final point on this matter, it is interesting to notice that Nakamura's mobilities for SF_5^- (Fig. 9) and the recent data (de Urquijo-Carmona et al., 1990) for SF_5^+ (Fig. 12) are nearly equal in the region around the maximum. On the one hand, this fact provides a stronger support to the assumptions of Brand and Jungblut (1983) regarding the identity of th positive ion, for which no mass-analyzed data were available by that time. On the other hand, previous evidence of equal mobilities for positive and negative ions in the neighborhood of E/N = 360 Td was established by Teich and Branston (1973) as early as 1973, using the PTM.

Longitudinal Diffusion Coefficients

The only sets of data for this parameter seem to be those of Nakamura (1988) for SF_5^-, SF_6^- and F^- over the E/N range 100-360 Td, which were derived from the same ATS he used to obtain the mobilities referred to above. Also, values of this parameter were reported by de Urquijo-Carmona et al. (1985b) over the range 909-5450 Td, from the PTM. These data, together with very recent ones obtained with a DTMS (de Urquijo-Carmona et al., 1990) are plotted in Fig. 13 as a function of E/N. The latter data correspond to SF_5^+.

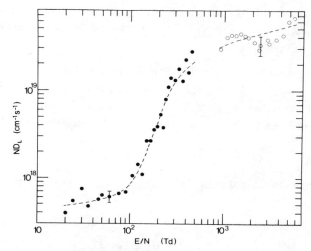

Fig. 13. Recent, density-normalized longitudinal diffusion
coefficients. Closed circles, mass-analyzed data (de
Urquijo-Carmona et al., 1990). Open circles from de
Urquijo-Carmona et al., 1985b.

Although there is a gap between these two sets of data, they would seem to
connect rather well through a typical S-shaped curve predicted by elementary
diffusion theories, although this mere fact is of no conclusive evidence.
The rather large scatter of the points at low E/N does not allow a precise
extrapolation to zero field in order to compare it with the predicted value
by Einstein's relation.

CONCLUSIONS

The above review on the determination of ion transport data of SF_6 ions
in their parent gas can be summarized as follows:

A. Attempts to use the pulsed Townsend method for the measurement of
ion transport data should be made with care, requiring either the assurance
that the models used for fitting the observed transients are complete and
correct and/or previous knowledge of the presence of a single, stable ion
present in the drift gap. On the other hand, the PTM still proves to be a
powerful technique for the determination of ionization and attachment
coefficients from observed ion transients.

B. There is a need for more mass-analyzed measurements of the mobility
in and around the maximum of the K_o versus E/N curve, so that a confirmation
of the only, recently measured mobilities of Nakamura for SF_5^-, SF_6^- and F^-
can be provided. Furthermore, it would be advantageous to provide at least
a second set of data over the E/N region covered by Patterson, since his
data are so far the only ones with mass analysis.

C. It is desirable to increase the range of measurement for the mobility of $SF_6^- SF_6$, for which no data are available in and above the predicted peak of Brand and Jungblut.

D. With the advent of a drift tube-mass spectrometer for the study of ion transport at high E/N (Stefansson et al. 1988), an adoption of this technique for the present case would be very useful in order to cover wider E/N ranges than have been hitherto possible with the pulsed Townsend method.

ACKNOWLEDGEMENT

This work was partially supported by CONACyT, project No. P228CCOX880316.

REFERENCES

Aschwanden, T., 1984, in "Gaseous Dielectrics IV," L. G. Christophorou, and M. O. Pace, (Eds.), 24-33, Pergamon, New York.
Aschwanden, T., 1985, DTW Thesis, ETH Zurich, unpublished.
Blevin, H. A., K. H. Nygaard, and K. R. Spriggs, 1981, J. Phys. D 14, 841.
Brand, K. P., and H. Jungblut, 1983, J. Chem. Phys. 78, 1999.
Christophorou, L. G., and S. R. Hunter, 1984, in "Electron-Molecule Interactions and their applications," 90-291, L. G. Christophorou, (Ed.), Academic Press, New York.
Cornell, M. C., and I. M. Littlewood, 1986, J. Phys. D 19, 1889.
Curran, R. K., 1961, J. Chem. Phys. 34, 1069.
de Urquijo-Carmona, J., 1980, Ph. D. Thesis, University of Manchester, England, unpublished.
de Urquijo-Carmona, J., 1983, J. Phys. D 16, 1603.
de Urquijo-Carmona, J., I. Alvarez, and C. Cisneros, 1985a, J. Phys. D 18, 29.
de Urquijo-Carmona, J., I. Alvarez, and C. Cisneros, 1986, J. Phys. D 19, L207.
de Urquijo-Carmona, J., I. Alvarez, C. Cisneros, and H. Martínez, 1990, J. Phys. D in press.
de Urquijo-Carmona, J., C. Cisneros, and I. Alvarez, 1985b, J. Phys. D 18, 2017.
Fehsenfeld, F. C., 1970, J. Chem. Phys. 53, 2000.
Fleming, I. A., and J. A. Rees, 1969, J. Phys. B 2, 777.
Gallagher, J. W., E. C. Beaty, J. Dutton, and L. C. Pitchford, 1983, J. Chem. Phys. Ref. Data 12, 109.
Gatland, I. R., 1984, in "Swarms of Ions and Electrons in Gases," W. Lindinger, T. D. Mark, and F. Howorka, (Eds.), 44-59, Springer-Verlag, Wien-New York.
Grünberg, R., 1969, Z. Naturforsch 24a, 1039.
Hayashi, M., and T. Nimura, 1984, J. Phys. D 17, 2215.
Hunter, S. R., J. G. Carter, and L. G. Christophorou, 1986, J. Appl. Phys. 60, 24.
Huxley, L. G. H., and R. W. Crompton, 1974, "The Diffusion and Drift of Electrons in Gases," Wiley, New York.
Itoh, H., Y. Miura, N. Ikuta, Y. Nakao, and H. Tagashira, 1988, J. Phys. D 21, 922.
Lindinger, W., T. D. Märk, and F. Howorka, (Eds.), 1984, "Swarms of Ions and Electrons in Gases," Springer-Verlag, Wien-New York.
Løvaas, T. H., H. R. Skullerud, O. -H. Christensen, and D. Linhjell, 1987, J. Phys. D 20, 1465.
McDaniel, E. W., and E. A. Mason, 1973, "The Mobility and Diffusion of Ions in Gases," Wiley, New York.

McGeehan, J. P., B. C. O'Neill, A. N. Prasad, and J. D. Craggs, 1975, J. Phys. D $\underline{8}$, 153.

Naidu, M. S., and A. N. Prasad, 1970, J. Phys. D $\underline{3}$, 951.

Nakamura, Y., 1988, J. Phys. D $\underline{21}$, 67.

Nakamura, Y., and T. Kizu, 1987, Proc. 5th Swarm Seminar, Birmingham, U.K.

O'Neill, B. C., and J. D. Craggs, 1973, J. Phys. B $\underline{6}$, 2634.

Patterson, P. L., 1970, J. Chem. Phys. $\underline{53}$, 696.

Phelps, A. V., and R. J. van Brunt, 1988, J. Appl. Phys. 64, 4269.

Picard, A., G. Turban, and B. Grolleau, 1986, J. Phys. D. $\underline{19}$, 991.

Raether, H., 1964, "Electron Avalanches and Breakdown in Gases," Butterworths, London.

Rees, J. A., 1974, Vacuum $\underline{24}$, 603.

Stefansson, T., T. Berge, R. Lausund, and H. R. Skullerud, 1988, J. Phys. D $\underline{21}$, 1359.

Teich, T. H., 1981, in "Electron and Ion Swarms," L. G. Christophorou, (Ed.), 241-50, Pergamon Press, New York.

Teich, T. H., and D. W. Branston, 1973, Nature $\underline{244}$, 504.

Teich, T. H., and W. S. Zaengl, 1979, in "Current Interruption in High Voltage Networks," Plenum Press, New York.

Thompson, B. E., K. D. Allen, A. D. Richards, and H. H. Sawin, 1986, J. Appl. Phys. $\underline{59}$, 1890.

COLLISIONAL ELECTRON-DETACHMENT AND ION-CONVERSION PROCESSES IN SF$_6$

J. K. Olthoff[a], R. J. Van Brunt[a], Yicheng Wang[b],
L. D. Doverspike[b], and R. L. Champion[b]

[a] National Institute of Standards and Technology
Gaithersburg, MD 20899

[b] Physics Department
College of William and Mary
Williamsburg, VA 23185

INTRODUCTION

It has long been hypothesized that collisional electron detachment from negative ions may be the cause of discharge initiation in SF$_6$-insulated high-voltage equipment. However, uncertainties continue to exist concerning the identities of the significant negative ions, and the magnitudes of the detachment cross sections and corresponding reaction coefficients. Discharge inception studies, which usually assume that SF$_6^-$ is the negative ion of interest, have predicted collisional-detachment cross sections of SF$_6^-$ in SF$_6$ with thresholds ranging from 1 to 8 eV (Kindersberger, 1986). Collisional-detachment coefficients determined from low-pressure, uniform-field drift-tube experiments (O'Neill and Craggs, 1973; Hansen et al., 1983) lie several orders of magnitude above those calculated from breakdown data (Kindersberger, 1986), while exhibiting an unexplained pressure dependence. Additionally, evidence exists that the destruction of negative ions in SF$_6$ is dominated by collisional dissociation into ionic fragments (O'Neill and Craggs, 1973; McAfee, Jr. and Edelson, 1963; Compton et al., 1971; McGeehan et al., 1975; Urquijo-Carmona et al., 1986; Nakamura and Kizu, 1987) or by charge-transfer processes rather than by the loss of an electron. However, the exact nature of these ion-conversion processes has not previously been determined.

In this report we summarize results from the first direct measurements of absolute cross sections for electron-detachment and ion-conversion processes involving interactions of SF$_6^-$, SF$_5^-$, and F$^-$ with SF$_6$ (Wang et al., 1989). These cross sections are used to calculate electron-detachment and ion-conversion reaction coefficients as functions of electric field-to-gas density ratios (E/N) for the reactions listed in Table 1 (Olthoff et al., 1989). We then discuss the relevance of these results to the interpretation of data from uniform-field drift-tube measurements and measurements of electrical-discharge initiation processes.

CROSS SECTIONS

Details of the experimental method used to measure the cross sections reported here are given elsewhere (Wang et al., 1989; White et al., 1984), and will not be covered here. We shall focus here only on the results obtained for collisions of SF_6^-, SF_5^-, and F^- on the SF_6 target gas, although measurements have also been made using rare gas targets (Wang et al., 1989), namely He, Ne, Ar, Kr, and Xe.

The measured center-of-mass energy dependencies of the collisional electron-detachment cross section, $\sigma_i(\varepsilon_{cm})$ for SF_6^-, SF_5^-, and F^- on an SF_6 target are shown in Fig. 1. The thresholds for the "prompt" collisional electron-detachment processes

$$SF_6^- + SF_6 \longrightarrow \text{neutral products} + e^-$$

$$SF_5^- + SF_6 \longrightarrow \text{neutral products} + e^-,$$

as indicated by the vertical arrow in Fig. 1, are seen to be quite high (approximately 90 eV), while the detachment process for F^-,

$$F^- + SF_6 \longrightarrow \text{neutral products} + e^-,$$

has a lower detachment threshold of about 8 eV. At center-of-mass collision energies below the indicated thresholds there is no evidence to suggest non-zero values for the $\sigma_i(\varepsilon_{cm})$'s. Similar high values for the collisional electron-detachment thresholds were found for collisions of SF_6^- and SF_5^- ions with rare gas targets (Wang et al., 1989). The implication of these results as shown below is that prompt electron detachment from either SF_6^- or SF_5^- cannot be important for electrical-discharge conditions because, even for the highest E/N values which could conceivably occur in a gas discharge gap, only an insignificant fraction of the negative ions could acquire a kinetic energy of 90 eV or more.

Cross sections for the ion-conversion processes (reactions (4)-(9)) listed in Table 1 are shown in Fig. 2 together with extrapolations down to the thresholds which are used later in the calculation of corresponding rate coefficients. The bases for these extrapolations are also discussed later. The solid symbols indicate cross sections due to collisional-induced-dissociation (CID) processes, and the open symbols indicate cross sections for charge-transfer processes (including charge-transfer decomposition). There is evidence that the charge-transfer process involving SF_6^- on SF_6 is predominately dissociative

Table 1. Collisional processes for which cross sections are
presented in the present work

cross section (σ_i)	reaction
σ_1	$SF_6^- + SF_6 \longrightarrow e^- + SF_6 + SF_6$
σ_2	$SF_5^- + SF_6 \longrightarrow e^- + SF_5 + SF_6$
σ_3	$F^- + SF_6 \longrightarrow e^- + F + SF_6$
σ_4	$SF_6^- + SF_6 \longrightarrow F^- + SF_5 + SF_6$
σ_5	$SF_5^- + SF_6 \longrightarrow F^- + SF_4 + SF_6$
σ_6	$SF_6^- + SF_6 \longrightarrow SF_5^- + F + SF_6$
σ_7	$SF_6^- + SF_6 \longrightarrow$ charge-transfer products $+ SF_6$
σ_8	$SF_5^- + SF_6 \longrightarrow$ charge-transfer products $+ SF_5$
σ_9	$F^- + SF_6 \longrightarrow F + SF_6^-$

(Lifshitz et al., 1973; Foster and Beauchamp, 1975), and that charge
transfer involving F^- on SF_6 may lead to both SF_6^- and SF_5^- (O'Neill and
Craggs, 1973; Lifshitz et al., 1973). It is interesting to note that the
ion-conversion processes predominate for ε_{cm} <200eV, i.e., upon collision
of an SF_6^- or SF_5^- ion with SF_6, ion conversion is much more likely than
prompt electron detachment.

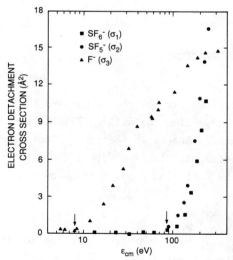

Fig. 1. Collisional electron-detachment cross sections for F^-, SF_5^-,
and SF_6^- on SF_6 target gas as a function of center-of-mass
energy.

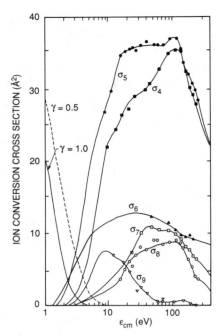

Fig. 2.   Measured cross sections for collision-induced ion-
conversion processes in SF_6 target gas: (●) F^- from SF_5^-;
(■) F^- from SF_6^-; (▲) SF_5^- from SF_6^-; (o) ions due to
charge-transfer reactions of SF_5^-; (□) ions due to charge-
transfer reactions of SF_6^-; and (∇) ions due to charge-
transfer reactions of F^-.   The two exponentially-decaying
curves on the left side of the figure represent the
kinetic-energy distribution of Eq. (5) for γ = 1.0 and γ =
0.5 scaled relative to each other.

REACTION RATES AND COEFFICIENTS

Calculations

     The analysis of rates for chemical processes in drift tubes or electrical
discharges requires expressing inelastic-collision probabilities in terms of
rate coefficients rather than cross sections. For a process where the pro-
jectiles have a velocity distribution, $f(v)$, the rate coefficient, k, is given
by an averaging of the collision cross section over all velocities, namely

$$k = \int_0^\infty \sigma(v)vf(v)dv, \tag{1}$$

were $\sigma(v)$ is the velocity-dependent cross section for the process and
$f(v)$ is subject to the normalization

$$\int_0^\infty f(v)dv = 1. \tag{2}$$

For measurements where charged particles are accelerated in a gas by an electric field (such as in a drift tube), the reaction coefficient, κ, is defined as the reaction probability per unit length in the direction of the electric field, and is related to the rate coefficient by

$$\frac{\kappa}{N} = \frac{k}{v_d},$$ (3)

where N is the target-gas number density and v_d is the charged particle drift velocity. For the specific case of collisional electron detachment from a negative ion, the reaction coefficient is referred to as the detachment coefficient, δ (i.e. $\kappa/N = \delta/N$ for collisional electron detachment).

Reaction coefficients for detachment and ion-conversion processes involving SF_6^-, SF_5^-, and F^- in SF_6 can be derived from the measured cross sections, $\sigma_i(\varepsilon_{cm})$, presented in Figs. 1 and 2 using

$$\frac{\kappa_i}{N} = \frac{1}{mv_d} \int_o^\infty \sigma_i(\varepsilon_L) f(\varepsilon_L) d\varepsilon_L,$$ (4)

where ε_L is the projectile energy in the lab frame, m is the mass of the negative ion, and $f(\varepsilon_L)$ and $\sigma(\varepsilon_L)$ are the ion kinetic-energy distribution and the process cross sections, respectively. The ion drift velocity is assumed to be given by $v_d = \mu E$, where μ is the ion mobility and E is the applied electric-field strength.

It should be noted that the determination of reaction coefficients from collisional cross sections suffers from certain difficulties, the greatest arising from the assumed form of the ion kinetic-energy distribution, $f(\varepsilon_L)$ (Albritton et al., 1977). While determination of ion kinetic-energy distributions has received considerable theoretical (Lin and Bardsley, 1977) and experimental (Khatri, 1984) attention, the theoretical work is hampered by the lack of detailed ion-atom (or molecule) potential-energy surfaces and the experimental work suffers from a lack of reliability (Albritton et al., 1977). Accurate direct measurements of ion-velocity distributions have been demonstrated in recent optical-probing experiments, (Dressler et al., 1988) but to date no experimental data are available for the kinetic-energy distributions of SF_6^-, SF_5^-, or F^- in SF_6.

In general, experimental work has indicated that ions with masses less than or equal to that of the molecules of the gas in which they are moving exhibit kinetic-energy distributions with high-energy tails (Albritton et al., 1977; Moruzzi and Harrison, 1974). Differences in the velocity distributions at high energies in Eq. (4) will obviously have a large effect upon the calculated reaction coefficients derived from cross sections with threshold energies considerably in excess of the average ion kinetic energies.

An example of the large differences in calculated values of reaction coefficients which can occur when different energy distributions are assumed is shown in Fig. 3, where the collisional-detachment coefficient for F^- in $SF_6(\delta_3)$ has been calculated as a function of E/N using the cross section data (σ_3) from Fig. 1 and several different indicated energy distributions. The solid line in Fig. 3 represents detachment coefficients calculated using the kinetic-energy distribution of Kagan and Perel (Kindersberger, 1986; Kagan and Perel, 1954),

$$f(\varepsilon_L) = \frac{\sqrt{6\gamma}}{\pi \, v_d} \exp\left(\frac{-\gamma\varepsilon_L}{\pi \, mv_d^2}\right),\tag{5}$$

which assumes that charge exchange is the dominant ion-molecule interaction. For the standard Kagan and Perel distribution, $\gamma = 1.0$. However, as will be discussed later, better agreement between ion-conversion reaction coefficients calculated here and those from analysis of drift-tube results is obtained by assuming $\gamma = 0.5$ which introduces a larger high-energy tail in the distribution.

The dashed line in Fig. 3 was obtained using a Maxwellian speed distribution (Kindersberger, 1986) of the form,

$$f(\varepsilon_L) = 3\sqrt{\frac{3m}{\pi}} \frac{\varepsilon_L}{\bar{\varepsilon}^{3/2}} \exp\left(\frac{-3\varepsilon_L}{2\bar{\varepsilon}}\right),\tag{6}$$

where the mean energy of the ion in the lab frame is given by (Wannier, 1951)

$$\bar{\varepsilon} = \frac{3kT}{2} + \frac{m}{2} v_d^2 + \frac{M}{2} v_d^2,\tag{7}$$

and M is the mass of the collision-gas molecules. This distribution has been used previously when analyzing discharge-inception data (Kindersberger, 1986) and is similar to a strongly anisotropic velocity distribution derived by Skullerud (1973) for ions drifting in a gas composed of molecules of the same mass as the ions under high electric fields.

Figure 3 clearly demonstrates how the use of different ion kinetic-energy distributions in Eq. (4) can produce reaction coefficients which differ by many orders of magnitude. This indicates the need for more accurate ion kinetic-energy distribution measurements.

In addition to the uncertainty associated with the assumed distribution function, another source of uncertainty in deriving reaction coefficients from cross-section data is the choice of experimentally determined values of ion mobilities (μ). Several previous mobility measurements (Brand and Jungblut, 1983; Morrow, 1986; Patterson, 1970; Nakamura, 1988) for SF_6^- and SF_5^- in SF_6 are not in complete agreement.

234

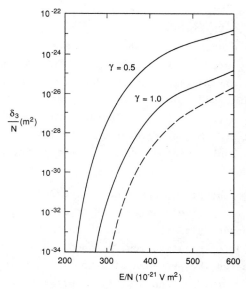

Fig. 3. Calculated collisional electron-detachment coefficients for
F^- on SF_6 gas using Eq. (4) and $\sigma_3(\varepsilon_L)$, and assuming
different ion kinetic-energy distributions: (——) Kagan and
Perel; (– – –) Maxwellian.

However, uncertainties in calculated reaction coefficients using different mobilities are significantly smaller than the uncertainties due to
the use of different energy distributions as discussed above. For the
remainder of this paper, the mobilities used for SF_6^- and SF_5^- are those
reported by Brand and Jungblut (1983) and for F^-, the values of Nakamura
(1988).

Detachment Coefficients

As expected, the extremely high apparent thresholds for prompt
collisional electron detachment from SF_6^- and SF_5^- yield detachment coefficients from Eq. (4) that are insignificantly small in the E/N range of
interest here. Simple estimates of these coefficients using a Kagan and
Perel velocity distribution in Eq. (4) indicate that the detachment
coefficients for SF_6^- and SF_5^- will be tens of orders of magnitude below
the detachment coefficients determined in drift-tube experiments and the
detachment coefficients calculated for F^- (Fig. 3). Thus one must conclude that prompt collisional electron-detachment processes for SF_6^- (and
SF_5^-) in SF_6 cannot be significant reactions for production of electrons
in discharge-inception processes, if, in fact, $\sigma_i(\varepsilon_{cm})$ is zero below the
apparent thresholds. It should be noted that this conclusion is independent of the assumed ion kinetic-energy distribution or the ion-
mobility data used.

It is clear from Fig. 3, however, that the collisional electron detachment from F^- in SF_6 is a significant process, thus suggesting that previously observed electron-detachment processes due to motion of negative ions in SF_6 are very likely from F^-. This agrees with earlier work (Eccles et al., 1967) which indicated that detachment in SF_6 was predominately from F^-, and also with recent reanalysis of pulsed-electron avalanche experiments (Teich, 1973). In fact, the observed threshold for electron detachment from F^- in SF_6 near 8 eV is consistent with the hypothesized thresholds predicted by some discharge-inception experiments (Kindersberger, 1986).

Other discharge-inception experiments (Van Brunt, 1986) have, however, suggested that negative cluster ions involving H_2O or HF may be responsible for discharge initiation in SF_6. In fact, recent measurements (Sauers et al., 1989) of negative ions produced in negative corona discharges, have indicated the presence of several types of cluster ions of the forms $F^-(HF)_n$, $OH^-(H_2O)_n$, and $SF_6^-(HF)$. Further investigations of the collisional detachment cross sections for these ions are necessary to determine their role in discharge initiation.

The conclusions drawn above depend upon the assumption that $\sigma(\varepsilon_L) = 0$ at energies below the apparent threshold marked in Fig. 1. If one assumes that $\sigma(\varepsilon_L) = 0.1\ \overset{\circ}{A}^2$ (i.e., the experimental uncertainty) for energies which extend down to the thermodynamic threshold for electron detachment, then detachment coefficients for SF_6^- and SF_5^- are found to be of the same order of magnitude as those determined by drift-tube experiments. However, detachment coefficients derived with such an assumption are not compatible with previously observed pressure dependencies observed in drift-tube experiments (O'Neill and Craggs, 1973; Hansen et al., 1983) as discussed in the next section.

Ion-Conversion Reaction Coefficients

In order to calculate the reaction coefficients for the ion-conversion processes listed in Table 1, it is necessary to extrapolate the measured cross sections down to assumed thresholds at lower energies. The extrapolations used for the subsequent calculations are shown in Fig. 2. These extrapolations were chosen to agree with known thermodynamic thresholds and to minimize the discrepancies with previously determined reaction coefficients as discussed below. The thresholds for production of F^- from SF_6^- and SF_5^- (σ_4 and σ_5) were determined to be 2.0 eV and 1.5 eV, respectively, by using the observed thresholds for F^- production from collisions of SF_6^- and SF_5^- with the rare gas targets (Wang et al., 1989). The cross sections for SF_5^- production (σ_6) and SF_6^- were extrapolated down to the thermodynamic threshold of 1.35 eV. For the charge-transfer reaction involving F^- and SF_6, the cross sections were extrapolated down the thermodynamic threshold of 2.25 eV under the assumption that the primary product is SF_6^-. The other charge-transfer cross sections (σ_7 and σ_8) were both extrapolated down to a threshold near 3 eV which corresponds to the thermodynamic threshold for a symmetric charge transfer between SF_6^- and SF_6 as suggested by Hay (1982). There is a large uncertainty in these last assumed thresholds since the identity of the charge-transfer products in these processes are indistinguishable in the present experiment.

The calculated reaction coefficients for processes 6, 5 and 9 of Table 1 are shown in Figs. 4, 5 and 6 respectively, along with reaction coefficients for the same reactions as determined by previous drift-tube experiments. The solid lines represent the coefficients calculated using the standard Kagan and Perel distribution (i.e. $\gamma = 1$) shown in Eq. (5). Note that these calculated reaction coefficients all fall substantially below those determined previously despite the fact that the extrapolated thresholds for these cross sections were all assumed to be thermodynamic thresholds. Any reasonable change in the assumptions concerning the reaction thresholds or the behavior of the cross sections near threshold would necessarily cause the reaction coefficients to be even smaller, thus implying that the discrepancies cannot be resolved by changing the assumed cross section thresholds or extrapolations.

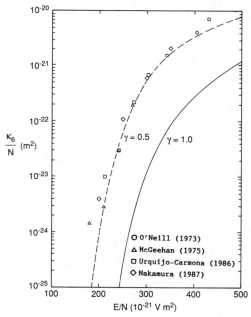

Fig. 4.   Calculated reaction coefficients for the reaction $SF_6^- + SF_6$ $\rightarrow SF_5^- + F + SF_6$ using measured cross-section data $\sigma_6(\varepsilon_L)$ and a Kagan and Perel energy distribution. The symbols are previously calculated reaction coefficients for the same process derived from uniform-field drift-tube data.

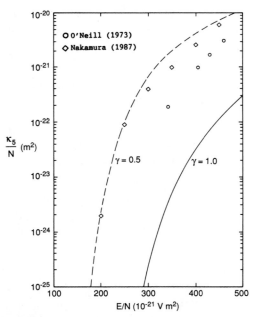

Fig. 5. Calculated reaction coefficients for the reaction $SF_5^- + SF_6^-$ $\rightarrow F^- + SF_4 + SF_6$ using measured cross-section data $\bar{\sigma}_5(\varepsilon_L)$ and a Kagan and Perel energy distribution. The symbols are previously calculated reaction coefficients for the same process derived from uniform-field drift-tube data.

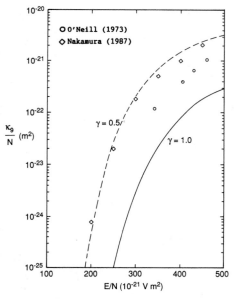

Fig. 6. Calculated reaction coefficients for the reaction $F^- + SF_6 \rightarrow F$ $+ SF_6^-$ using measured cross-section data $\sigma_9(\varepsilon_L)$ and a Kagan and Perel ion kinetic-energy distribution. The symbols are pre- viously calculated reaction coefficients for the same process derived from uniform-field drift-tube data.

Despite the fact that the Kagan and Perel distribution produces the largest coefficients of any of the commonly used ion kinetic-energy distributions, better agreement can be obtained between our calculated coefficients and those from previous experiments if one assumes that the kinetic-energy distribution has a longer high-energy tail, in agreement with the previous discussion of kinetic-energy distributions. The Kagan and Perel distribution in Eq. (5) can be conveniently altered by allowing γ to vary between 0 and 1. The dashed lines in Figs. 4 to 6 represent reaction coefficients calculated with $\gamma = 0.5$, while in Fig. 2 the relative magnitudes of the two distributions ($\gamma = 1$ and $\gamma = 0.5$) are shown for comparison.

Obviously, the curves in Figs. 4-6 with $\gamma = 0.5$ are in better agreement with the previously reported coefficients than are the curves calculated using $\gamma = 1$. This may indicate that previously assumed energy distributions need to be modified (Kindersberger, 1986; Hansen et al., 1983). However, one must note that the reaction coefficients derived from drift-tube data are model-dependent and that only reactions 5, 6 and 9 (Table 1) are assumed in the previous analysis of data from drift tubes. Thus discrepancies may also arise because this commonly used model does not consider the collision-induced dissociation of SF_6^- into $F^- + SF_5$ (reaction 4). This reaction is found here to be significant (see Fig. 2) and its omission may produce errors in the reaction rates derived from drift-tube data.

The calculated reaction coefficients for reaction 4 (and for reactions 7 and 8) are shown in Fig. 7 using Kagan and Perel distributions with $\gamma = 1.0$ and 0.5. As stated before, no previously determined coefficients for these processes exist for comparison.

MODEL FOR ELECTRON DETACHMENT FROM ION DRIFT IN SF_6

A different interpretation of the processes which lead to detachment coefficients derived from drift-tube data (O'Neill and Craggs, 1973; Hansen et al., 1983) and their observed pressure dependence can be obtained if one assumes that electron production in a drift tube is not due primarily to prompt electron detachment from SF_6^-, but arises from either detachment from F^- or from collisionally-excited, energetically-unstable $(SF_6^-)^*$ ions as suggested by the previously presented data. A model can then be developed which assumes that under drift-tube conditions, prompt detachment from SF_5^- and SF_6^- is insignificant, and that a steady state condition exists for intermediate products (i.e. $d[SF_5^-]/dt = d[F^-]/dt = d[(SF_6^-)^*]/dt = 0$). If one analyzes the drift-tube data as done previously (O'Neill and Craggs, 1973; Hansen et al., 1983), assuming that all ejected electrons come from SF_6^- ions, then the measured electron-production rate can be written in terms of an "effective" detachment coefficient (δ_{eff}) according to the expression

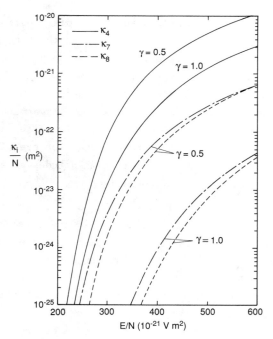

Fig. 7. Calculated reaction coefficients using a Kagan and Perel
ion kinetic-energy distribution for the following
reactions: (———) $SF_6^- + SF_6 \rightarrow F^- + SF_5 + SF_6$; (— — —) $SF_6^- + SF_6 \rightarrow$ charge-transfer products; and (— · —) $SF_5^- + SF_6 \rightarrow$ charge-transfer products.

$$\frac{d[e]}{dt} = v_d \left(\frac{\delta_{eff}}{N}\right) N[SF_6^-], \tag{8}$$

where $[e]$ is the number of free electrons produced per unit volume which
can be detected, v_d refers specifically to the drift velocity of SF_6^- in
SF_6, and δ_{eff}/N is determined by analysis of drift-tube data (O'Neill and
Craggs, 1973; Hansen et al., 1983). Assuming that the simplified set of
processes indicated in Table 2 dominates in a drift tube, an expression
for δ_{eff}/N, can be found in terms of the relevant rate coefficients by
solving the set of coupled rate equations. The result is

$$k_{eff} = v_d \left(\frac{\delta_{eff}}{N}\right) = k_3 \left(\frac{k_6 + k_4}{k_9 + k_3}\right) + k_{12} \left(\frac{k_{10}}{k_{12} + k_{11}N}\right). \tag{9}$$

The effective detachment coefficient given by Eq. (9) is seen to consist
of two terms, a pressure independent term which depends upon various ion-
conversion and direct-detachment processes involving F^-, and a pressure-
dependent term which depends upon the rates for collisional relaxation,
excitation and auto-detachment of $(SF_6^-)^*$. This expression is more complex
than those previously derived (Kindersberger, 1986; O'Neill and Craggs, 1973;

Table 2. Proposed ion-molecule reactions for drift-tubes containing SF_6

$SF_6^- + SF_6 \longrightarrow SF_5^- + F + SF_6$	k_6	⎫
$SF_6^- + SF_6 \longrightarrow F^- + SF_5 + SF_6$	k_4	⎬ Dissociative Ion Conversion
$SF_5^- + SF_6 \longrightarrow F^- + SF_4 + SF_6$	k_5	⎭
$F^- + SF_6 \longrightarrow F + SF_6^-$	k_9	Charge Transfer
$F^- + SF_6 \longrightarrow$ neutrals $+ e^-$	k_3	e^- Detachment
$SF_6^- + SF_6 \longrightarrow (SF_6^-)^* + SF_6$	k_{10}	Excitation
$(SF_6^-)^* + SF_6 \longrightarrow SF_6^- + SF_6$	k_{11}	De-excitation
$(SF_6^-)^* \longrightarrow SF_6 + e^-$	k_{12}	Auto-detachment

Hansen et al., 1983) which assume that direct detachment from SF_6^- was the sole source of electrons.

If $k_{11}N$ is approximately the collision frequency of SF_6^- in SF_6 (e.g., ~ 10^8/sec at 1 kPa) and k_{12} is on the order of the inverse of the excited-state lifetime (τ ~ 10μs to 2 ms) (Odom, et al., 1975), then $k_{11}N \gg k_{12}$ and the model predicts an inverse pressure dependence for δ_{eff}/N at low N. This inverse pressure dependence is consistent with previous drift-tube measurements (O'Neill and Craggs, 1973; Hansen et al., 1983). O'Neill and Craggs (1973) also report no detachment from F^- or SF_5^- at low pressures in agreement with the dominance of the second term of Eq. (9) for smaller N. The model proposed here is consistent with the observed (Hansen et al., 1983) variations in δ_{eff}/N with the "age" of the SF_6^- ions if there is a substantial fraction of SF_6 anions which are initially in excited or autodetaching states.

At higher pressures, the first term on the right side of Eq. (9) dominates, giving a δ_{eff}/N that is essentially pressure independent. In this pressure regime, electron production involves mainly process (3) with a threshold of 8 eV. These results are consistent with: 1) the lack of pressure dependence (Kindersberger, 1986; Teich, 1973) for δ_{eff}/N suggested from the analysis of high-pressure electrical-discharge initiation data, and 2) previously discussed results (Eccles et al., 1967) suggesting detachment in SF_6 is predominately from F^-.

Ideally, one would like to calculate the effective detachment coefficients using Eq. (9) to compare with the previously determined drift-tube measurements. However, values for k_{10}, k_{11} and k_{12} are not available, so only the contribution to δ_{eff}/N from the first term of Eq. (9) can be calculated using the reaction coefficients derived above. The solid curve in Fig. 8 shows the contribution from the first term of Eq. (9) using a Kagan and Perel distribution with $\gamma = 1.0$. Note that the

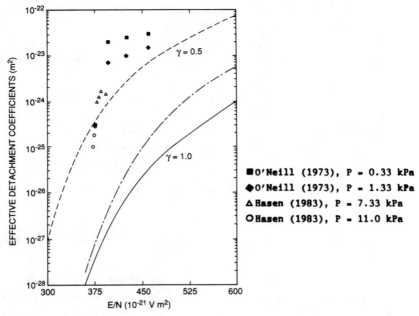

Fig. 8.   Effective detachment coefficients determined from drift-tube
           experiments at different pressures are shown as identified by the
           symbols in the key. Effective detachment coefficients calculated
           from discharge-inception data from (Kindersberger, 1986) are also
           shown (—— · ——). The contribution to the effective detachment
           coefficients from the first term of Eq. (9) are shown for $\gamma = 1.0$
           (——) and for $\gamma = 0.5$ (- - -).

magnitude of the solid curve is similar to that of the effective detachment coefficient derived from discharge-inception experiments (Kindersberger, 1986) (dot-dashed curve) but is substantially smaller than the
coefficients derived from drift-tube experiments (O'Neill and Craggs,
1973; Hansen et al., 1983) (symbols). If one uses the reaction coefficients derived using a Kagan and Perel energy distribution with $\gamma = 0.5$
(dashed curve) then the δ_{eff}/N derived from the first term of Eq. (9)
becomes of the same order of magnitude as the coefficients derived from
the highest pressure drift-tube experiments. The fact that the dashed
curve actually lies above the smallest measured drift-tube values
indicates that taking $\gamma = 0.5$ may overestimate the high-energy tail for
the kinetic-energy distribution.

CONCLUSION

     The measured cross sections for collisional electron detachment and
ion conversion of negative ions in SF_6 have been used in a theoretical
model which invokes detachment from long-lived, energetically-unstable
states of collisionally excited SF_6^- to explain the pressure dependence of

242

previously measured detachment coefficients and the high apparent detachment thresholds implied by analysis of breakdown-probability data for SF_6. The model suggests that measured effective detachment coefficients depend upon many different reaction rates, thus implying that detachment processes in SF_6 are more complex than previously assumed. At high pressures, measured detachment coefficients appear to depend primarily upon the rates for ion-conversion and direct-detachment processes involving F^-, consistent with earlier suggestions. Also, calculation of ion-conversion rates for SF_6^-, SF_5^-, and F^- in SF_6 indicate the need to reexamine both reaction coefficients derived from drift tube experiments and ion kinetic-energy distributions assumed in interpreting the cross section data presented here.

REFERENCES

Albritton, D. L., I. Dotan, W. Lindinger, M. McFarland, J. Tellinghuisen, and F. C. Fehsenfeld, 1977, J. Chem. Phys. 66, 410.

Brand, K. P., and H. Jungblut, 1983, J. Chem. Phys. 78, 1999.

Compton, R. N., D. R. Nelson, and P. W. Reinhardt, 1971, Int. J. Mass Spectrom. Ion Phys. 6, 117.

Dressler, R. A., J. P. M. Beijers, H. Meyer, S. M. Penn, V. M. Bierbaum, and S. R. Leone, 1988, J. Chem. Phys. 89, 4707.

Eccles, M., A. N. Prasad, and J. D. Craggs, 1967, Electronics Lett. 3, 410; B. H. Crichton, G. C. Crichton, and D. J. Tedford, 1972, "Proceedings of 2nd International Conference on Gas Discharges," London, p. 385.

Foster, M. S., and J. L. Beauchamp, 1975, Chem. Phys. Lett. 31, 482.

Hansen, D., H. Jungblut, and W. F. Schmidt, 1983, J. Phys. D: Appl. Phys. 16, 1623.

Hay, P. J., 1982, J. Chem. Phys. 76, 502.

Kagan, I. M., and V. I. Perel, 1954, Doklady Akad. Nauk S.S.S.R. 98, 575.

Khatri, M. H., 1984, J. Phys. D.: Appl. Phys. 17, 273; P. P. Ong, and M. J. Hogan, 1985, J. Phys. B: At. Mol. Phys. 18, 1897; H. A. Fhadil, A. T. Numan, T. Shuttleworth, and J. B. Hasted, 1985, Int. J. Mass Spectrom. Ion Process 65, 307; T. Makabe, and H. Shinada, 1985, J. Phys. D: Appl. Phys. 18, 2385.

Kindersberger, J., 1986, Ph.D. Thesis, Technical University of Munich, Munich, W. Germany, unpublished; N. Wiegart, L. Niemeyer, F. Pinnekamp, W. Boeck, J. Kindersberger, R. Morrow, W. Zaengl, M. Zwicky, I. Gallimberti, and S. A. Boggs, 1987, IEEE/PES 1987 Winter Meeting, New Orleans, LA; N. Wiegart, 1985, IEEE Trans. Electr. Insul. EI-20, 587.

Lifshitz, C., T. O. Tiernan, and B. M. Hughes, 1973, J. Chem. Phys. 59, 3182.

Lin, S. L., and J. N. Bardsley, 1977, J. Chem. Phys. 66, 435; J. H. Whealton, and S. B. Woo, 1971, Phys. Rev. A 6, 2319.

McAfee, Jr., K. B., and D. Edelson, 1963, Proc. "Phys. Soc.," London, 81, 382.

McGeehan, J. P., B. C. O'Neill, A. N. Prasad, and J. D. Craggs, J. Phys. D: Appl. Phys. 8, 153.

Morrow, R., 1986, IEEE Trans. Plasma Sci. PS-14, 234.

Moruzzi, J. L., and L. Harrison, 1974, Int. J. Mass spectrum. Ion Phys. 13, 163.

Nakamura, Y., 1988, J. Phys. D: Appl. Phys. 21, 67.

Nakamura, Y., and T. Kizu, 1987, "Proceedings of 5th International Swarm Seminar," Birmingham, U.K., 126.

Odom, R. W., D. L. Smith, and H. H. Futrell, 1975, J. Phys. B: Atom.
    Molec. Phys. 8, 1349; J. E. Delmore, and L. D. Appelhans, 1986, J.
    Chem. Phys. 84, 6238.
Olthoff, J. K., R. J. Van Brunt, Y. Wang, R. L. Champion, and L. D.
    Doverspike, 1989, J. Chem. Phys. 91, 2261.
O'Neill, B. C., and J. D. Craggs, 1973, J. Phys. B: At. Mol. Phys. 6,
    2634.
Patterson, P. L., 1970, J. Chem. Phys. 53, 696.
Sauers, I., M. C. Siddagangappa, and G. Harman, 1989, "Proc. 6th Int.
    Symposium High Voltage Eng.," New Orleans.
Skullerud, H. R., 1973, J. Phys. B: Atom. Molec. Phys. 6, 728.
Teich, T. H., private communication; D. W. Branston, 1973, Thesis,
    University of Manchester, UK.
Urquijo-Carmona, J., I. Alvarez, and C. Cisneros, 1986, J. Phys. D:
    Appl. Phys. 19, L207.
Van Brunt, R. J., 1986, J. Appl. Phys. 59, 2314.
Wang, Y., R. L. Champion, L. D. Doverspike, J. K. Olthoff, and R. J. Van
    Brunt, 1989, J. Chem. Phys. 91, 2254.
Wannier, G. H., 1951, Phys. Rev. 83, 281; 1952, Phys. Rev. 87, 795.
White, N. R., D. Scott, M. S. Huq, L. D. Doverspike, and R. L. Champion,
    1984, J. Chem. Phys. 80, 1108.

A CLOSE ENCOUNTER BETWEEN THEORY AND EXPERIMENT IN ELECTRON-ION COLLISIONS

M. A. Hayes

SERC Daresbury Laboratory
Warrington, WA4 4AD, UK

INTRODUCTION

In the calculation of electron impact excitation cross-sections of ions, only for a relatively few simple cases can we make a reasonable assessment of the accuracy of the cross-sections. By simple cases, I mean for targets for which there are relatively few strongly interacting terms, e.g., Li-like ions, and for electron energies either low enough that all possible scattering channels can be included in the calculation or high enough that perturbation methods, i.e., distorted wave or Coulomb-Born type methods, can be applied. Although the numerical accuracy of the calculations can be checked, for most theoretical results we have to depend on comparison with experimental data to help us judge the reliability of our data. Even for simple cases, there are some surprising discrepancies between theoretical and experimental results.

Of course, theory has the advantage that you can calculate data for systems at present inaccessible to experiment, and it can also be useful in determining what are the important interactions in a collision.

In this paper, I shall look at some examples of systems for which both theoretical and experimental cross-sections or rate-coefficients are available, to see what can be learned about the current state of the theory. This study does not claim to be a full survey - I have selected only a few sample cases - and I hope that more information will have been gained from discussions at the seminar.

THEORETICAL METHODS

The calculations that I have selected, while not necessarily being 'state of the art' are never-the-less fairly sophisticated. They employ multi-configuration descriptions of the target. The close-coupling (CC) method is used for evaluating the scattering cross-sections, i.e., the full (non-relativistic) interaction is included between the target and electron, but because of the complexity of the calculation, only a fairly small number of terms are included in the wavefunction expansion, which is of the form

$$\Psi = \Sigma_i A \psi_i \theta_i + \Sigma_j c_j \phi_j$$

where ψ are the target wavefunctions, θ are the free-electron functions to be determined, ϕ are known 'bound state' functions of the target plus electron, c_j are coefficients to be determined and A is an operator for antisymmetrisation.

The cross-sections can only be as good as the target functions and hence much effort should be spent in obtaining an accurate description of the target. A large part of the following discussion is centered around the effect of the neglected terms in the above expansion. All the examples are in LS coupling.

A "TYPICAL" PROBLEM

I was asked to calculate rate-coefficients for transitions among the $n = 2$ states of boron-like neon, Ne^{5+}, and for the excitation of some of the low-lying $n = 3$ states, to compare with laboratory plasma measurements of Lang (1989). These rate-coefficients will also be useful in analyzing ultra-violet spectra from the sun to determine electron temperatures and densities. Accuracies of 10 – 20% in the rates are usually required for such analyses. Although I am aware that there will be deficiencies in the calculation of the $n = 3$ rates, because of pressure to obtain these rate-coefficients and because there was free time on the Cray XMP at Rutherford Appleton Laboratory during its commissioning, I agreed to attempt the calculation. A partial term scheme for Ne^{5+} is shown in Fig. 1.

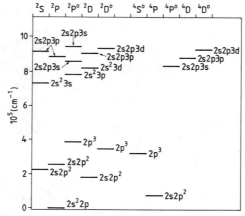

Fig. 1. A partial term scheme for Ne^{5+}.

In the target description, I included configurations of the form $2^3, 2^2 3, 2^2 4$, where $2^2 3$ stands for all configurations containing 2 electrons in the n = 2 shell and 1 in the n = 3 shell, etc. The n = 4 orbitals (s, p and d only) were non-physical pseudo-orbitals which attempt to mock up the effect of the omitted configurations in the target expansion. For the terms of interest, this gave energies with respect to the ground term in agreement with experimental energies to better than 1%, except for $2s2p^2 \; ^4P$ and ^2S where the difference is 3%.

For the scattering problem, all the target terms up to $2s2p(^3P°)3p$ were included which led to 32 scattering channels in the calculation. Adding the $2s2p(^3P°)3d$ terms would have contributed up to 16 channels which would have made the calculation too large. (A channel is defined by the target state and orbital angular momentum of the scattered electron.)

For the electron temperature of interest (9×10^5 K), a large part of the rate coefficient comes from electron energies greater than the ionization potential where there is not only an infinite number of bound target states but also continuum states which are energetically accessible. What is the effect on the cross-section of the severe truncation of the CC expansion to a few terms containing low-lying target states? In Fig. 2 is a section of the collision-strength for excitation of the metastable first excited state showing the complicated resonance structure, but how accurate is the structure?

COMPARISON OF RESULTS FROM THEORY AND SOME CROSSED-BEAM EXPERIMENTS

As said in the introduction, for simple systems we have a fair idea of the accuracy of the theoretical cross-sections, but there are still some unaccountable discrepancies. One example which has been discussed in the

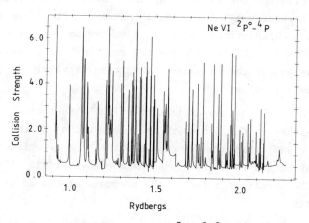

Fig. 2.   Collision strength for the $2s^2 2p \; ^2P°$ - $2s2p^2 \; ^4P$ transition in Ne^{5+}.

literature is the excitation of the resonance 2s - 2p transition in Be^+, where there is an 18% difference at the threshold between the measurements of Taylor et al. (1980) and the CC calculations of Hayes et al. (1977) which included up to the n = 3 target states. In fact, the difference between experiment and theory is roughly constant at 20% over the electron energy range 4 -740 ev. The experimental uncertainty is on average 8%, and near threshold the theoretical results were thought to be accurate to about 5%.

Similar comparisons have been carried out for N^{4+} and C^{3+}, see Gregory et al. (1979) and references therein. In these cases, agreement is well within the 17% experimental uncertainty. We note that theoretically, Be^+ is harder to treat than higher ions in the iso-electronic sequence, as the electron-electron interactions are relatively more important than the central coulomb potential.

The effect of the neglected channels in the CC expansion for Be^+ was investigated by Henry et al. (1978) by using n = 3 pseudo states rather than the real n = 3 states. They obtained a decrease in the cross-section of 3 - 4% over the energy range 26 - 110 ev. One would expect that a bigger close-coupling calculation would decrease the cross-section further. However, below the n = 3 thresholds, the decrease is expected to be smaller as all the open (energetically accessible) channels have been included. I have also looked at the effect of resonances on this transition, but it was small and tended to enhance the cross-section, increasing the discrepancy with experiment.

Parpia et al. (1986) used a much more sophisticated target, concentrating on the effect of short-range electron-electron interactions and allowing ab initio for the polarization of the $1s^2$ core by the valence electron, rather than semi-empirically as was done by Hayes et al. They found agreement to within 1% with the earlier calculation for the total cross-section although for the partial cross-section, for individual partial waves, the difference was 3 - 4%. Note that even with a sophisticated 44-configuration calculation, the 2s - 2p energy separation obtained by Parpia et al., is 1.1% different from the experimental energy. The calculated f-value for the resonance transition is within 4% of the most recent experimental value.

We shall now look at the excitation of the resonance 4s - 4p transition in the alkali-like ion Ca^+ for which there are recent CC calculations of Mitroy et al. (1988) which can be compared with the experimental cross-sections of Taylor and Dunn (1973) and Zapesochnyi et al. (1976). This case is slightly more complicated due to the presence of the 3d state between the 4s and 4p states.

In the calculation emphasis was put on obtaining a good target description including polarization of the n = 3 core by the valence electron. The interaction of the target polarization with the scattered electron was also taken into account. A semi-empirical polarization potential was used; from the above experience with Be$^+$ this should yield an adequate description of the target. One can compare the theoretical and experimental f-values to estimate the effect of short-comings in the target model on the cross-section. There is a spread of about 20% in the f-values for the resonance transition derived from observation. The calculated f-value is consistent with all of the experimental results and agrees with the most accurate of these which has an error of 3%. It was found that the inclusion of polarization decreased the cross-section by about 20%, illustrating the importance of core polarization for singly charged ions and the need for a good model for the target.

In Figs. 3 and 4, taken from Mitroy et al. (1988), are given various theoretical values for the cross-section and the measurements from the two experiments. The CC calculations are; CCV3 - including the 4s, 3d and 4p states; CCV6 - as CCV3 plus the 5s, 4d and 5p states. (The V indicates inclusion of the polarization potential.) The error bars on the measurements of Taylor and Dunn, denoted by crosses, are approximately 10%. Good agreement, in general within the error bars, is obtained for incident energies below the 5p threshold.

Above the 5s threshold, cascades have to be taken into account. Values of the cross-sections for excitation of the 5s, 4d and 5p states are obtained in the CCV6 calculation, and although they are not expected to be as accurate as for the resonance transition, they allow an estimate of the

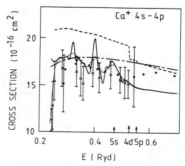

Fig. 3. Figure 1 of Mitroy et al. (1988). Absolute emission cross-section for electron impact excitation of the Ca$^+$ 4p$_{3/2}$ state as a function of incident energy. The following calculations are depicted: CCV6 ——, CCV3 — — — —, UCBAV6 — — —, and CCV6 convoluted with an energy resolution function of full width 0.023 Ryd ······. The experimental data of (Taylor and Dunn, 1973; Zapesochnyi et al., 1976) are depicted as x and •, respectively.

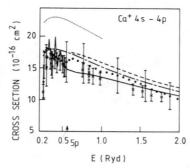

Fig. 4.   Figure 2 of Mitroy et al. (1988). Absolute emission cross-
section for electron impact excitation of the Ca$^+$ 4p$_{3/2}$
state as a function of incident energy. The following
calculations are depicted: CCV6 ———, CCV3 – —— – —, the
3 state CC calculations of Burke and Moores, (1968) ·····,
and CCV6 with cascade corrections – – –. The experimental
data of Taylor and Dunn (1973) and Zapesochnyi et al.
(1976) are depicted as x and •, respectively.

effect of cascading to be performed. A 20% increase caused by cascading
from these levels is obtained just above the 5p threshold. In this region,
theory and experiment disagree by 20%. The cause of this discrepancy is
probably two-fold. The major contribution to cascading comes from the decay
of the 4d state. The measurements of Zapesochnyi et al. (1976) indicate
that the calculated 4s – 4d cross-section is too large by a factor 4. There
is very little evidence of an increase in the 4p emission cross-section of
Taylor and Dunn at the 4d threshold, and a small effect is discernible in
the 4p emission data of Zapesochnyi et al. (1976). Hence it would seem that
cascading is overestimated. One might also expect a decrease in the 4s – 4p
cross-section as channels based on higher target states are added to the CC
expansion.

From the above, we can probably put an upper limit of 20% on the error
of resonance transitions with no change in principal quantum number over the
whole energy range. One would have expected the accuracy to be greater near
threshold when all the open channels are included in the CC expansion; this
would seem to be the case for Ca$^+$, but a problem remains with Be$^+$. There is
considerable difficulty for the forbidden 4s – 4d transition in Ca$^+$. For
forbidden transitions, interactions with target states other than the
initial and final states of the transition are more important than for
allowed transitions, and more terms should be added to the CC expansion.

COMPARISON BETWEEN THEORY AND PLASMA MEASUREMENTS OF RATE COEFFICIENTS

We turn to spectra of laboratory plasmas to investigate the rate-
coefficients for more complicated systems. I have selected two examples.

250

The first looks at transitions among the n = 2 terms of the Be-like ion Ne^{6+}. The second looks again at Li-like ions but at transitions with a change of principal quantum number.

The Ne^{6+} measurements were carried out by Lang (1983) on a Θ pinch plasma whose electron temperature and density are measured from laser scattering to be 10^6K and 2×10^{16} cm^{-3}, respectively. The data consist of emission line intensity ratios as the number density of the Ne^{6+} ions in the filling gas was not known accurately enough to find absolute intensities. A very detailed account of the analysis is given in Lang (1983).

Theoretical cross-sections and rate-coefficients have been calculated by Kingston et al. (1985). The six n=2 terms arising from the configurations $2s^2$, $2s2p$ and $2p^2$ were included in the CC expansion. The populations of the ten n = 2 fine-structure levels were obtained using the resulting rates in a collisional-radiative model, including estimates of cascading from the n = 3 terms and heavy particle collisions among the triplet fine-structure levels.

The intensities of three lines from the n = 2 terms, with respect to the resonance transition intensity, are considered. The first is for the transition $2s2p\,^3P^{\circ} - 2p^2\,^3P$ where two components of the multiplet are measured, for both He and H_2 filling gases. The theoretical ratios are higher than experimental. For three out of the four measured ratios, the theoretical values are just outside the experimental error bars of approximately 40%; for the fourth measurement the theoretical result is within the error bars. Most of the population of the $2p^2\,^3P$ term arises from excitation from the metastable $2s2p\,^3P^{\circ}$ state. The metastable state is mainly depopulated by collisions rather than radiative decay at this electron density, and hence the multiplet intensity depends mainly on the rate-coefficients for $2s2p\,^3P^{\circ} - 2p^2\,^3P$ and $2s^2\,^1S - 2s2p\,^3P^{\circ}$. Decreasing these rates with respect to that of the resonance transition would improve the agreement with the experimental results.

The second ratio is for $2s2p\,^1P^{\circ} - 2p^2\,^1D$ with respect to the resonance transition. Here again the theoretical ratios are larger; the average of the observations for the two filling gases is 58% of the theoretical ratio. From the level population calculations, we find that 72% of the population of the upper state comes from the $2s2p\,^1P^{\circ}$ term, and the error in this rate coefficient is the largest source of uncertainty in the intensity ratio calculation.

Finally, the intercombination multiplet $2s^2\,^1S - 2s2p\,^3P^{\circ}$ is studied. In this case, theory yields a lower result; a decrease of 1.5 in the experimental ratio is required to bring theory within the error bars and a decrease of 2.8 to bring coincidence with the theoretical value. Lang

251

(1983) notes the relative intensity of the intercombination multiplet depends directly on the value of the radiative lifetime for the C III $2p^2$ ^1D - 2s3p $^1P^o$ line used to calibrate the spectrometer in the wavelength region of the intercombination multiplet. This lifetime is difficult to calculate and could be the source of the discrepancy in this intensity ratio.

Kingston et al. (1985) discussed the sensitivity of the intensity ratios to the atomic data. Even changing the relevant parameters by their quoted uncertainty (10% and 20%, respectively, for allowed and forbidden rate coefficients) in such a way to improve agreement between theory and experiment did not resolve the problems for the any of the ratios. The effect of higher channels in the close-coupling expansion needs to be investigated, both for the increase caused by resonances converging to higher states and the decrease due to loss of flux into the higher scattering channels. If we assume the experimental results are correct, then we need to revise our estimates of the accuracy of the calculated data.

From the comparison of theory and crossed-beam results for the 2s - 2p transitions in Li-like ions, it seems that the neglect of channels with n greater than 3 would not lead to an error of more than 20%, even for energies above the ionization potential where one is neglecting the continuum channels in the calculations. How large is the error for transitions where the principle quantum number changes?

I have looked at the convergence of the CC expansion in He^+ at the first ionization threshold (Hayes, 1987). The results for the individual transitions with n less than or equal to 3 are given in Table 1 and in Table 2 and are the sums of the transitions for different n - n' for n' up to 5. The various CC approximations are X1, including up to the n = 5 target states; NX1, as X1 but excluding exchange; NX2, including up to the n = 6 target states but excluding exchange; NX3 as NX2 but with n = 6 pseudo-states instead of real states; HS calculations of Hummer and Storey (1987) with exchange and including up to the n = 3 states. One can see that for the 2 - 3 collision strengths, there is a large decrease in the collision strength as more channels are added to the CC expansion.

One would expect the convergence of the CC expansion to be better for more highly ionized systems. We shall look at the 2-3 transitions in Ne^{7+} for which there have been a number of plasma experiments (Chang et al., 1984; Kunze and Johnston, 1971; Haddad and McWhirter, 1973) for plasmas of differing electron temperature, ranging from 75ev to 260ev. For these temperatures roughly 30-40% of the rate-coefficient will come from energies between 1 and 2 times the ionization threshold. (The ionization potential of Ne^{7+} is 239 eV.) The results are summarized in Chang et al. (1984). Agreement between the theoretical rates of van Wyngaarden and Henry (1976)

Table 1. Collision Strengths for He^+

	X1	HS
1s–2s	0.119	0.143
1s–2p	0.510	0.594
1s–3s	0.376^{-1}	0.712^{-1}
1s–3p	0.911^{-1}	0.162
1s–3d	0.323^{-1}	0.505^{-1}
2s–3s	0.921	1.41
2s–3p	1.99	3.00
2s–3d	2.25	4.01
2p–3s	0.935	1.56
2p–3p	4.20	7.21
2p–3d	15.3	24.8

See text for a description of
the calculations.

is well within the experimental error of 50% of Chang et al. (1984) and
Haddad and McWhirter (1973) and a factor 2 of Kunze and Johnston (1971),
except for the lowest electron temperature where the 2s–3s theoretical rate
is just within the error bars. The calculations included only the n = 2 and
n = 3 states of Ne^{7+}. Note that in these experiments, the 2p level is
strongly populated and the theoretical rates were used to estimate the
relative contribution to the n = 3 populations from the ground and 2p terms.

Table 2. Collision Strengths for He^+ from n to n'

n–n'	X1	NX1	NX2	NX3
1–2	0.628	0.943	0.934	0.894
1–3	0.161	0.251	0.244	0.221
1–4	0.905^{-1}	0.124	0.115	0.975^{-1}
1–5	0.759^{-1}	0.887^{-1}	0.724^{-1}	0.590^{-1}
2–3	25.8	33.3	31.7	28.2
2–4	9.31	12.4	11.0	8.60
2–5	7.81	8.72	6.64	4.38
3–4	220	250	214	175
3–5	131	149	99.0	61.1
4–5	1450	1512	848	632

See text for a description of the approximations.

In passing, we note agreement within 22% between similar calculations by van Wyngaarden and Henry (1976a) for N^{4+} and the rate coefficients measured by Kunze and Johnston (1971).

Thus we can see that for more highly ionized systems, the neglect of higher channels probably leads to an error of less than 50% in the cross-section at intermediate energies. It would be good to try to tighten this estimate.

PHOTOIONIZATION

Photoionization experiments can also be used to estimate the accuracy of scattering calculations. In the work described below, both the wavefunctions for the initial target atom and the final state of ion plus electron were obtained by solving the CC equations for the electron plus ion scattering problem. For the target states, the equations are solved for negative electron energies, yielding an eigen-value problem for the bound states. The function for the ejected free electron is found as before. The photoionization cross-section is then obtained from the matrix of the dipole operator between the initial atomic state and the final ion plus electron state.

Photoionization experiments have the advantage that in general the energy resolution of the incident photon beam is better than that of an electron beam, and hence, experiments with photons can better map out resonance structure. There are also fewer final states to disentangle because of the selection rules. I have chosen two examples where we can look at the resonance structure in the photoionization cross-section to see how well our close-coupling model can deal with resonant states where we have 2 electrons in excited orbitals and one would expect their motion to be correlated.

For photoionization of He there is a wealth of experimental and theoretical data. Hayes and Scott (1988) looked at the resonance structure below the n = 3 and n = 4 ion thresholds in the photoionization cross-section of the ground state. We solved the e + He^+ scattering problem including all target states of He^+ up to n = 4. Solving for negative energies, we obtained an ionization potential for the $1s^2$ state of He of 23.85 eV compared to the observed energy of 24.58 eV.

In Fig. 5, taken from Hayes and Scott (1988), we compare some recent calculated and measured data (Solomonson et al., 1985; Woodruff and Samson, 1980; Lindle et al., 1985) for ionization to the n = 2 states. As can be seen, the two theoretical curves lie above experiment, the results of Solomonson et al. (1985) being in better agreement with experiment. At 68.88 eV, we have compared our total cross-section with the value from the

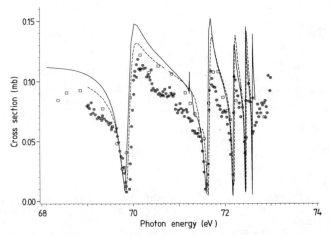

Fig. 5. Figure 2 of Hayes and Scott (1988). Partial cross-section
for the photoionization of He to the n = 2 states of He$^+$
against photon energy. ——— theoretical results of Hayes and
Scott (1988), − − − theoretical results of (Solomonson et al.,
1985), □ experimental results of Lindle et al. (1985),
o experimental results of Woodruff and Samson (1980).

compilation of Marr and West (1976) to which the two experiments are normalized. Our value was 7% too high which is a reflection of the inaccuracy
of our wave functions. The more recent experimental values, from Lindle et
al. (1985), have an uncertainty of (10-15)% and lie 20% lower than ours.

Turning our attention to the resonance structure, there is good agreement between theory and experiment for the resonance position and shapes as
can be seen from Fig. 5. There has been a recent high resolution experiment
on the synchrotron at Daresbury by Zubek et al. (1989). Embarrassingly good
agreement is obtained for the position and width of the first large resonance from this experiment and our calculation. We can have some confidence
that for simple systems, we can reproduce the resonance structure correctly.

Zubek et al. have in fact measured the differential cross-sections for
the photo-electron ejected along the direction of the major polarization
axis of the radiation. They obtained results for photoionization to the
n = 2, n = 3 and n = 4 states of He$^+$. A photon bandpass of 0.1 eV was used
for the n = 3 measurements.

One would expect the calculated cross-sections just below the n = 4
thresholds to be less accurate than those below the n = 3 thresholds, as we
do not have states higher than n = 4 in our calculation. Also, the cross-
sections to the n = 3 states will be less accurate than for the n = 2. I
have convoluted the differential cross-section of Hayes and Scott with a
Gaussian of width 0.1 eV and compared the result with the non-absolute
experimental data of Zubek et al. This is shown in Fig. 6. Fair agreement

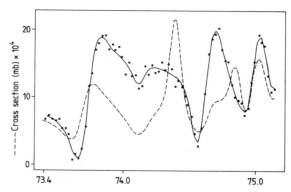

Fig. 6. Partial cross-section for the photoionization of He to the
         n = 3 states of He$^+$ against photon energy. The experi-
         mental points and solid curve are from Fig. 3 of Zubek et
         al. (submitted), the curve being a least squares fit to the
         data including seven resonances. The dashed curve is from
         the results of Hayes and Scott (1988) convoluted with a
         gaussian of full width 0.1 eV.

is achieved for the position of the structure in the cross-section. There
are extremely sharp resonances in the calculated data from 74.4 eV upwards
[see Fig. 6 of Hayes and Scott (1988)]. The energy mesh used in the
calculation is too coarse to represent these adequately, and hence the
heights of the peaks in the smoothed cross-section could be in error.

Of course, for He$^+$, the target functions are known "exactly". We look
now at a harder problem, photoionization of O I. Bell et al. (submitted)
performed a close-coupling calculation for the photoionization from the
three terms in the ground configuration $2s^2\, 2p^4$. In solving the O$^+$ scattering problem, all terms arising from the configurations $2s^2\, 2p^3$, $2s\, 2p^4$, and
$2s^2\, 2p^2\, 3s$ were included in the CC expansion. The description of the O$^+$
target also contained configurations with 3p and 3d pseudo-orbitals
determined by optimising the ground state energy and static polarizability.
Bound state functions of O$^+$ plus electron containing 3p and 3d pseudo-
orbitals will occur in the second summation in the CC expansion. For their
target model, they obtained O$^+$ energies within 3% of experimental values.
In solving the scattering problem for negative energies to obtain the oxygen
functions, they obtained ionization energies of the ground configuration
terms to within 1% of observed energies.

Figure 7, taken from Bell et al. (submitted), shows their results,
those of other CC calculations and from experiment. The calculations
increase in complexity from Taylor and Burke (1976), Pradhan (1978) and Bell
et al., and one can see the convergence of their results.

The most complete set of experiments is of Samson and Pareek (1985) and
Angel and Samson (1988), the latter being normalized to the former. There

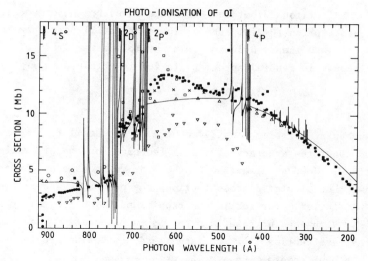

PHOTO-IONISATION OF OI

Fig. 7.   Figure 1 of Bell et al. (submitted). Photoionization
          cross-section of the ^{3}P state of atomic oxygen. Theory:
          ——— from Bell et al. (submitted), △ Pradhan (1978), X
          Taylor and Burke (1976). Experiment: ∇ Comes et al.
          (1968), o Kohl et al. (1978), □ Hussein et al. (1985),
          ■ Samson and Pareek (1985), ● Angel and Samson (1988).
          Thresholds correspond to experimental values.

are two regions of discrepancy between theory and these experiments. The
first is just above the ionization threshold where theory is up to 50%
larger. No explanation is available. The second is above the ^{2}P^o ion
threshold, where a maximum in the experimental results is not seen in the
theoretical ones. It is possible that this could be caused by the presence
of excited states of O in the interaction region. The calculations of Bell
et al. show a large resonance in the photoionization cross-section out of
the excited ground configuration terms at this energy. The presence of
molecular oxygen could also cause difficulty.

If we look at the resonance structure below the ^{2}P^o threshold, we see
good agreement between theory and experiment, except that the calculated
resonance positions are a little too high. For resonances converging to the
^{4}P threshold, Angel and Samson have not estimated cross-sections from their
experimental results. However, they do give the trace of their O$^+$ ion
yield, and good agreement is obtained for the resonance positions and shapes
when the energy scales are adjusted so that the ^{4}P threshold energies
coincide.

From the above, it is seen that we can have reasonable confidence in
resonance profiles for second row ions, where a fairly accurate description
of the target ion can be found. This may not be the case for third row
systems. For example, in the calculation of electron scattering cross-

sections for S^{2+}, I found that the resonance structure depended strongly on the target description (Hayes, 1986). There are still some puzzling differences between theory and experiment as to the magnitude of the results, although in general agreement to 20% is obtained.

SUMMARY

The number of examples considered here is too small to draw any general conclusions about the accuracy of calculated electron impact excitation cross-sections. However attention has been drawn to some problems and I believe that the calculated cross-sections may not be as accurate as we had supposed, particularly for forbidden transitions, even for fairly simple target ions. With the increase in super-computer power, we can include more terms in the wavefunctions for the target and scattering problem and hopefully sort out some of these problems.

There is obviously a need for more comparisons between theory and experiment. In particular, we need experimental data of greater accuracy to help us tie down the uncertainties in the calculations.

ACKNOWLEDGEMENTS

I wish to thank K.L. Bell, P.G. Burke, G.C. King, M. Zubek, P.M. Rutter, F.H. Read, A. Hibbert and A.E. Kingston for allowing me to use their results before publication.

REFERENCES

Angel, G. C., and J. A. R. Samson, 1988, Phys. Rev. A 38, 5578.
Bell, K. L., P. G. Burke, A. Hibbert, and A. E. Kingston, 1989, J. Phys. B (submitted).
Burke, P. G., and D. L. Moores, 1968, J. Phys. B 1, 575.
Chang, C. C., P. Greve, K-H. Kolk, and H-J. Kunze, 1984, Physica Scripta 29, 132.
Comes, F. J., F. Speier, and A. Elzer, 1968, Z. Naturforsch 23a, 125.
Gregory, D, G. H. Dunn, R. A. Phaneuf, and D. H. Crandall, 1979, Phys. Rev. A 20, 410.
Haddad, G. N., and R. W. P. McWhirter, 1973, J. Phys. B. 6, 715.
Hayes, M. A., 1986, J. Phys. B 19, 1853.
Hayes. M. A., 1987, "Proceeding of the Atomic Data Workshop at St. Catherine's College," Oxford, 1-2 August, 1987, Edited by W. Eissner and A. E. Kingston.
Hayes, M. A., D. W. Norcross, J. B. Mann, and W. D. Robb, 1977, J Phys. B 11, L429.
Hayes, M. A., and M. P. Scott, 1988, J. Phys. B 21, 1499.
Henry, R. J. W., W-L. van Wyngaarden, and J. J. Matese, 1978, Phys Rev. A 17, 798.
Hummer, D. G., and P. J. Storey, 1987, Mon. Not. R. Astron. Soc. 224, 801.
Hussein, M. I. A., D. M. P. Holland, K. Codling, P. R. Woodruff, and E. Ishiguro, 1985, J. Phys. B 18, 2827.

Kingston, A. E., P. L. Dufton, J. G. Doyle, and J. Lang, 1985, J. Phys. B. 18, 2561.

Kohl, J. L., G. P. Lafyatis, H. P. Palenius, and W. H. Parkinson, 1978, Phys. Rev. A 18, 571.

Kunze, H-J., and W. D. Johnston III, 1971, Phys Rev. A 3, 1384.

Lang, J., 1989, private communication.

Lang, J., 1983, J. Phys. B 16, 3907.

Lindle, D. W., T. A. Ferret, U. Becker, P. H. Kobrin, C. M. Truesdale, H. G. Kerkoff, and D. A. Shirley, 1985, Phys. Rev. A 31, 714.

Marr, G. V., and J. B. West, 1976, At. Data Nucl. Data Tables, 18, 497.

Mitroy, J., D. C. Griffin, D. W. Norcross, and M. S. Pindzola, 1988, Phys. Rev. A 38, 3339.

Parpia, F.A., D. W. Norcross, and F. J. da Paixao, 1986, Phys Rev A 34, 4777.

Pradhan, A. K., 1978, J. Phys. B 11, L729.

Samson, J. A. R., and P. N. Pareek, 1985, Phys. Rev. A 31, 1470.

Solomonson, S., S. L. Carter, and H. P. Kelly, 1985, J. Phys. B 18, L149.

Taylor, K. T., and P. G. Burke, 1976, J. Phys. B 9, L353.

Taylor, P. O., and G. H. Dunn, 1973, Phys. Rev. A 8, 2304.

Taylor, P. O., R. A. Phaneuf, and G. H. Dunn, , 1980, Phys. Rev. A 22, 435.

van Wyngaarden, W-L., and R. J. W. Henry, 1976, Can J. Phys. 54, 2019.

van Wyngaarden, W-L., and R. J. W. Henry, 1976a, J. Phys. B 9, 1461.

Woodruff, P. R., and J. A. R. Samson, 1980, Phys. Rev. Lett. 45, 110.

Zapesochnyi, I. P., V. A. Kel'man, A. I. Imre, A. I. Dashchenko, and F. F. Danch, 1976, Sov. Phys. JETP 42, 989.

Zubek, M., G. C. King, P. M. Rutter, and F. H. Read, 1989, J. Phys. B (submitted).

259

ELECTRON-ION, ION-ION, AND ION-NEUTRAL INTERACTIONS

Rainer Johnsen

University of Pittsburgh

INTRODUCTION

In my talk, I will present an overview of ongoing work in the area of electron-ion, ion-ion, and ion-neutral reactions. Clearly, this is a very wide field. Rather than trying to cover the whole field, I will focus my discussion on effects which are of importance in high-pressure plasmas, where the "high pressure" regime is defined as that in which collision processes are significantly influenced by third-body effects. In the absence of experimental data on collision processes in high-pressure plasmas, models of such plasmas, for instance, electrical discharge or laser plasmas, often make use of rate coefficients that were obtained in low-pressure experiments. This may be justified in many cases, but the problem deserves some further scrutiny.

I will begin by looking at three-body association of atomic ions in atomic gases, including the case where the association produces an excited, usually radiating, state of the product molecular ion.

My second topic deals with recombination of positive and negative ions in the presence of an ambient gas. I will present some recent data on recombination of molecular ions, in which case mutual neutralization in addition to three-body recombination has to be considered as a mechanism leading to charge neutralization, and I will then proceed to ion-ion recombination of atomic ions (Xe^+ and F^-) into excimer states.

My last topic deals with electron-ion recombination in high-pressure plasmas. We are presently carrying out some work in this area and have obtained some preliminary data. I will conclude my talk with some speculative remarks on diffusion limits to electron-ion recombination and how it may be observed in cryogenic plasmas at very high densities.

Nonequilibrium Effects in Ion and Electron Transport
Edited by J. W. Gallagher *et al.*, Plenum Press, New York, 1990

The subjects have some common aspects, and it may be of interest to look at them from a more general perspective. Consider a general "reaction" between two particles A and B, which may be ions, electrons, or neutrals. An ambient gas, consisting of atoms or molecules M is present. In general, M is different from A or B, but the case M=B may be thought of as being included. We are interested in the rate at which particles A and B or converted to other species. We assume that an attractive potential exists between A and B, as is usually the case in the reactions of interest. The rate at which A and B react may be limited by the rate at which the two particles enter the sphere of mutual attraction by diffusion or drift. We will postpone discussion of the diffusion limit until the end of the talk but assume for the time being that the diffusion limited rate far exceeds other relevant rates.

As a result of collisions of A and B with the ambient atoms M, a population of weakly bound pairs AB^{**} will be formed very rapidly. By "weakly bound" we mean that their total energy (the sum of their kinetic and potential energies) is negative by an amount of the order of thermal energy; i.e., these pairs will be destroyed and created at a high rate. It is, in part, the existence of a quasi-equilibrium population of these weakly bound pairs that distinguishes reactive collisions in an ambient gas from the true binary collisions that one studies, for instance, in a beam experiment.

Figure 1 shows a potential energy diagram of A and B; the weakly bound states may correspond to highly excited vibrational molecular levels, if A and B are heavy particles, or Rydberg states of electrons with a core ion in the case of electron-ion recombination. Now, let us consider the further fate of AB^{**} pairs. We distinguish two mechanisms: (1) Additional removal of energy by collisions with atoms M or (2) an internal conversion of AB^{**} without further interaction with atoms M. Internal conversion, for instance, may be a chemical reaction of A and B producing new particles, or a radiative transition to a lower state of AB. In Fig. 1, these internal conversion mechanisms are indicated by the horizontal arrow, while the energy removal is indicated by the arrow along the potential curve. In the latter

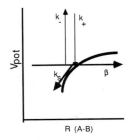

Fig. 1.

case, the ion pair eventually becomes stable; i.e., its collisional redis-
sociation becomes negligible. It is important to realize that the two
mechanisms are in competition with each other and with the collisional
destruction of AB^{**} into A and B. A solution to the set of rate equations
shows that the dependence of the overall loss rate of A and B on the density
of M is not a simple linear function, but can be more complicated. Denoting
the rate coefficients as k_+ (the three-body rate coefficient for formation
of AB^{**}), k_- (the two-body rate coefficient for destruction of AB^{**}), k_s
(the net rate of collisional stabilization of AB^{**}), and β (the unimolecular
rate of internal conversion), the overall reaction rate k_{eff} for the
reaction of A with B becomes:

$$k_{eff} = k_o + (k_+/k_-) \ (\beta + k_s \ N) \ [1 + (\beta + k_s \ N)/k_- N]^{-1} \qquad (1)$$

where N denotes the density of the ambient gas M and k_0 denotes the rate
coefficient in the absence of an ambient gas. The function k_{eff} vs. N has
the shape shown in Fig. 2. In the low-density limit the slope tends to k_+,
at higher densities it approaches the ratio $k_+ k_s/k_-$.

It should also be kept in mind that for simplicity we have assumed that
only one type of AB^{**} pairs exist. A more realistic model should allow for
the existence of a distribution of ion pairs of different energies and with
different rates of stabilization and internal conversion. There is experimental evidence in the case of ion-ion recombination that supports the
complicated density dependence of Eq. (1). A single three-body rate
coefficient cannot be defined in that case. Usually, however, one observes
only a simple, linear dependence of the rate coefficient on gas density.

ION-ATOM ASSOCIATION

The perhaps simplest type of a three-body reaction is given by the
association of ions in atomic gases

$$A^+ + B + M \longrightarrow AB^+ + M \qquad (2)$$

where it is assumed that the ionization potential of A is less than that of B.

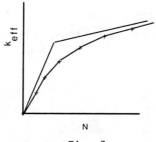

Fig. 2.

Numerous experiments have been performed and a considerable body of data on rate coefficients and their variation with gas temperature exists. I will not discuss the experimental techniques (flowing afterglow and ion drift tubes) (Liu and Conway, 1974; Johnsen et al., 1980), since they are well known and have been described extensively in the literature. Somewhat surprisingly, comparatively little theoretical work (Dickinson et al., 1972) has been done on association in atomic gases, while the case of molecular ions or molecular gases has been treated extensively. The case where the ionization potential of A is larger than that of B is of interest in this context, especially if fast charge transfer between A^+ and B does not occur. The association

$$A^+ + B + M \longrightarrow AB^{+*} + M \tag{3}$$

then results in an electronically excited state which subsequently decays by a radiative transition to the ground state that may be bound or unstable against dissociation:

$$AB^{+*} \longrightarrow AB^+ + h\nu \tag{4a}$$

$$\longrightarrow A + B^+ + h\nu. \tag{4b}$$

The second process may be called three-body radiative charge transfer, if AB^+ dissociates into A and B^+. Rates for several reactions of the type have been measured (Johnsen and Biondi, 1978). Rigorous theories of three-body association are difficult to construct. For the purpose of estimation of rate coefficients we have developed a simple formula, to be given below, which reproduces experimental data fairly well. We restrict ourselves to the case where the third body atom M is equal to the reactant atom B. The formula is based on the following, rather crude derivation. The basic idea is that ion-atom association is essentially an energy loss process. As a result of the attractive induced-dipole interaction between the ion and the atom, the kinetic energy of the ion during the collision is increased. Dissipation of this excess energy in collisions with a third atom then leads to a bound ion-atom pair which is stabilized in further collisions. The potential energy of an ion or charge Z at a distance r from the atom is given by

$$U(r) = - (Ze)^2 \alpha / 8\pi\varepsilon_o r^4 \tag{5}$$

where α is the atom's polarizability, and ε_o is the permittivity of the vacuum. In the presence of N atoms/unit volume, randomly distributed, the average potential energy is given by

$$\langle U(r) \rangle = N \int_{r_c}^{\infty} U(r) \, 4\pi r^2 dr = -N \, (Ze)^2 \, \alpha / 2 \, \varepsilon_o \, r_c \tag{6}$$

where r_c is a cutoff radius to be defined later. If we assume that the center of mass of the ion/atom system is (on average) at rest in the laboratory frame of reference, the average excess kinetic energy of the ion during a collision is given by

$$\langle \varepsilon \rangle = \langle U \rangle \, m/m_i \tag{7}$$

Here, m is the reduced mass of the ion and the atom and m_i is the mass of the ion. In the next step, we make the crude assumption that the rate at which the energy $\langle \varepsilon \rangle$ is lost can be estimated from the collisional energy loss rate of ions drifting through a gas under the influence of an electric field with ionic mobility μ. It is shown easily, using Wannier's formula for the average field energy of drifting ions, that the collisional energy loss in that case is given by

$$d\langle \varepsilon \rangle/dt = 2 \, \langle \varepsilon \rangle \, e/\mu \, (m+M) \tag{8}$$

By combining equations and choosing the cutoff radius r_c to be the radius for which $\langle \varepsilon \rangle = 3kT/2$, we obtain an expression for the energy loss rate of an ion in the average potential due to the atoms B. The three-body rate coefficient for association reaction should be related to the energy loss rate by an expression of the form

$$k_3 = (3kT/2)^{-1} \, N^{-2} \, d\langle \varepsilon \rangle/dt \tag{9}$$

We have assumed here that a loss of an energy of the amount $3kT/2$ renders an ion-atom pair stable, which is of course somewhat arbitrary, but factors on the order of unity are not of interest here. The final expression for k_3 then becomes

$$k_3 = (64\pi)^{1/4} \, (Ze)^{5/2} \, (\alpha \, m/Tm_i)^{3/4} \, /N_L \, (3\varepsilon_o k)^{3/4} \, \mu_o \, (m_i + M) \tag{10}$$

where N_L (Loschmidt's number) enters when we introduce the reduced mobility μ_o. Expressed in the customary units, the three-body rate coefficient is given by

$$k_3 \, (cm^6/s) = 3.27 \times 10^{-27} \, Z^{5/2} \, (\alpha \, m/Tm_i)^{3/4} \, /\mu_o \, (m_i+M) \tag{11}$$

In this formula α should be inserted in units of 10^{-24} cm^3, T in Kelvin, all masses in atomic units, and the reduced mobility of the ion in cm^2/Vs. Clearly, the derivation of this formula is crude, but it seems to reproduce experimental data quite well, as may be seen in Fig. 3, where the rate coefficients have been calculated using Eq. (11) and are compared to experimental data for a variety of three-body association processes. Several of the data points refer to simple association reactions of ions in the parent gases, some are for three-body radiative charge transfer of doubly-charged rare gas ions in their parent gases, and finally, some for ions in different gases. In several cases, the reactions have been measured at different temperatures (usually 77 K and 300 K). Both values are plotted and it may be seen that the theoretical formula describes the temperature variation quite well. References to the numerous sources of data will be

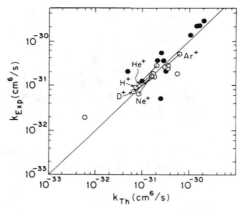

Fig. 3. Comparison of experimental data on ion-atom association
with results of Eq. (11). Open circles: Ion-atom
association. Solid circles: Three-body radiative charge
transfer. Points marked H^+ and D^+: Association of those
ions to helium. Points marked by rare-gas symbols:
Association of ions in their parent gases. Points
connected by lines refer to the same reaction at different
temperatures (300 K and 77 K).

given in a more detailed publication (Chatterjee and Johnsen). Obviously,
to make use of this formula, one needs to know ionic mobilities either from
measurements or one may have to estimate them using the Langevin theory,
which works quite well in most gases except for helium.

I do not claim that the "theory" presented here is rigorous in any way,
but it may be of some use in obtaining quick estimates for association
coefficients. Very elaborate theoretical work on the three-body association
of helium ions in helium has recently been published by Russel (Russel,
1985). His results are in excellent agreement with measured rate
coefficients for this process.

ION-ION RECOMBINATION

Ion-Ion Recombination

$$A^+ + B^- \longrightarrow \text{neutrals} \tag{12}$$

is often the dominant deionization mechanism in plasmas containing electronegative gases, and it also provides an important mechanism of forming the
upper laser level in excimer lasers. The theoretical work on ion-ion recombination is extensive. Numerous contributions have been made by Bates
and Flannery, 1968, and by Flannery, 1982, and Monte Carlo simulations have
been carried out also by Morgan et al., 1982. The state of the art of
experiments is far less satisfactory. An important series of measurements
have been performed by Smith and Adams, 1982, using the flowing afterglow

technique, but these measurements were limited to low ambient gas pressures (less than 8 Torr). All of the earlier work at higher pressures was done without mass analysis of the recombining plasma, so that the identity of the ions was essentially unknown. We have recently developed an experimental technique for the study of ion-ion recombination at near atmospheric pressures, which employs mass analysis of the recombining ions. In addition, spectroscopic observations are made to ascertain the nature of the recombining ions. This work is now beginning to give some interesting results which I would like to present here.

The experiment makes use of afterglows of photoionized plasmas. The experimental arrangement is shown in Fig. 4. A fuller description of the technique has been given elsewhere (Lee and Johnson, 1989). We use an array of spark gaps to produce a burst of ultraviolet light which then ionizes the gas contained in a probe volume, defined by the space between two concentric cylindrical electrodes. The ion density in the decaying plasma (typically on the order of 10^9 to 10^{10} cm^{-3}) is obtained by measuring the ionic conductivity of the plasma. Radio frequency voltages (frequencies of 455 kHz to 15 MHz) are applied to one electrode, and the current to the other electrode is measured using a rf-bridge method to cancel the displacement current between the electrodes. In addition, ions from the plasma are sampled into a mass spectrometer to monitor the plasma ion composition. In some cases, ion-ion recombination results in an electronically excited excimer state which then decays radiatively. The fluorescence light is dispersed by a monochromator and then detected by a photomultiplier.

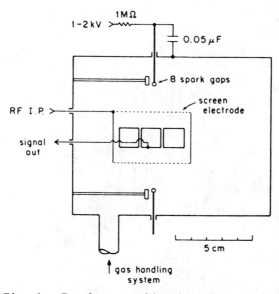

Fig. 4. Ion-ion recombination apparatus.

Rates of recombination are obtained from the decay of the ion density. It follows the usual recombination decay

$$1/n_i(t) = 1/n_i(t=0) + \alpha t \qquad (13)$$

where α is the ion-ion recombination coefficient. For a good measurement of α, the linear dependence of the reciprocal ion density on a afterglow time should be observed over a significant range of ion densities (at least a factor of 8). At the high pressures used in this work, diffusion of ions is very slow and need not be considered.

We have applied this technique to two examples of ion-ion recombination processes, one involving molecular ions and one involving atomic ions. The first process

$$H_3O^+ (H_2O)_n + NO_3^- + M \longrightarrow \text{neutrals } (n=2,3) \qquad (14)$$

has been studied for M=He and M=Ar at pressures up to about 1 at. This work has been published (Lee and Johnsen, 1989). The second process

$$Xe^+ + F^- + He \longrightarrow XeF^* + He \qquad (15)$$

is an example of excimer formation by ion-ion recombination which is thought to be an important reaction in the XeF* excimer laser. This B-X fluorescence (two features at 350 nm) was observable in the experiment (see Fig. 5) showing that an excimer is actually being formed.

A comparison of the observed recombination rates for the two different reactions (see Figs. 6 and 7) shows that the magnitudes of the coefficients

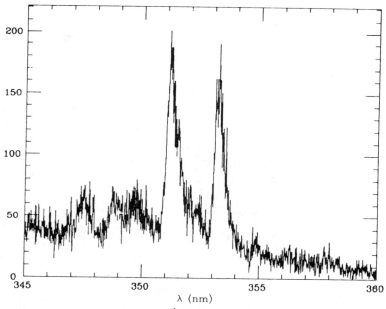

Fig. 5. XeF* emission spectrum.

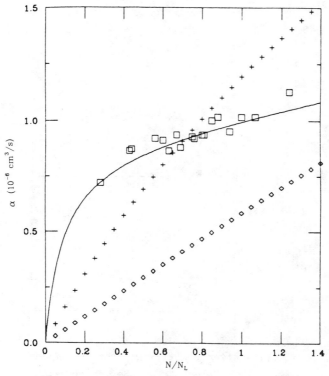

Fig. 6. Recombination coefficients for the reaction,
$H_3O^+(H_2O)_n + NO_3^- + M \longrightarrow$ neutrals, in helium.

are not too different, but the dependence on helium pressure is quite dis-
similar in the two cases. A linear extrapolation of the data in Fig. 6 to
low He pressures would yield a very high recombination coefficient which
would be incompatible with the low-pressure results of Smith and Adams for
the same reaction. We are forced to conclude that a linear extrapolation is
invalid and that the recombination coefficients depend on pressure in a more
complicated way. The curve drawn through the data points is based on a
model which involves both mutual neutralization and three-body recombination, as discussed in the introduction. In this case, mutual neutralization
most likely occurs by proton transfer from the water cluster ion to the NO_3^-
ion. (See Lee and Johnson, 1989, for a fuller discussion and parameters
used in the fitting). Figure 7 also shows theoretical recombination
coefficients calculated from the theories of Natanson, 1959, and Bates and
Mendas, 1982. In these theories, the ions are modeled as structureless
point charges and effects of mutual neutralization are not included.

The realization that the pressure dependence of ion-ion recombination
is complicated by mutual neutralization is due to Bates and Mendas. Our
model is based on their ideas. There is direct experimental evidence (Smith

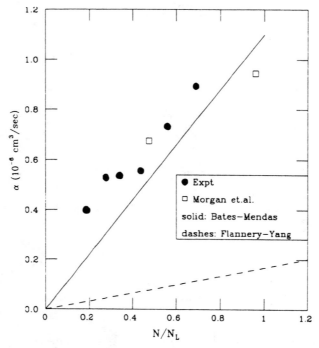

Fig. 7. Recombination coefficients for the reaction,
$Xe^+ + F^- + He \longrightarrow XeF^* + He$, in helium.

and Adams, 1982) that a sharp rise of α with gas density occurs at low
pressures (up to 8 Torr).

In the case of the $Xe^+ + F^-$ recombination, the dependence of α on
density is well described by a simple linear function. An extrapolation of
the data to zero pressure yields a small value which may not be significant
in view of the uncertainty of the data. Mutual neutralization by electron
transfer appears to be very slow, which is not surprising since the only
available crossing between potential curves occurs at a very large internuclear separation. Figure 7 also indicates results of several calculations. We have used the theory of Bates and Mendas, 1982, to calculate the
line indicated in the figure. Recombination coefficients have also been
calculated by Flannery and Yang (1978) using the modified Natanson formula
for the very similar process of $Kr^+ + F^-$ and Monte Carlo simulations have
been carried out by Morgan et al. (1982), also for $Kr^+ + F^-$. The difference
in the masses of Xe and Kr should not be important. It is clear that the
theory of Bates and Mendas and the Monte Carlo results are in far better
agreement with our experimental data than those of Flannery and Yang. We
are unable to offer an explanation why the theoretical results are so different. It should be noted that the Monte Carlo simulations were carried
out for ion densities near 10^{15} cm^{-3} under which conditions the shielding of

the Coulomb field leads to a slight reduction of the recombination coefficient. Thus, a Monte Carlo calculation for low ion densities, appropiate to our conditions, should yield somewhat larger coefficients. This work was carried out by H.S. Lee (to be published).

The results of our experiments show that recombination coefficients can exhibit quite complicated pressure dependences if mutual neutralization is present, but that a simple linear rise with pressure is observed in the absence of mutual neutralization. The agreement between experiment and theory is good, but the theory of Flannery and Yang appears to give rate coefficients that are too small.

Electron-ion Recombination at High Gas Densities

The effect of the ambient gas on recombination of electrons with ions

$$A^+ + e + M \longrightarrow neutrals \tag{16}$$

is considerably smaller than the effect on ion-ion recombination, the principal cause being that an electron loses only a small fraction of its energy in a collision with the much heavier atom. To measure the pressure dependence of electron-ion recombination in atomic gases one needs to measure recombination rates at very high gas densities (corresponding to many atmospheres). In molecular gases inelastic collisions of electrons with molecules lead to stronger pressure dependences, which have been measured at below atmospheric pressures (Warman et al., 1980).

We have used an experimental method for studies of electron-ion recombination at high densities which is similar to that used for measuring ion-ion recombination rates. A plasma is created by the ultraviolet light produced by a spark in helium. The subsequent decay of the plasma is monitored by measuring the conductivity of the afterglow plasma using radio-frequency probes. The experimental arrangement (Fig. 8) is simple and compact and can easily be housed in a cryogenic dewar for cooling.

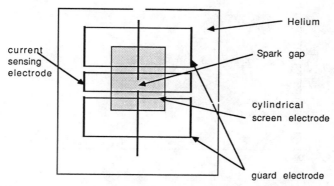

Fig. 8. Experimental apparatus used in the electron-ion recombination studies. The pressure chamber is housed in a liquid-helium dewar.

The probe volume is that enclosed by the inner cylindrical screen electrode and the central segment of the outer cylindrical electrode. The decay of the electron density during the afterglow is analyzed in the usual manner to obtain the recombination coefficient. Experimental observations indicated a recombination controlled decay over a factor of at least 8 in electron density, sufficient for a reasonably accurate determinations of recombination coefficients. Some preliminary results of measurements in helium are shown in Fig. 9. It is clear that the recombination coefficient increases linearly with helium density. An extrapolation of the data to zero helium density yields a coefficient of about 4×10^{-7} cm^3/s, which is fairly typical for simple ions. The question of the identify of the ion is probably not very important here, since the three-body enhancement of the recombination rate should not dependent much on the nature of the ion. One might ask, however, if the observed increase of the recombination rate with gas density results from a shift of the ion population to heavier, faster recombining ions. The question cannot be answered unambiguously at the present time.

The three-body rate coefficient for electron-ion recombination, derived from the data in Fig. 9 has a value of 4.7×10^{-27} cm^6/s. There are few experimental data available for comparison. Gousset et al., 1977, measured the three-body recombination coefficient for Cs$^+$ ions in helium at a temperature of 625 K and obtained a coefficient of 4×10^{-29} cm^6/s. If we assume that the temperature dependence of the coefficient is given reasonably well by the Thomson model (Bates, 1980) ($T^{-5/2}$) their data, scaled to 77 K, would yield a coefficient of 7.5×10^{-27} cm^6/s, compatible with our result. The theoretical results of Bates and Khare, 1965, would give higher (by a factor of about 3) values than our experiment. To make this comparison, we used the data calculated by Bates and Khare for He$^+$ ions at 125 K and have scaled

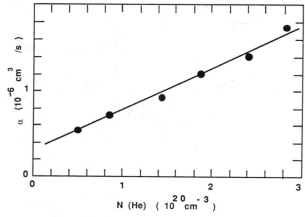

Fig. 9. Measured recombination rate in ambient helium at 77 K as a function of helium density.

their results to 77 K by using a $T^{-5/2}$ dependence. According to those authors, the mass of the ion enters the recombination coefficient only weakly through an inverse dependence on the reduced mass of the ion/gas system. We have assumed that the ions in our experiments were considerably heavier than helium and have reduced the theoretical results accordingly by a factor of two.

We are in the process of extending our experiments to different temperatures in order to obtain the variation of the three-body coefficient recombination coefficient with temperature.

As one goal of this work, we wish to investigate the recombination in plasmas in which the electrons are "localized". It is known that at very low temperatures (near 4 K) and densities above about 1×10^{21} cm^{-3}, electrons in helium exhibit a drastic (by 4 to 5 orders of magnitude) reduction in mobility and diffusion constant (Levine and Sanders, 1966). The effect may be related to "Anderson localization" in random scattering media. In a helium plasma at 4 K, recombination of electrons should be very large at low gas densities, but it should be strongly reduced by the slow rate of electron diffusion if the gas density exceeds a critical density of about 1×10^{21} cm^{-3}. One should also expect that the recombination rate should increase with increasing temperature in this regime, since increasing the electron temperature leads to delocalization of electrons. It will be interesting to explore this density region which is intermediate between the gas and liquid phases.

ACKNOWLEDGMENTS

The author acknowledges many important contributions to this work by his graduate students B.K. Chatterjee, H.S. Lee, and Y. Cao. This work was supported, in part, by the US Army Research Office and NASA.

REFERENCES

Bates, D. R., 1980, J. Phys. 13, 2587.
Bates, D. R., and M. R. Flannery, 1968, Proc. Roy. Soc. London, Ser. A, 69, 910.
Bates, D. R., and S. P. Khare, 1965, Proc. Phys. Soc. 85, 231.
Bates, D. R., and I. Mendas, 1982, Proc. Roy. Soc., Ser. A, 304, 447.
Chatterjee, B. K., and R. Johnsen, to be published.
Dickinson, A. S., R. E. Robert, and R. B. Bernstein, 1972, J. Phys. B 5, 355.
Gousset, G., B. Sayer, and J. Berlande, 1977, Phys. Rev. A 16, 1070.
Flannery, M. R., 1976, "Atomic Processes and Applications," Burke, P.G. and Moisewitsch, B.L., (Eds.), North Holland, Amsterdam; Flannery, M.R., 1982, "Applied Atomic Collision Physics," Academic Press, New York, 3, 141.
Flannery, M. R., and T. P. Yang, 1978, Appl. Phys. Lett. 33, 574.
Johnsen, R., and M. A. Biondi, 1978, Phys. Rev. A 18, 989.
Johnsen, R., A. K. Chen, and M. A. Bioindi, 1980, J. Chem. Phys. 73, 1717.
Lee, H. S., and R. Johnsen, 1989, J. Chem. Phys. 90, 6328.

Lee, H. S., and R. Johnsen, to be published.

Levine, J. L., and T. M. Sanders, 1966, Phys. Rev. 154, 138.

Liu, W. F., and D. C. Conway, 1974, J. Chem. Phys. 60, 784.

Morgan, W. L., J. N. Bardsley, J. Lin, and B. L. Whitten, 1982, Phys. Rev. A 26, 1696.

Natanson, G. L., 1959, Sov. Physics-Tech. Phys. 4, 1263.

Russel, J. E., 1985, J. Chem. Phys. 83, 3363.

Smith, D., and N. G. Adams, 1982, "Physics of Ion-Ion and Electron-Ion Collisions," Edited by Brouillard, F. and McGowan J. Wm., (Plenum, New York) p. 105.

Warman, J. M., E. S. Sennhauser, and D. A. Armstrong, 1980, J. Chem. Phys. 15, 479.

ELECTRON-ION RECOMBINATION IN DENSE MOLECULAR MEDIA

Kyoji Shinsaka[a] and Yoshihiko Hatano[b]

[a]Department of Electronics
Kanazawa Institute of Technology
Nonoichi-machi, Ishikawa 921, Japan

[b]Department of Chemistry
Tokyo Institute of Technology
Meguro-ku, Tokyo 152, Japan

INTRODUCTION

The behavior of excess electrons such as electron transport and reactivity in dense media has recently attracted great interest (Freeman, 1981; Hatano and Shimamori, 1981; Warman, 1982; Christophorou and Siomos, 1984; Johnsen and Lee, 1985; Morgan, 1985; Hatano, 1986; Marsolais and Sanche, 1988). Excess electron mobilities μ_e have been measured extensively in a variety of nonpolar organic compounds in the liquid and the dense gas phase, and several theoretical models have been proposed for the transport mechanism of excess electrons in these media. In the solid phase of these compounds, however, there are few investigations of electron mobilities. Excess electron reactivity with solute molecules or positive ions is closely related to the transport phenomena.

It is of great importance to investigate the electron-ion recombination in dense molecular media in the density range from one atm. gases to liquids and solids (Tezuka et al., 1983; Nakamura et al., 1983; Shinsaka et al. 1988; Shinsaka et al., 1989).

In the gas phase at the pressures of one to several atm., it has been shown (Warman et al., 1979; Sennhauser et al., 1980) that electron-ion recombination is overall a three-body process and that an observed recombination rate constant k_r is expressed by

$$k_r = k_2 + k_3\,n, \tag{1}$$

where k_2 and k_3 are the two-body and the three-body coefficients, respectively, and n is the density of the medium.

In the liquid phase, it has been shown (Tezuka et al., 1983) that the observed electron-ion recombination rate constant k_r in a variety of nonpolar media are in good agreement with the calculated values k_D by the reduced Debye equation,

$$k_D = \frac{4\pi e}{\varepsilon} \mu_e ,$$ (2)

in the range of μ_e from about 0.09 to 160 cm^2/Vs, where e is the electron charge and ε is the dielectric constant of the medium, and thus the reaction in these media was concluded to be diffusion controlled.

In liquid and high pressure gaseous methane, however, it has been found recently (Nakamura et al., 1983) that the observed k_r values are much lower than k_D. This experiment has stimulated the several theoretical investigations of the recombination process in dense media. Since the magnitude of the electron mobility in argon is as high as in methane, it is interesting to investigate the electron-ion recombination in argon and to compare further the result with theoretical ones. Electron drift mobilities μ_e and electron-ion recombination rate constants k_r in gaseous, liquid, and solid argon have been recently measured and compared with Eq. (2) (Shinsaka et al., 1988). The observed k_r values in both liquid and gas phases are again found to be much smaller than those calculated by the reduced Debye equation. The deviation, which is bigger in gas than in liquid, has been compared with theoretical studies. It has been concluded that the recombination in liquid and gaseous argon as well as in methane is not a usual diffusion-controlled reaction. In the solid phase, however, the observed k_r values are almost in agreement with those calculated by Eq. (2). The effect of an external DC electric field strength on both k_r and μ_e values has been also measured.

This paper summarizes in the following, the experimental studies of electron-ion recombination in both methane and argon media and surveys briefly the related theoretical studies.

EXPERIMENTAL

A research grade methane or argon was purified in a grease-free vacuum system which was pumped down to a pressure less than 10^{-6} Torr, passed through columns of "Gas Clean" which is an oxygen absorber, KOH pellets, and molecular sieves 5A in this order, bubbled through Na-K liquid alloy, transferred to the conductivity cell which is shown in Fig. 1, and finally sealed in the cell.

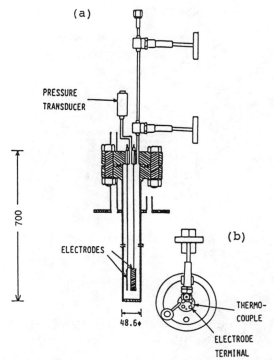

Fig. 1. Conductivity cell. (a) Cross sectional view, (b) top view
(Shinsaka et al., 1988).

The conductivity cell made of stainless steel can be operated up to
the inner pressure of 130 atm. The cell has two parallel rectangular
electrodes with a rectangular guard electrode.

The experimental setup for the measurement of transient electron
current is shown in Fig. 2. An x-ray pulse of a few nanoseconds was
obtained by impinging an electron beam pulse of 0.6 MeV from a Febetron
706 on a 0.1 mm thick tungsten foil at the front of an electron-pulse
emission window. The pulse dose was controlled by setting lead plates of
various thickness behind the tungsten foil. The dose was calibrated by a
thermoluminescent dosimeter.

The lower half of the conductivity cell containing the sample was
set in a Dewar vessel in an electromagnetic shielding box. Temperature
was controlled by controlling the flow rate of cold nitrogen gas into the
Dewar vessel using a temperature control system. Temperature of the
sample was monitored by the two copper-constantan thermo-couples; one of
them was set near the upper side of the guard electrode and the other was
set near the lower side of the guard electrode. The difference in both
temperature indications was less than one degree.

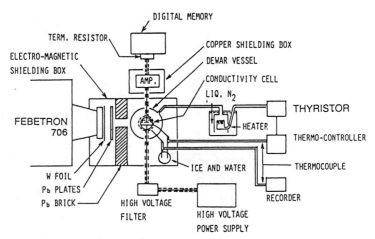

Fig. 2.   Setup for measurement of transient electron current
          (Shinsaka et al., 1988)

When thermal equilibrium was established, a negative high voltage
from a dc power supply through a low-pass filter (8 kHz) was applied to
the high voltage electrode of the cell, and then the sample was irradiated with an x-ray pulse. Under the influence of an electric field, a
transient electron current was induced in the external circuit of the
cell.

The values of μ_e and k_r were determined by the decay curve analysis
method (Tezuka et al., 1983; Nakamura et al., 1983; Shinsaka et al.,
1988; Shinsaka et al., 1989; Wada et al., 1977). In short, μ_e and k_{tr}
which is a rate constant for electron trapping due to the medium itself
or residual trace of electron attaching impurity are determined by
fitting Eq. (3) to the observed decay of electron current under the
condition of a low x-ray pulse dose, where the contribution of
electron-ion recombination can be neglected,

$$I(t) = I_0 \, (1-\mu_e Et/d)\exp(-k_{tr}t). \tag{3}$$

Using the obtained μ_e and k_{tr}, k_r was determined by fitting Eq. (4)
to the observed decay of electron current at a high x-ray pulse dose
where the contribution of electron-ion recombination is taken into
account,

$$I(t) = \frac{I_0 \, (1-\mu_e Et/d)}{1 + n_0 \, k_r \, t} \, \exp(-k_{tr}t). \tag{4}$$

In Eqs. (3) and (4), I(t) is the electron current at the time t
after the x-ray pulse, I_0 the peak current at the end of the pulse, E the

electric field strength, d the distance between the electrodes, and n_o the initial concentration of excess electrons. The error limits of the obtained values of μ_e and k_r were 10% and 15%, respectively.

The densities of the sample were evaluated by the measured pressure and temperature. Temperature dependence of the dielectric constant of the sample was calculated from the Clausius-Mossotti equation using the evaluated density.

RESULTS AND DISCUSSION

Methane (Nakamura et al., 1983; Shinsaka et al., 1989)

Figure 3 shows the temperature dependence of μ_e in liquid and solid methane together with earlier results in the fluid state for comparison.

Figure 4 shows the density dependence of the electron mobility in gaseous, liquid, and solid methane together with earlier results in the gas phase.

On the phase transition from liquid to solid, μ_e abruptly increases as shown in Fig. 3. Similar phenomena have been observed in neopentane, tetramethylsilane, and heavy rare gases such as Ar, Kr, and Xe, and explained well by the Cohen-Lekner model (Cohen and Lekner, 1967; Lekner, 1967), in which μ_e is given by

$$\mu_e = \frac{2}{3} \left(\frac{2}{\pi m k T} \right)^{1/2} \frac{e}{4\pi a^2 n^2 k T \chi^T} \qquad (5)$$

where m is the electron mass, a the electron scattering length of medium molecules, χ^T the isothermal compressibility of the medium, k the Boltz-mann constant, and T the absolute temperature. It has been concluded, therefore, that the increase of μ_e on the phase transition from liquid to solid is due to the decrease in the value of χ^T.

Recent electron transport models (Berlin et al., 1978; Basak and Cohen, 1979), except the Cohen-Lekner model, have shown that the latter model is not valid in the critical or the dense gas region in nonpolar compounds and that electron scattering by the fluctuation of the conduc-tion band energy ΔV_o is important in such density region.

The temperature dependence of k_r from the critical region to solid phase and the density dependence of k_r in gaseous, liquid, and solid methane are shown in Figs. 5 and 6, respectively. In the solid phase, the k_r value is surprisingly large.

In our previous work (Tezuka et al., 1983) it has been shown that the observed k_r values in several liquid and solid nonpolar organic

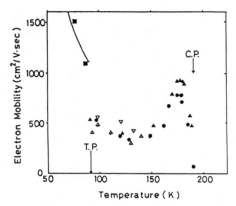

Fig. 3. Temperature dependence of μ_e in liquid ( ) and solid (■) phase of methane. T.P. and C.P. denote the triple point and the critical point, respectively. Solid line shows the $T^{-3/2}$ dependence of μ_e in the solid phase. Earlier results (▲, Δ, and ▽) are shown for comparison (Nakamura et al., 1983).

compounds with μ_e lower than about 160 cm^2/ Vs are in good agreement with those estimated from Eq. (2), and that, therefore, the electron-ion recombination in those media is diffusion controlled.

In the present work, k_r/k_D values in liquid and gaseous methane are much smaller than unity except for low μ_e region, while the k_r/k_D value in the solid phase is nearly equal to unity as shown in Fig. 7. The

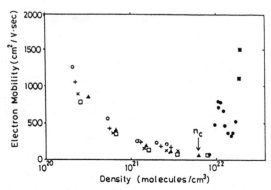

Fig. 4. Density dependence of μ_e in methane, gas (O, x, +, and □), liquid (●), and solid (■). n_c means the critical density (Nakamura et al., 1983).

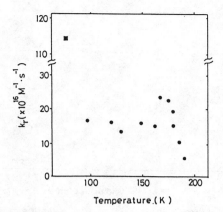

Fig. 5. Temperature dependence of k_r in liquid (●) and solid (■) phase of methane (Nakamura et al., 1983)

variation of k_r/k_D value over a much wider range of μ_e is shown in Fig. 8, where the larger deviation of k_r from k_D is observed in media with the larger μ_e value.

Argon (Shinsaka et al., 1988)

Figure 9 shows the temperature dependence of μ_e in liquid and solid argon together with earlier results for comparison. A large increase in μ_e on the phase change from liquid to solid is again explained quantitatively by the Cohen-Lekner theory.

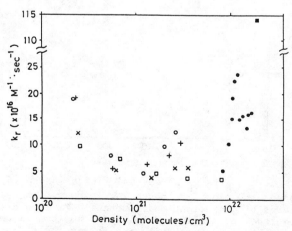

Fig. 6. Density dependence of k_r in methane. Symbols are the same as Fig. 4 (Nakamura et al., 1983).

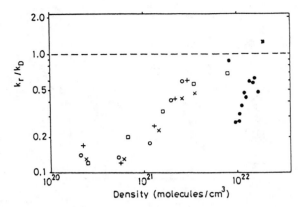

Fig. 7. Density dependence of k_r/k_D in methane. Symbols are the same as Fig. 4 (Nakamura et al., 1983).

In regard to the variation of electron mobility with temperature in liquid argon, several theoretical models have been proposed and they are rather far from giving theoretical values which are in good agreement with the experimental values over the whole liquid region. However, Ascarelli's treatment (Ascarelli, 1986) considering phonons and density fluctuations is interesting because it gives rather good calculated

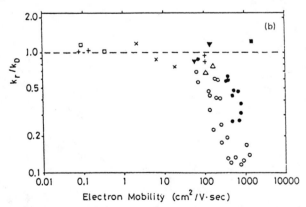

Fig. 8. Plots of k_r/k_D vs μ_e in several fluids and solids; ■, solid methane; ●, liquid methane; O, gaseous methane; x, liquid mixtures of neopentane and n-hexane; ▼, liquid and solid neopentane; Δ, liquid and solid tetramethylsilane; +, liquid n-pentane, n-hexane, and tetramethylsilane; □ , liquid n-hexane and cyclohexane (Tezuka et al., 1983; Nakamura et al., 1983; Shinsaka et al., 1988).

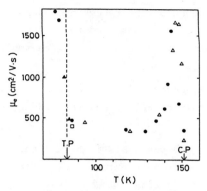

Fig. 9. Temperature dependence of electron mobility in liquid and
solid argon. Earlier results (▲, Δ, and □) are shown for
comparison (Shinsaka et al., 1988).

values of electron mobility in good agreement with the experimental
values if effective electron mass is used as a fitting parameter with
density.

Electric field dependences of density-normalized electron mobility
in liquid and gas phases are shown in Fig. 10. The electron mobilities
were constant in the region of low electric field in both gas and liquid
and decreased gradually with further increasing electric field strength
in the region above a certain critical electric field strength in the
liquid phase. The gradual decrease in electron mobility in higher
electrical field is due to heating of excess electrons by the external
electric field.

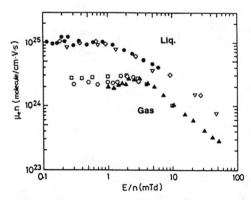

Fig. 10. Electric field dependence of density-normalized electron
mobility in liquid (●) and gaseous (○ and ▲) argon.
Earlier results (◊, ▽, and □) are shown for comparison
(Shinsaka et al., 1988).

Temperature dependence of the recombination rate constants in liquid argon is shown in Fig. 11. The observed recombination rate constants in both liquid and gas phases were found to be much smaller than those calculated by the reduced Debye equation.

In the solid phase, however, the observed recombination rate constant was close to that calculated by the reduced Debye equation (Fig. 12).

The recombination rate constants in both liquid and gas phases increase roughly in proportion to external electric field strength in the region of low electric field (Figs. 13 and 14). They form a peak around the critical electric fields which are almost the same magnitude as those for electron mobility and then decrease with further increase in external electric field.

The external electric field dependence of the ratios of the observed recombination rate constant k_r to that calculated from the reduced Debye equation k_D are shown in Figs. 15 and 16.

A large deviation of k_r/k_D from 1 in both liquid and gas phases means electron-ion recombination is not a usual diffusion-controlled reaction. The difference in the values of k_r/k_D in gaseous (~ 0.01), liquid (~ 0.1 corrected for electric field dependence), and solid argon (~ 1) may be mainly caused by the difference of the medium density; the more easily the electrons lose their energy, the more smoothly the recombination can occur.

It should be noted that in solid argon which is known as a plastic crystal, the ratio k_r/k_D is ~ 1 in spite of higher electron mobility than in liquid argon. In this case, the lattice of the plastic crystal may effectively work as an energy sink for the kinetic energy of excess electrons in collision with the electrons.

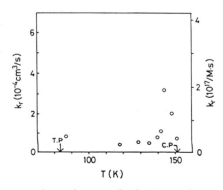

Fig. 11.   Temperature dependence of electron-ion recombination rate constant in liquid argon (Shinsaka et al., 1988)

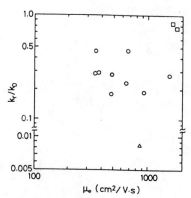

Fig. 12. Ratio of observed recombination rate constant k_r to that
calculated by the reduced Debye equation k_D vs electron
mobility μ_e in gaseous (Δ), liquid (O), and solid ($\square$)
argon (Shinsaka et al., 1988)

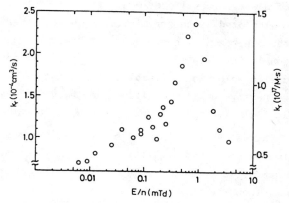

Fig. 13. Electric field dependence of electron-ion recombination
rate constant k_r in liquid argon at 87 K (Shinsaka et al.,
1988).

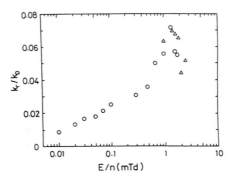

Fig. 16.   Variation of k_r/k_D with electric field strength in gaseous
argon (Shinsaka et al., 1988).

In regard to the problem that the ratio of k_r/k_D becomes < 1, Warman
(1983) proposed a theoretical formula [Eq. (6)] of the recombination rate
constant, taking the diffusion-controlled recombination rate constant and
the energy exchange controlled recombination rate constant
into account:

$$k_r/k_D = (1+55T^4\varepsilon^2 /10^4 E_{10}^2)^{-1}, \qquad\qquad (6)$$

where E_{10} is the critical field strength at which the drift velocity
deviates 10% from μ_{th} E.   In spite of many assumptions for the derivation
of Eq. (6), the calculated value for liquid argon is rather in good
agreement with the experimental one.   However, the calculated value for a
dense gas of argon is far from the experimental one.   In case of solid
argon, the calculated value is much smaller than that of experiment.   The
difference may be partly due to uncertainty in the determination of
critical electric field for solid argon reported in an earlier paper.

Recently, Tachiya (1986; 1987a; 1987b) pointed out that if l/a where
l is the mean free path of reacting particles and a the reaction radius,
namely the distance between two particles at which they react immediately
is not negligible, then the rate constant of bulk recombination will be
affected and deviated from k_D in the recombination for neutral particles
and for an electron-ion recombination by use of Monte Carlo techniques.
It is very interesting that our experimental values of recombination rate
constant for liquid methane locate nearly on the theoretical curve of
this simulation which uses only l/a as a parameter.

The scatter of experimental points for liquid argon which is shown
in Fig. 12 is not an experimental error but reflects subtle differences
in conditions of the measurement and the medium.   Even if we take account

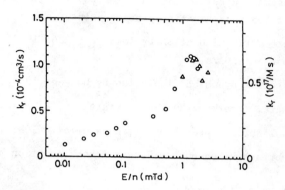

Fig. 14. Electric field dependence of electron-ion recombination
rate constant k_r in gaseous argon (Shinsaka et al., 1988).

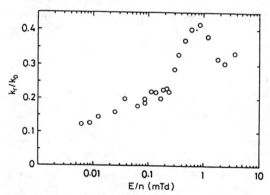

Fig. 15. Variation of k_r/k_D with electric field strength in liquid
argon at 87 K (Shinsaka et al., 1988).

of such a complexity in the recombination processes, it seems to be true that a large $1/\underline{a}$ is an important factor for the deviation of k_r from k_D.

In the following theoretical investigations which have been stimulated by the present experimental results of $k_r/k_D \ll 1$ for dense methane and argon are summarized.

(1) A semi-empirical treatment by Warman (1983) taking both the diffusion-controlled and the electron-energy-exchange controlled recombination processes into account,

(2) A Monte Carlo simulation by Tachiya (1986; 1987a; 1987b) with the parameter $1/\underline{a}$,

(3) A molecular dynamics simulation by Morgan (1985; 1986),

(4) A fractal treatment by Lopez/Quintela et al. (1988),

(5) A gas kinetic approach by Kitahara et al. (1988), and

(6) An approach based on the Fokker-Planck equation by Sceats (1989).

ACKNOWLEDGEMENTS

The authors thank T. Wada-Yamazaki, H. Namba, Y. Nakamura, T. Tezuka, M. Chiba, S. Yano, M. Yamamoto, M. Codama, T. Srithanratana, K. Serizawa, and K. Endou for their excellent collaboration, and also thank M. Tachiya, K. Kitahara, K. Kaneko and J.M. Warman for helpful discussions.

REFERENCES

Ascarelli, G., 1986, Phys. Rev. B 33, 5825.
Basak, S., and M. H. Cohen, 1979, Phys. Rev. B 20, 3034.
Berlin, Yu. A., L. Nyikos, and R. Shiller, 1978, J. Chem. Phys. 69, 2401.
Christophorou, L. G., and K. Siomos, 1984, "Electron-Molecule Interactions and Their Applications," Edited by L. G. Christophorou (Academic Press) Vol. 2, p. 221.
Cohen, M. H., and J. Lekner, 1967, Phys. Rev. 158, 305.
Freeman, G. R., 1981, "Electron and Ion Swarms," Edited by, L. G. Christophorou (Pergamon Press) p. 93.
Hatano, Y., 1986, "Electronic and Atomic Collisions," Edited by D. C. Lorents, W. E. Meyerhof, and J. R. Peterson, (Elsevier) p. 153.
Hatano, Y., and H. Shimamori, 1981, "Electron and Ion Swarms," Edited by, L. G. Christophorou (Pergamon Press) p. 103.
Johnsen, R., and H. S. Lee, 1985, "Swarm Studies and Inelastic Electron-Molecule Collisions," Edited by L. C. Pitchford, B. V. McKoy, A. Chutjian, and S. Trajmar (Springer-Verlag) p.23.
Kaneko, K., Y. Usami, and K. Kitahara, 1988, J. Chem. Phys. 89, 6420.
Lekner, J., 1967, Phys. Rev. 158, 130.
Lopes-Quintela, M. A., M. C. Bujan-Nunes and J. C. Perez-Moure, 1988, J. Chem. Phys. 88, 7478.
Marsolais, R. M., and L. Sanche, 1988, Phys. Rev. B 38, 11118.
Morgan, W. L., 1986, J. Chem. Phys. 84, 2298.

Morgan, W. L., 1985, "Swarm Studies and Inelastic Electron-Molecule Collisions," Edited by L. C. Pitchford, B. V. McKoy, A. Chutjian, and S. Trajmar (Springer-Verlag) p. 43.

Nakamura, Y., K. Shinsaka, and Y. Hatano, 1983, J. Chem. Phys. 78, 5820.

Sceats, M. G., 1989, J. Chem. Phys. 90, 2666.

Sennhauser, E. S., D. A. Armstrong, and J. M. Warman, 1980, Radiat. Phys. Chem. 15, 479.

Shinsaka, K., M. Codama, T. Srithanratana, M. Yamamoto, and Y. Hatano, 1988, J. Chem. Phys. 88, 7529.

Shinsaka, K., M. Codama, Y. Nakamura, K. Serizawa, and Y. Hatano, 1989, Radiat. Phys. Chem. 34, 519.

Tachiya, M., 1986, J. Chem. Phys. 84, 6178.

Tachiya, M., 1987a, J. Chem. Phys. 87, 4108.

Tachiya, M., 1987b, J. Chem. Phys. 87, 4622.

Tezuka, T., H. Namba, Y. Nakamura, M. Chiba, K. Shinsaka, and Y. Hatano, 1983, Radiat. Phys. Chem. 21, 197.

Wada, T., K. Shinsaka, H. Namba, and Y. Hatano, 1977, Can. J. Chem. 55, 2144.

Warman, J. M., 1982, "The Study of Fast Processes and Transient Species by Electron Pulse Radiolysis," Edited by J. H. Baxendale and F. Busi (D. Reidel Publishing Co.) p. 433.

Warman, J. M., 1983, J. Phys. Chem. 87, 4353.

Warman, J. M., E. S. Sennhauser, and D. A. Armstrong, 1979, J. Chem. Phys. 70, 995.

THE MOBILITY OF ELECTRONS IN LIQUID ARGON; SOME DIFFERENCES AND SOME SIMILARITIES WITH THE MOTION OF ELECTRONS IN CRYSTALS AND GASES

G. Ascarelli

Department of Physics
Purdue University
West Lafayette, Indiana   47907

ABSTRACT

After a brief description of the Basak and Cohen's and the Vertes' calculations of electron mobility, my calculation is presented in which I consider that the electron is scattered by both acoustic phonons and large density fluctuations localized over a region whose dimension is of the order of half an electron wavelength. Both the dimensions of these density fluctuations, as well as the magnitude of the density fluctuation themselves, are varied over a wide range in order to calculate the mean free path of electrons whose energy is varied between 1/20 kT and 10 kT. The resulting mobility is compared with recent Hall mobility results as well as with the measured ratio of the Hall and drift mobilities. In order to get a reasonable agreement between experiment and theory, it is found that an effective mass that varies linearly with temperature is necessary. If we extrapolate both this density-dependent effective mass as well as the density dependence of V_o to the density of the solid at the triple point, the calculated phonon-limited mobility is in excellent agreement with experiment. An anomalous dependence of the Hall angle on the magnetic field near the maximum of the electron mobility is ascribed to weak localization. A brief review of the weak localization phenomenon is given, and the calculated results are compared with experiment.

In this talk I would like to describe experiments and calculations aimed at the elucidation of the mechanisms determining the mobility of electrons in rare gas liquids.

Most measurements of mobility employ the so called time of flight (TOF) technique in which a pulse of ionizing radiation creates electrons at time t=0 and the experimenter measures the time it takes these charges to move a specified distance.

Even in a perfectly pure liquid, as opposed to a perfect crystalline solid, disorder permits the existence of two types of states: localized and delocalized. Only the latter, whose wave function extends over the whole sample, can contribute to electronic transport. In the language used in the study of glasses, the localized states are states below the mobility edge.

In a TOF experiment the transit time of the electrons is determined both by the time spent in states above and below the mobility edge respectively. By way of contrast, in the course of a Hall mobility measurement, localized states give no contribution to the measured signal. In general the Hall and the drift mobility are expected to differ by a factor of the order of 1 that depends on the energy dependence of the scattering time. A calculation of the electron mean free path is, however, insufficient for the calculation of the electron mobility. The latter depends not only on the scattering processes but also on the stationary energy levels of an electron in the fluid. A comparison of the Hall and the TOF mobility can give us an indication on whether the TOF mobility corresponds to the drift mobility. If it does, when the latter is compared with the Hall mobility, it can provide an additional test for model calculations of the scattering mechanisms, since the ratio of the mobilities is nearly independent of the electronic band parameters.

A comparison of TOF and Hall measurements gives conflicting results in the case of high mobility organic liquids. As an example, in the case of tetramethyl silane (Munoz and Ascarelli, 1983b; Munoz and Holroyd, 1987) and 22 dimethyl butane (Itoh et al., 1989), the Hall and the TOF mobilities were the same within the experimental errors. By way of contrast, in the case of neopentane (22 dimethyl propane) (Munoz and Ascarelli, 1983a), the difference between the two measurements suggested the possibility of a significant contribution by localized states (see Fig. 1).

However, there are insufficient data on the thermodynamic properties of these substances, and frequently data on the value of the energy corresponding to the bottom of the conduction band are unavailable, making a comparison between experiment and a model calculation difficult.

In the case of rare gases, particularly Ar and Xe, there is much more available thermodynamic data, and it is possible to make such a comparison. A rotation of the atoms cannot affect the potential seen by the electrons; thus disorder and the resulting scattering associated with the relative orientation of the molecules is absent. The mobilities are large, reaching about 6000 cm^2/Vs.

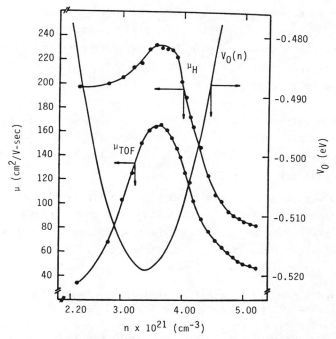

Fig. 1. Hall mobility (μ_H) and time of flight mobility (μ_{TOF})
         (Munoz and Ascarelli, 1983) and position of the conduction
         band (V_0) (Holroyd and Cipollini, 1978) in neopentane.

The first calculation of electron mobility that approximately reproduced the experimental data was carried out by Basak and Cohen (1979).

These authors assumed that an electron moving in liquid Ar can be
imagined as moving in a band similar to what would be expected in the case
of the motion of an electron in a crystalline solid. The bottom of this
band is at an energy V_0 from the vacuum. The electron is bound and can be
represented by a wave packet whose extent is large compared with the average
interatomic distance. Its momentum can be specified in the same way as in
the case of electrons in a crystalline solid. Implicit in such a picture is
the idea that the electron is not scattered by individual atoms of the fluid
but only by those atoms that are not in their equilibrium positions. This
is in contrast to what happens in a gas where each atom or molecule gives
rise to a scattering potential.

The energy dispersion of an electron in the fluid is assumed to be

$$E = V_o + (\hbar k)^2 / 2m^*\tag{1}$$

where m^* is the so called effective mass. It is a parameter that reflects
the fact that the electron is not free but is shared by a large number of atoms
of the fluid (in principle all). The quantity $\hbar k$ that behaves like a momentum
is associated in the case of the solid with the periodicity of the crystal.

It must be stressed that $\hbar k$ is not the real momentum of the electron in the crystal. Indeed it is called the pseudo momentum of the electron. Although strictly speaking translational symmetry does not exist in a fluid, our experience indicates that volumes containing a large number of atoms can be superposed by an appropriate translation. What must still be found out is whether the absence of the perfect superposition that exists in a liquid can be viewed as a perturbation in a way that resembles the case of a solid in which the atoms are not frozen in place. We shall thus assume that over large distances, or equivalently for small values of the wave vector k, a liquid behaves as if it would be periodic.

Such an approach is certainly appropriate for wavelengths long compared with the interatomic distance as when light and sound interact and are detected in light scattering experiments. These experiments indicate there is conservation of momentum of the interacting waves; i.e. that the medium behaves as if it would be periodic on this length scale.

Basak and Cohen (1979) realized that $-n \frac{dV_o}{dn}$ is precisely the so called deformation potential appearing in the calculation of the electron acoustic phonon coupling used in the theory of the mobility of electrons in solids. The main term in their calculation arises in a way that resembles the calculation of the scattering of electrons by acoustic phonons in a crystal.

These authors considered the change of potential produced by a density fluctuation Δn. For the case when Δn is small, they expanded the scattering potential $V_o(\bar{n}+\Delta n) - V_o(\bar{n})$ in series of powers of Δn. In the remainder of the calculation Δn was expanded in a Fourier series and the calculation of the main terms of the result was carried out along the lines used in calculating the phonon-limited mobility; e.g., in the case of semiconductors. The calculation is also very similar to the Ornstein-Zernike theory of critical opalescence (1914).

Shortly after this calculation was published Reininger et al. (1983) published measurements of V_o as a function of density for Ar, Kr, Xe (Fig. 2) and compared the mobilities obtained with Basak and Cohen's calculation to experimental data (Fig. 3).

Overall the agreement between experiment and theory was extremely encouraging. It provided the correct order of magnitude of the mobility with $m^*=m_o$, the free electron mass. A mobility maximum appeared in the vicinity of the density where $dV_o/dn=0$ as had been surmised by Holroyd and Cipollini (1978) from data on $V_o(n)$ and mobility in organic liquids. However, an inescapable prediction of the theory is that, as a consequence of the divergence of the isothermal compressibility, the electron mobility is zero at the critical point, $T\sim150.75K$, $n \sim 0.8x10^{22}cm^{-3}$. The calculated peak of the mobility maximum is much narrower than the experimental result.

294

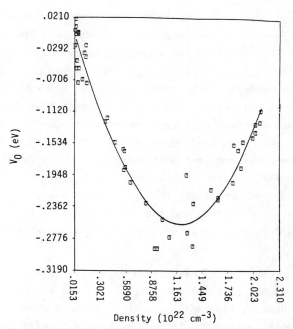

Fig. 2. Experimental (Reininger et al., 1983) $V_0(n)$. The full line is a fit of the expression of V_o suggested in Basak and Cohen, 1979.

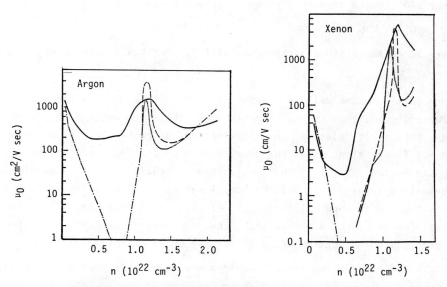

Fig. 3. Zero field mobilities in argon and xenon from Reininger et al., 1983, ... simplified calculation according to Basak and Cohen, 1979, --- full calculation, ·—·— simplified and full calculation coincide. _____ experimental TOF data used in Reininger et al., 1983.

The density fluctuations considered by Basak and Cohen are necessarily small in order for the series expansion to converge. They are certainly appropriate for phonon scattering where the phonon wave vector is of the order of the electron wave vector k appearing in Eq. (1). In the case of a phonon, the corresponding frequencies are of the order of $10^{12}s^{-1}$. In such a time an atom can only move a fraction of a nearest neighbor distance. Such fluctuations are very well represented by the series expansion and the subsequent Fourier series used by Basak and Cohen. Similarly fluctuations are also small when we consider scattering of light. The important volumes have dimensions $\sim\lambda/2$, where λ is the wavelength of light. In the case of a thermal electron, its wavelength is $\sim100\overset{\circ}{A}$. Within a volume whose diameter is $50\overset{\circ}{A}$, there are ~1000 atoms and the corresponding probability of large density fluctuations is much larger than in a volume whose radius is 50 times larger.

Although not explicitly recognized by Vertes (1983; 1984), this must have been part of his rationale when he fitted the experimental mobility assuming large density fluctuations over a volume Ω whose radius was left as a fitting parameter. From statistical mechanics (Landau and Lifchitz, 1967), the probability of a density fluctuation Δn

$$P(\Delta n) \ \alpha \ \exp \ - \ \frac{\Omega(\Delta n)^2}{2\bar{n}S(0)} \tag{2}$$

where $S(0)$ is the structure factor for zero momentum transfer.

$$S(0)=\bar{n}k_BT\chi_T \tag{3}$$

where χ_T is the isothermal compressibility. The scattering potential is

$$\Delta V_0 = V_0(\bar{n} + \Delta n) - V_0(\bar{n}) \tag{4}$$

represents a square well or barrier over the volume Ω. The radii of the volumes necessary to fit the experimental mobilities were of the order of $20\overset{\circ}{A}$. In a fluid volume of these dimensions the probability of a large density fluctuation, where the density is e.g. zero or equal to the density of the crystal, is not negligible. Clearly, if such a density fluctuation develops, it will be long lived compared with times $\sim10^{-12}s$. The latter is an upper limit for the electron mean free time. From the point of view of the electron such density fluctuations can be considered as static and the corresponding scattering purely elastic.

These potential wells and barriers are similar to what was described by Coker et al. (1987) in their Monte Carlo study of the behavior of excess electrons in fluid Xe.

In the calculation of the mobility there are therefore two types of scattering: scattering by phonons and scattering by large, localized density fluctuations. Phonon scattering is inelastic, while scattering by large density fluctuations is elastic. An alternative view is that phonons are either created or destroyed when they scatter an electron. By way of contrast, static density fluctuations remain unaltered during the scattering process.

Phonon scattering has been extensively studied e.g. in the case of semiconductors. The corresponding mobility (Brooks, 1955) is

$$\mu_p = \frac{2(2\pi)^{1/2}e\hbar^4}{3k_B^{3/2}m_o^{5/2}} \times \frac{\rho\, C_s^2}{T^{3/2}\Xi^2 \left(\frac{m^*}{m_o}\right)^{5/2}} \qquad (5)$$

Here ρ is the density of the fluid (g/cm^3), C_s is the velocity of sound (cm/s), T the temperature in K, $\Xi = -n\dfrac{dV_o}{dn}$ is the deformation potential, m^* and m_o are respectively the effective mass and the free electron mass. With parameters appropriate to Ar we get:

$$\mu_p = \frac{2\times10^{-27}C_s^2}{nT^{3/2}\left(\frac{dV_o}{dn}\right)^2 \left(\frac{m^*}{m_o}\right)^{5/2}} \qquad (6)$$

This expression is appropriate every where except when either $\dfrac{dV_o}{dn} \sim 0$ or near the critical point where $C_s \to 0$. In these vicinities we must consider that phonon scattering will be changed within a static density fluctuation and the above expression for μ_p must be corrected by considering their effect (Ascarelli, 1986a). Fig. 4 shows the density dependence of the phonon limited mobility in the case of a constant effective mass. The value of V_o and it derivatives is taken from Reininger et al, 1983. It is a well known result from semiconductor theory that the mean free path associated with acoustic phonon scattering is independent of the electron energy.

We must now extend the Vertes (1983) calculation so as to take into account all possible volumes Ω in which there might be a density fluctuation.

The probability of finding a constant density in a volume Ω whose radius is r is:

$$P(r) \propto e^{-r/\xi} \qquad (7)$$

where ξ is the correlation length.

An order parameter ξ much larger than the dimensions of the volume we are considering cannot influence the existence of order within that volume. Therefore, if we consider volumes whose radii are much smaller than ξ, P(r) must be constant independent of ξ. The necessary behavior of P(r) may be described by substituting ξ in Eq. (7) with a quantity ξ_e that is a function of both ξ and r, whose limit is ξ where r>>ξ, and r when r<<ξ.

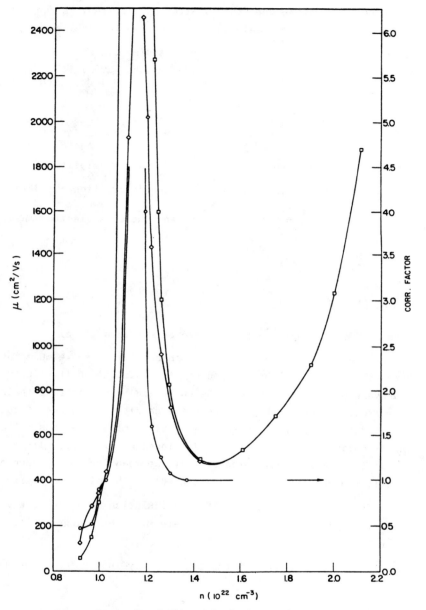

Fig. 4. Phonon-limited mobility of electrons in argon (Ascarelli,
1986) assuming $m^* = m_0$. □ mobility according to Eq. (6),●
mobility taking into account density fluctuations: $\mu_c = \mu_p/X$
where X is labeled as the correction factor in the figure
(○) assuming density fluctuations whose dimensions are
$$\left[\frac{2\pi\, m^* k_B T}{h^2} \right]^{-3/2}$$

298

On account of the isotropy of the fluid, P(r) cannot change when x is changed into -x, y into -y or z into -z. Therefore, r must appear as some power of r^2. The simplest rational function that has this property as well as the appropriate limits is:

$$\xi_e^{-2} = \xi^{-2} + r^{-2} \tag{8}$$

The number of atoms that are normally within the volume Ω must be relatively large (of the order of hundreds?); otherwise neither the effective mass approximation we are using nor statistical mechanics are applicable. This limits the smallest volume that can be used in the calculation. The values of Δn must also be limited. The number of atoms in the volume Ω cannot be either negative or larger than in the crystalline solid. With both these limitations in mind, we can write for the mean free path for momentum transfer:

$$\Lambda^{-1} = \frac{\displaystyle\sum_{\Delta n}\sum_{r} e^{-\frac{\Omega(\Delta n)^2}{2\bar{n}S(0)}}\, e^{-\frac{r}{\xi_e}}\left[\frac{2\pi}{\Omega}\int_0^{\pi}\sigma(\theta,\Delta n,\Omega,k)\,(1-\cos\theta)\,\sin\theta\,d\theta\right]}{\displaystyle\sum_{\Delta n}\sum_{r} e^{-\frac{\Omega(\Delta n)^2}{2\bar{n}S(0)}}\, e^{-\frac{r}{\xi_e}}} \tag{9}$$

where $\sigma(\Delta n,\Omega,k)$ is the cross section for scattering of an electron by a square well:

$$\sigma(\Delta n,\Omega,k) = 2\pi\int_0^{\pi}\sigma(\theta,\Delta n,k)\,(1-\cos\theta)d\theta \tag{10}$$

$$= \frac{4\pi}{k^2}\left[\sin^2\delta_0 + 3\sin^2\delta_1 + 5\sin^2\delta_2 + 7\sin^2\delta_3 + 9\sin^2\delta_4\right.$$

$$- 2\sin\delta_0\sin\delta_1\cos(\delta_1-\delta_0) - 8\sin\delta_1\sin\delta_2\cos(\delta_2-\delta_1)$$

$$\left. - 12\sin\delta_2\sin\delta_3\cos(\delta_3-\delta_2) - 16\sin\delta_3\sin\delta_4\cos(\delta_4-\delta_3)\right]$$

Here s to g waves have been considered. I limited the calculation to g waves because I found that they did not give a large contribution to the mean free path. The spikes that are seen in the energy dependence of the mean free path when Ω contains more than 128 atoms (Figs. 5, 6, 7) are primarily due to p and d waves. They are the consequence of the choice of a square well potential.

In order to do the machine calculation in a reasonable time, I have limited the values of r to the case when $r > r_{min}$ and $e^{-r/\xi_e} \geq 10^{-6}$. The value of Ω was then progressively doubled up to when the upper limit of r was reached. Similarly the maximum and minimum value of Δn are limited by sections were calculated for each electron momentum and each value of Ω at 100 equally spaced values of Δn.

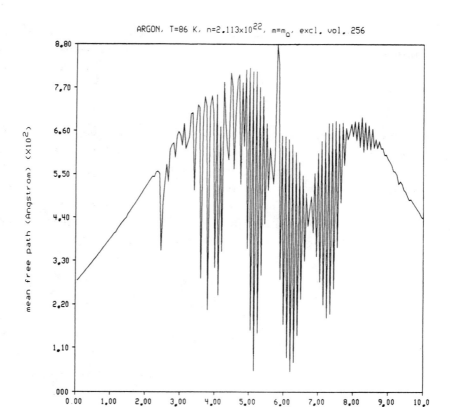

Fig. 5. Static density fluctuation limited mean free path for electrons with $m^* = m_o$ in argon at 86K and density $n=2.11 \times 10^{22} cm^{-3}$. On average, Ω_{min} contains 256 atoms.

The values of the electron momentum that were considered correspond to energies from $k_B T/20$ to $10 k_B T$ in intervals of $k_B T/20$.

Knowing the energy dependence of the mean free path for scattering by static density fluctuations as well as the energy independent mean free path arising from phonon scattering, the drift mobility, the Hall mobility, as well as their ratios can be calculated for whatever electron distribution is desired. I considered only the case of a Boltzmann distribution. The calculation was carried out along the liquid vapor coexistence line.

In order to measure the Hall mobility of an insulator we cannot use the same techniques used in semiconductors. We also must be careful that polarization is not set up so as to distort our knowledge of the applied fields. We would like then to make pulse measurements with few charges over times that are short compared with the time it takes an electron to traverse the sample under the effect of the applied fields. ''The Redfield Technique'',

300

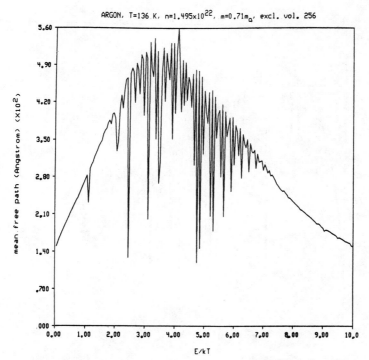

Fig. 6. Same as Fig. 5 for $T=136K$, $n=1.495\times10^{22}cm^{-3}$, $m^{*}=0.71m_o$.

initially developed for the measurement of the Hall mobility in ionic crystals (Redfield, 1954), is ideally suited for this purpose.

The sample holder that has been designed produces a uniform axial field in the active volume, even when it is inserted in a metallic container that would tend to severely distort the electric field distribution (Fig. 8).

A narrow well-collimated x-ray beam produces electron hole pairs that drift in the applied electric field. Adjustment of one of the potentiometers connected across the resistive plates D and D' rotates the equipotentials and nulls any differential signal detected by the amplifier.

The application of a magnetic field unbalances the system. By changing again one of the potentiometers a transverse electric field E_y is created to balance the $\vec{v}\times\vec{B}$ force created by the magnetic field. The ratio of the longitudinal and transverse electric fields is equal to the tangent of the Hall angle

$$\frac{E_y}{E_z} = \tan\theta_0 = -\frac{\left\langle \dfrac{\omega_c \tau^2}{1 + (\omega_c \tau)^2} \right\rangle}{\left\langle \dfrac{\tau}{1 + (\omega_c \tau)^2} \right\rangle} \tag{11}$$

When $\omega_c \tau$ is small, $\tan\theta_0 = \mu_H B \sim \omega_c \tau$.

301

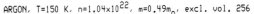

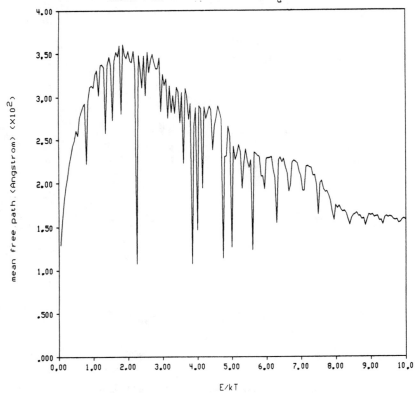

Fig. 7. Same as Fig. 5 for T=150K, n=1.04x10^{22}cm^{-3}, m*=0.49m$_o$.

The result of the Hall mobility measurements (Ascarelli, 1989) is shown in Fig. 9. Most of the data have been taken along the liquid vapor coexistence line while some of the Hall and TOF mobility data have also been taken along the 50 atmosphere isobar.

The calculated drift mobility is shown in Fig. 10 for the case of a constant effective mass m* = m$_o$. It is clear that although it has some features of the experimental results, quantitative agreement is poor. A much better agreement between experiment and theory is found with an effective mass that decreases linearly with density. Clearly such a density dependence of the effective mass cannot be extrapolated to the case of the dilute gas. It may be instead reasonable to extrapolate it to the density of the solid (Fig. 11).

In the case of xenon (Ascarelli, 1986b) the effective mass in the liquid at the triple point has been estimated from exciton data. With such a value of m* the calculated drift mobility and the experimental TOF mobility agree with no adjustable parameters (Fig. 12).

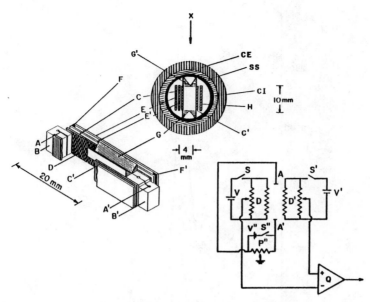

Fig. 8.   Electrode configuration used for Hall measurements. D, D',
G, G' are thick-film resistors printed on either alumina or
Macor. E, E' is a parallel plate capacitor used for
measuring the density of the fluid. A, A' electrodes
connected to the top of resistors G, G' printed on Macor
spacers B, B'. These electrodes are enclosed in a
stainless steel cell lined inside and outside with copper.

   In order to find values of the effective mass in argon obtained from
other experiments we must go to the case of crystalline argon. The effective
mass that is obtained (Zimmerer, 1987; Resca et al., 1978a; Resca et al.,
1978b; Reininger et al., 1984; Perluzzo et al., 1985) varies between
approximately 0.5 m_o and 0.7 m_o, depending both on how it is calculated and on
the original experimental data used. However, since this mass is in most
cases estimated from spectra of a relatively small radius exciton (Zimmerer,
1987; Resca et al., 1978a; Resca et al., 1978b; Reininger et al., 1984), there
are important corrections that must be introduced in the Rydberg formula that
was used. These corrections are not necessarily negligible compared with the
exciton binding energy, even in the case of the n=2 exciton state. Their
estimate is difficult because they have different signs and, in the case of
the lowest exciton state, similar magnitudes (Hermanson and Phillips, 1966;
Hermanson, 1966).

   It is appropriate to ask whether the density dependent effective mass
that is obtained in the fit of the mobility calculation to experiment is
just a parameter or whether it represents an approximation of reality.

   As pointed out above, the ratio of the Hall and the drift mobility has
been calculated. It is nearly independent of the choice of effective mass.

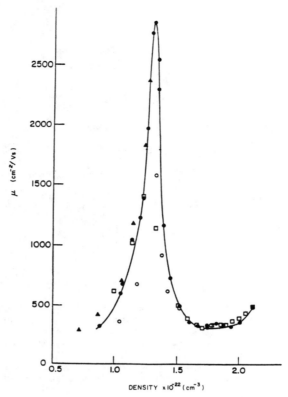

Fig. 9. Results of mobility measurements. ● Hall mobility along the
liquid vapor coexistence line, ▲ Hall mobility along the 50
atmosphere isobar, □ TOF mobility along the 50 atmosphere
isobar (Jahnke et al., 1971) and ○ TOF mobility along the
liquid vapor coexistence line (Shinsaka et al., 1988).

It varies from approximately 1 at a density much above the mobility maximum
to 1.5 at the mobility maximum, falling again to 1 at the critical point.
The result corresponding to $m^* = m_o$ as well as to a variable effective mass
are shown in Figs. 10 and 11. A comparison with the drift and Hall mobility
given in Fig. 9 indicates that there is good agreement between the ratio of
the mobilities and the result of the calculation.

If both the values of V_o and of the effective mass measured in the
liquid are extrapolated to the solid at the triple point, the mobility can
be calculated with no adjustable parameters. Only phonon scattering should
be important in a sufficiently perfect crystal and the ratio of the Hall to
the drift mobility should be $3\pi/8$. The calculated Hall mobility is 2770
cm^2/Vs to be compared with an experimental mobility of 2820 cm^2/Vs!

A more detailed discussion of some small discrepancies between
experiment and the calculation are given in Ascarelli, 1989.

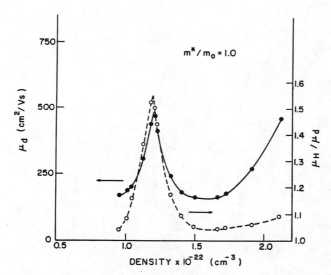

Fig. 10. Calculated drift mobility and ratio μ_H/μ_D along the liquid–vapor coexistence line; $M^* = m_o$. On average, Ω_{min} contains 256 atoms. Scattering is due to both phonons and static density fluctuations.

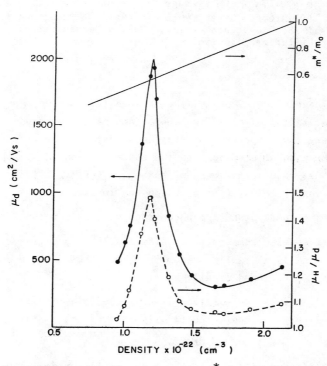

Fig. 11. Same as Fig. 10 except that m^* varies linearly with density as shown.

305

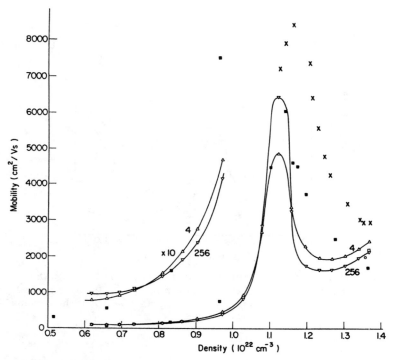

Fig. 12. Calculated drift mobility of electrons in xenon using
$m^* = 0.27 m_0$. On average Ω_{min} contains either 4 (Δ) or 256
(∇) atoms. Experimental TOF data: ■ Huang and Freeman,
1978, X Gushchin et al., 1982, □Yoshino et al., 1976, O
Miller et al., 1968.

Before concluding, I would like to speculate on the possible existence
of weak localization effects like those that have been found in metallic
alloys at very low temperatures (Bergman, 1984; Chakravarty and Schmid,
1986), in the case of electrons on the surface of solid hydrogen (Adams and
Paalanen, 1987), and in the case of scattering of light (John, 1988) by a
turbid solution of polystyrene microspheres in water.

In order to understand the phenomenon we should refer to the Fig. 13.

Consider an electron scattered elastically at position $\vec{r}_1$, then at $\vec{r}_2$,
$\vec{r}_3$, . . . , finally at $\vec{r}_N$. Consider the same electron wave scattered
at $\vec{r}_N$, then $\vec{r}_{N-1}$ .... and finally at $\vec{r}_1$.

If at all points the electron is scattered elastically, the phase
difference of the wave scattered finally from $\vec{r}_N$ and $\vec{r}_1$ is:

$$(\vec{k}_f - \vec{k}_1) \cdot (\vec{r}_N - \vec{r}_1) = \vec{q} \cdot (\vec{r}_N - \vec{r}_1) \ . \tag{12}$$

This phase difference is zero if the electron is incident and is scattered
back in the direction perpendicular to $(\vec{r}_N - \vec{r}_1)$.

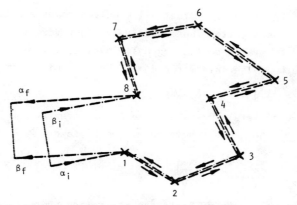

Fig. 13. Path L considered in the multiple scattering leading to
weak localization. Example with N=8. Figure from
Chakravarty and Schmid, 1986.

In the absence of a magnetic field the waves being scattered along
either the clockwise or the anti-clockwise paths gain the same phase. The
two waves interfere. Inelastic scattering will destroy the coherence and,
therefore, the interference of the waves scattered along the clockwise and
anti-clockwise path.

Considering the electron as a classical particle executing a random
walk starting at $\vec{r}_1$ and lasting a time τ, if the electron is to return to
$\vec{r}_N$ near $\vec{r}_1$, this implies:

$$\left|\vec{r}_N - \vec{r}_1\right| \sim (D\tau)^{1/2} = \left(\frac{\Lambda L}{3}\right)^{1/2} , \tag{13}$$

where D is the diffusion constant for electrons of momentum k. When the
electron is considered quantum mechanically, the correspondence principle
indicates that the most probable paths are those calculated classically.
The same is e.g. valid in the case of light. However, on account of the
wavelike nature of the electron when we calculate probabilities we must add
the amplitude of the waves traversing the path L in the clockwise and the
anticlockwise directions. As a result of interference, the probability of
finding the electron at the origin is twice what would be expected for a
classical particle (Bergman, 1984).

The total path L is expected to be of the order of the inelastic mean
free path, e.g. the mean free path determined by phonon scattering.

If $k\Lambda \ll 1$ the electrons are localized. This is the well-known Joffe-
Regel criterion. It just indicates that if the mean free path is short
compared with the wavelength, it is not meaningful to use the concepts
associated with a plane wave. Clearly, localized electrons cannot
contribute to transport.

From Eq. (12) it is clear that all electron waves scattered in the backwards direction interfere constructively. However, for large values of $\vec{k}$ this enhanced backwards scattering only involves few electrons because the angular width of the interference maximum is limited by the condition $\left| \vec{q} \cdot (\vec{r}_N - \vec{r}_1) \right| \sim \pi$.

In the case of metals where $\vec{k}_F$, the Fermi momentum, is large compared with $\left| \vec{r}_N - \vec{r}_1 \right|^{-1}$ the angle θ between $\vec{k}_f$ and $\vec{k}_i$ is $q = k_F \theta$, therefore:

$$\theta \sim \frac{\pi}{k_F \left| \vec{r}_N - \vec{r}_1 \right|} = \frac{\pi}{k_F} \left(\frac{3}{\Lambda L} \right)^{1/2} . \tag{14}$$

This is not so in the case of Boltzmann statistics where k is not always large. As an example, in the case of electrons in Ar near the mobility maximum the calculated $\Lambda \sim 700\overset{\circ}{A}$. The value of the phonon limited mean free path is calculated to be L $\sim 25 \times 10^{3}\overset{\circ}{A}$. The value of k $\sim \pi/\Lambda$ is much below the value that corresponds to the most probable energy of the electron distribution. For these electrons the lobe of the constructive backwards interference can be very wide and a large fraction of the electrons of momentum k $\lesssim \left(\frac{3\pi^2}{\Lambda L} \right)^{1/2}$ do not contribute to electrical transport.

We must now consider the effect of a magnetic field. According to the rules of quantum mechanics, we must substitute $\vec{p}$ by $(\vec{p} - e\vec{A})$ in the Hamiltonian of our "free" electrons; $-e$ is the charge of the electron and $\vec{A}$ is the vector potential. Therefore the phase difference of the waves traversing the path L in the clockwise and anticlockwise direction is

$$\frac{2}{\hbar} \oint_L e\vec{A} \cdot \vec{d1} = \frac{2e}{\hbar} \int_S (\nabla \times \vec{A}) \cdot \vec{ds} = \frac{2e}{\hbar} \Phi_S . \tag{15}$$

The factor 2 arises because $\vec{d1}$ has the opposite signs when the direction in which the path is traversed is reversed. Φ_S is the flux of magnetic field through the closed path that encloses the surface S. (The portion $\int_{\vec{r}_N}^{\vec{r}_1} \cdots$ missing in the phase integral can be added and subtracted along one of the paths. One of the terms is used closing the clockwise path, the other the anticlockwise path).

The area enclosed by L is of the order of Λ^2 so that $\Phi_S \sim B\Lambda^2/3$; the factor 3 arises because the average projection of the area S on three mutually orthogonal planes is the same.

We can now ask what happens to the condition that the lobe of the backward interference maximum is broad when the magnetic field is increased. We take $\theta \sim \pi/4$ as an indication of such wide lobe.

$$\left|q_0\right| \left|\vec{r}_N - \vec{r}_1\right|\cos\theta = \left|q\right|\left|\vec{r}_N - \vec{r}_1\right|\cos\theta - \frac{2eBS}{3\hbar} = \pi \qquad (16)$$

Since $\dfrac{2BSe}{3\hbar} >> q_0 \left|\vec{r}_N - \vec{r}_1\right| \sim q_0(L\sqrt{3})^{1/2}$ for fields of a few

kilogauss, it is clear that:

$$q \sim \frac{2BSe}{3\hbar \left|\vec{r}_N - \vec{r}_1\right|\cos\theta} + q_0 \sim \frac{2BSe}{3\hbar \left|\vec{r}_N - \vec{r}_1\right|\cos\theta} \qquad (17)$$

Because we wish to have large angle scattering, we must also have $q \sim k$.

This implies that the magnetic field ''scans a hole'' in the electron distribution. For calculations e.g. of the current, the "hole" that must be considered of the order of

$$C\left[\frac{\tau}{1 + (\omega_c\tau)^2}\right] p^4 \, e^{-\frac{p^2}{2m^*k_BT}} \Delta p \qquad (18)$$

Here Δp is a measure of its width and C reflects both the solid angle over which incident electrons contribute to the constructive interference and the normalization of the electron distribution.

In a liquid rare gas, weak localization will be most important near the mobility maximum because in that vicinity the mean free path associated with inelastic scattering is much longer than that associated with elastic scattering. If weak localization gives rise to a ''hole'' in the electron distribution that is ''scanned'' by the magnetic field over different values of the energy, we would expect that the tangent of the Hall angle is:

$$\tan\theta = \frac{\left\langle\dfrac{\omega_c\tau^2}{1 + (\omega_c\tau)^2}\right\rangle - C\left[\dfrac{\omega_c\tau^2}{1 + (\omega_c\tau)^2}\right] p^4 \, e^{-\frac{p^2}{2m^*k_BT}} \Delta p}{\left\langle\dfrac{\tau}{1 + (\omega_c\tau)^2}\right\rangle - C\left[\dfrac{\tau}{1 + (\omega_c\tau)^2}\right] p^4 \, e^{-\frac{p^2}{2m^*k_BT}} \Delta p} \qquad (19)$$

Rearranging terms and expanding the denominator in series, keeping only the terms first order in Δp, we get:

$$\tan\theta \cong \tan\theta_0 - Cp^4 e^{-\frac{p^2}{2m^*k_BT}}\left[\omega_c\tau - \tan\theta_0\right]\frac{\dfrac{\tau}{1 + (\omega_c\tau)^2}}{\left\langle\dfrac{\tau}{1 + (\omega_c\tau)^2}\right\rangle}\Delta p \qquad (20)$$

Here $\tan\theta_0$ is the tangent of the Hall angle in the absence of weak localization (Eq. 11).

In the case of argon the calculated average mean free path at the mobility maximum is 700 Å. The calculated phonon limited mean free path is 25×10^3Å. From an inspection of Fig. 14, it is seen that the term

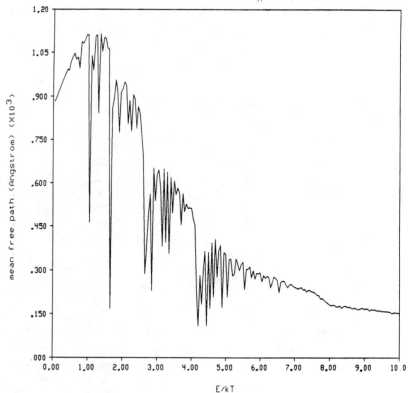

 T=148 K, n=1.17×10^{22}, m=0.55 m$_n$, excl. vol. 256

Fig. 14.   Calculated mean free path due to scattering by static
density fluctuations near the mobility maximum.

proportional to Δp in Eq. (20) is always positive because for the magnetic
fields available (1 T), $p < p_{max}$.

We may now compare the theoretical model with experiment. Contrary
to what is expected from the Boltzmann equation, in the present case when
$\omega_c \tau \lesssim 0.3$, $\tan \theta$ is not a linear function of B (Fig. 15). If weak local
ization is responsible for the deviation we should substitute p by

$\dfrac{BSe}{(3\Lambda L)^{1/2}}$ in Eq. (20) and get

$$\Delta = \tan\theta - \tan\theta_0 \ \alpha \ B^4 \exp - \frac{B^2 S^2 e^2}{6m^* \Lambda L k_B T} \qquad (21)$$

or $\log (\Delta/B^4) \ \alpha \ B^2$.

Despite the paucity of the data, this seems to be the case (Fig. 16).
The straight line was obtained by reading out 7 values of Δ from Fig. 14.

310

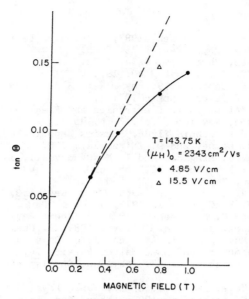

Fig. 15. Experimental magnetic field dependence of the Hall angle near the maximum of the mobility. The broken line is what is expected from Boltzmann equation.

From it, assuming that the path L is a regular polygon of side Λ, the estimate of L turns out to be of the order of 7Λ while from the mobility calculation I would have estimated $\sim 20\Lambda$.

I conclude with a question: is this an indication of weak localization in liquid argon?

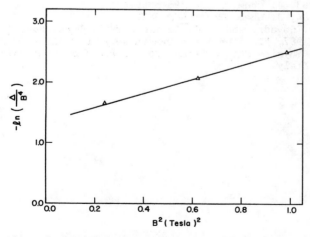

Fig. 16. Plot of the value of Δ from Fig. 15 to put in evidence the predictions of the weak localization model (Eq. 21). The straight line is obtained by reading 7 pairs of values of Δ and B from Fig. 15.

REFERENCES

Adams, P. W., and M. A. Paalanen, 1987, Phys. Rev. Lett. 58, 2106.
Ascarelli, G., 1986a, Phys. Rev. B 33, 5825.
Ascarelli, G., 1986b, Phys. Rev. B 34, 4278.
Ascarelli, G., 1989, Phys. Rev. B 40, 1871.
Basak, S., and M. H. Cohen, 1979, Phys. Rev. B 20, 3404.
Bergman, G., 1984, Physics Reports 107, 1.
Brooks, H., 1955, "Advances in Electronics and Electron Physics," L. Morton
    (Ed.), Academic, New York, Vol. 7, p. 85.
Chakravarty S., and A. Schmid, 1986, Physics Reports 140, 193.
Coker, D. F., B. J. Berne, and D. Thirumalai, 1987, J. Chem. Phys. 86, 5689.
Gushchin, E. M., A. A. Kruglov, and I. M. Obodovskii, 1982, Zh. Eksp. Teor.
    Fiz. 82, 1114; Sov. Phys. JETP 55, 650.
Hermanson, J., 1966, Phys. Rev. 150, 660.
Hermanson J., and J. C. Phillips, 1966, Phys. Rev. 150, 652.
Holroyd, R., and N. E. Cipollini, 1978, J. Chem. Phys. 69, 502.
Huang, S. S. S., and G. R. Freeman, 1978, J. Chem. Phys. 68, 1355.
Itoh, K., R. C. Munoz, and R. A. Holroyd, 1989, J. Chem. Phys. 90, 1128.
Jahnke, J. A., L. Meyer, and S. A. Rice, 1971, Phys. Rev. A 3, 734.
John, S., 1988, Comments Condensed Matter Physics 14, 193.
Landau, L., and E. Lifchitz, 1967, "Physique Statistique," Mir, Moscow,
    p.39.
Miller, L. S., S. Howe, and W. E. Spear, 1968, Phys. Rev. 166, 871.
Munoz, R. C., and G. Ascarelli, 1983a, Phys. Rev. Lett. 51, 215.
Munoz, R. C., and G. Ascarelli, 1983b, Chem. Phys. Lett. 94, 235.
Munoz, R. C., and R. A. Holroyd, 1987, Chem. Phys. Lett. 137, 250.
Ornstein, L. S., and F. Zernike, 1914, Proc. Acad. Sci. Amsterdam 17, 793.
Perluzzo, G., G. Bader, L. G. Caron, and L. Sanche, 1985, Phys. Rev. Lett.
    55, 545.
Redfield, A. G., 1954, Phys. Rev. 94, 526.
Reininger, R., U. Asaf, I. T. Steinberger, and S. Basak, 1983, Phys. Rev. B
    28, 4426.
Reininger, R., I. T. Steinberger, S. Bernstorff, and P. Laporte, 1984, Chem.
    Phys. 86, 189.
Resca, L., R. Resta, and S. Rodriguez, 1978a, Phys. Rev. B 18, 696.
Resca, L., R. Resta, and S. Rodriguez, 1978b, Phys. Rev. B 18, 702.
Shinsaka, K., M. Codoma, T. Srithanratana, M. Yamamoto, and Y. Hatano, 1988,
    J. Chem. Phys. 88, 7529.
Vertes, A., 1983, J. Chem. Phys. 79, 5558.
Vertes, A., 1984, J. Chem. Phys. 88, 3722.
Yoshino, K., U. Sowada, and W. F. Schmidt, 1976, Phys. Rev. A 14, 438.
Zimmerer, G., 1987, "Excited State Spectroscopy in Solids," Proc. Int.
    School of Physics Enrico Fermi, Course 96, Italian Physical Society,
    Bologna, Italy.

ULTRAFAST AND ULTRASENSITIVE DIELECTRIC LIQUIDS/MIXTURES: BASIC
MEASUREMENTS AND APPLICATIONS

L. G. Christophorou, H. Faidas, and D. L. McCorkle

Atomic, Molecular, and High Voltage Physics Group
Health and Safety Research Division
Oak Ridge National Laboratory
Oak Ridge, Tennessee 37831

The Department of Physics
University of Tennessee
Knoxville, Tennessee 37996

Basic properties of cryogenic and room temperature dielectric
liquids/mixtures with high electron yields (under irradiation by ionizing
particles) and high excess electron drift velocities are discussed. A
number of ultrafast and ultrasensitive liquid media--appropriate for
possible use in liquid-filled radiation detectors and other
applications--are identified.

INTRODUCTION

Contrary to low-pressure gases where electrons are "free," in
liquids electrons are quasifree or localized; e.g., see Christophorou and
Siomos (1984), Christophorou (1988), Schmidt (1984), Freeman (1987),
Kunhardt et al. (1988). Excess electrons are quasifree in liquids for
which the electron ground-state energy, V_o, is negative (< 0 eV). Such
liquids can be cryogenic or "warm" (room temperature). Examples are
given in Table 1. It is in such liquids (and their mixtures with
appropriate additives) that the search for ultrafast and ultrasensitive
liquid media is being focused.

ELECTRON DRIFT VELOCITIES, ELECTRON ENERGIES, AND ELECTRON ATTACHMENT TO
MOLECULES IN DIELECTRIC LIQUIDS WITH V_o < 0 eV

In liquids/mixtures with V_o < 0 eV (such as those in Table 1),
quasifree electrons drift fast, have energies in excess of thermal at
high applied electric fields E, and attach to molecules as in gases (but
with notable changes in the energy position, cross section, and lifetime
of the negative ion state(s) involved); e.g., see Christophorou and

Table 1. Examples of Liquids with Negative V_0

Liquid	$V_0(eV)$[a]
Cryogenic	
Ar(87 K)	−0.20
Xe(165 K)	−0.61
Room Temperature	
Neopentane $[C(CH_3)_4]$	−0.43
Tetramethylsilane $[Si(CH_3)_4]$	−0.57
Tetramethylgermanium $[Ge(CH_3)_4]$	−0.64
Tetramethyltin $[Sn(CH_3)_4]$	−0.75
2,2,4,4-Tetramethylpentane $[(CH_3)_3CCH_2C(CH_3)_3]$	−0.36

[a] See Christophorou and Siomos (1984), Christophorou (1988), Schmidt (1984), Freeman (1987), Kunhardt et al. (1988), Allen (1976) for original sources of V_0 data.

Siomos (1984), Christophorou (1988), Schmidt (1984), Freeman (1987), Kunhardt et al. (1988), Allen (1976), Christophorou (1985), Christophorou et al. (1989). Figs. 1 through 6 exemplify these properties for cryogenic liquids.

In Fig. 1 the electron drift velocity (w) as a function of the density-reduced electric field E/N is shown for gaseous Ar (Christohphorou, 1971; Robertson, 1977) and Xe (Christophorou, 1971; Hunter et al., 1988) and for liquid Ar and Xe (Miller et al., 1968). (w_L and w_G refer to the w in the liquid and the gas, respectively.) At any value of E/N, the w_L in the liquid far exceeds the w_G in the low-pressure gas, especially for Xe at low E/N, reflecting the profound changes in the momentum transfer cross section $\sigma_m(\varepsilon)$ between the gas, $(\sigma_m)_G(\varepsilon)$, and liquid, $(\sigma_m)_L(\varepsilon)$ (Christophorou, 1988; Christophorou et al., 1989).

In Fig. 2 the characteristic energy $\left(\frac{3}{2} e \frac{D_T}{\mu}\right)_L$ versus E/N (where

D_T is the transverse electron diffusion coefficient and μ is the electron mobility) is shown for liquid Ar (Shibamura et al., 1979) (T = 87 K) and liquid Xe (Kubota et al., 1982) (T = 165 K). For comparison, the

calculated (Christophorou et al., 1989; Hunter et al., 1988) $\left(\frac{3}{2} e \frac{D_T}{\mu}\right)_G$

versus E/N for gaseous Ar and gaseous Xe at room temperature (T = 300 K) and at T = 165 K for Xe and at T = 87 K for Ar are also shown in the

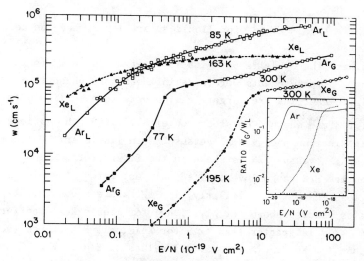

Fig. 1. w versus E/N in gaseous Ar [■,□ ; see Christophorou (1971), Robertson (1977)] and Xe [●, ○ ; see Christophorou (1971), Hunter et al. (1988)] and liquid Ar [□; see Miller et al. (1968)] and Xe [▲, see Miller et al. (1968)]. Inset: Ratio w_G/w_L versus E/N for Ar and Xe.

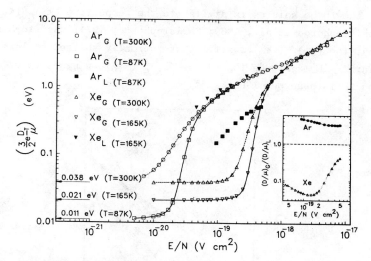

Fig. 2. Calculated (Christophorou et al., 1989; Hunter et al., 1988) $\left(\frac{3}{2} e \frac{D_T}{\mu}\right)_G$ versus E/N for gaseous Ar at T = 300 K (o) and T = 87 K (□) and gaseous Xe at T = 300 K (Δ) and 165 K (∇). The experimental $\left(\frac{3}{2} e \frac{D_T}{\mu}\right)_L$ versus E/N for liquid Ar [■: see Shibamura et al. (1979)] and for liquid Xe [▼: see Kubota et al. (1982)] respectively at 87 and 165 K. Inset: Ratio $\left(\frac{D_T}{\mu}\right)_G \Big/ \left(\frac{D_T}{\mu}\right)_L$ versus E/N for Ar (●) and Xe (Δ).

figure. While the $\left(\dfrac{3}{2}\,e\,\dfrac{D_T}{\mu}\right)_L$ for liquid Ar are lower than the corresponding

gaseous values, the opposite behavior is observed for Xe. In the E/N range over which D_T/μ measurements were made in the liquid phase (Fig. 2), the characteristic energies are larger in liquid Xe than in liquid Ar.

In Fig. 3 are shown (Christophorou et al., 1989) the various estimates of the mean electron energy $\langle\varepsilon\rangle_L$ versus E/N in liquid Ar. For comparison, the gaseous $\langle\varepsilon\rangle_G$ versus E/N is shown in the figure for T = 300 K and 87 K. Over the entire E/N range investigated, all estimates give—for a fixed E/N—$\langle\varepsilon\rangle_L$ less than $\langle\varepsilon\rangle_G$. Similarly, in Fig. 4 are shown the $\langle\varepsilon\rangle_L$ versus E/N estimates (Christophorou et al., 1989) for liquid Xe. Contrary to the case of Ar, $\langle\varepsilon\rangle_L$ is greater than $\langle\varepsilon\rangle_G$ for Xe at all E/N values (see Fig. 4). The fact that for Xe $\langle\varepsilon\rangle_L > \langle\varepsilon\rangle_G$ at all values of E/N in Fig. 2 is consistent with the result $w_L > w_G$ (Fig. 1) and indicates $(\sigma_m)_L \ll (\sigma_m)_G$ for $\varepsilon < 1$ eV.

In Fig. 5 is plotted the rate constant $(k_a)_L$ for electron attachment to SF_6, N_2O, and O_2 in liquid Ar (Bakale et al., 1976) as a function of $\langle\varepsilon\rangle_L$ (Christophorou et al., 1989; see Fig. 3). While the $\langle\varepsilon\rangle_L$ dependence of $(k_a)_L$ for the three solutes is similar to the respective $(k_a)_G$ ($\langle\varepsilon\rangle$) in gases (Christophorou et al., 1989; Christophorou, 1971; Christophorou et al., 1984), the $(k_a)_L$ ($\langle\varepsilon\rangle$) functions are normally shifted to lower energies and are larger in magnitude than those $(k_a)_G$ ($\langle\varepsilon\rangle$) in gases (e.g., see Fig. 6 and Christophorou, 1988; 1985).

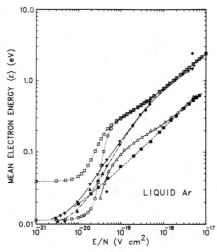

Fig. 3. Calculated $\langle\varepsilon\rangle_L$ versus E/N for liquid Ar, ▼: Nakamura et al. (1986); ■: Christophorou (1985); ▲: Gushchin et al. (1982); ◆: Lekner (1967) in comparison with the calculated values of Christophorou et al. (1989) in liquid Ar at T = 87 K (Δ) and gaseous Ar at T = 87 K (o) and 300 K (□).

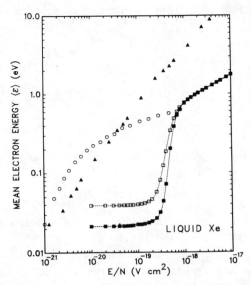

Fig. 4. Calculated $\langle\varepsilon\rangle_L$ versus E/N for liquid Xe, ▲: Gushchin et
al. (1982) in comparison with the calculated values of
Christophorou et al. (1989) in liquid Xe at T = 165 K (o)
and gaseous Xe at T = 165 K (■) and 300 K (□).

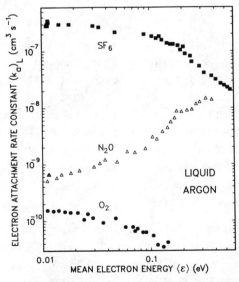

Fig. 5. Rate constant, $(k_a)_L$, for electron attachment to SF_6, N_2O
and O_2 measured in liquid Ar [Bakale et al. (1976)],
plotted versus $\langle\varepsilon\rangle_L$ [using the $\langle\varepsilon\rangle_L$ versus E/N estimates of
Christophorou et al. (1989)].

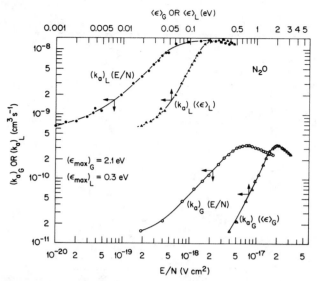

Fig. 6. Electron attachment rate constant for N_2O in gaseous, $(k_a)_G$, and liquid, $(k_a)_L$, argon plotted versus E/N and $\langle\varepsilon\rangle_G$ or $\langle\varepsilon\rangle_L$, see Christophorou (1985). The attachment is due to the reaction $e + N_2O \longrightarrow N_2O^{-*} \longrightarrow O^- + N_2$. Note the shift of the resonance to lower energies— and the increase in the rate constant—in liquid.

For room-temperature dielectric liquids, w(E) has been measured by many authors especially at low E [see Christophorou and Siomos (1984), Christophorou (1988), Schmidt (1984), Freeman (1987), Kunhardt et al. (1988), Allen (1976), Christophorou et al. (1989), Döldissen and Schmidt (1979), Faidas et al. (1989a)]. In Fig. 7 are shown our (Faidas et al., 1989a) recent measurements of w(E) in tetramethylsilane (TMS) and 2,2,4,4-tetramethylpentane (TMP). These measurements extend to $E \geq 1.2 \times 10^5$ V cm^{-1} and were made using a new technique (see below). For both TMS and TMP, the mobility μ (=w/E) decreases with E (Fig. 8) indicating that the $\langle\varepsilon\rangle_L$ exceeds 1.5 kT, see Bakale and Beck (1986).

The technique employed for the w(E) measurements in Fig. 7 is especially suitable for accurate measurements of w(E) in fast dielectric liquids; it is being developed (through the use of a subnanosecond laser pulse) to also measure longitudinal electron diffusion coefficients in dielectric liquids. The principle of the technique is shown in Fig. 9. After extensive purification (Faidas and Christophorou, 1987), the liquid under study was contained in a cell which consisted of a six-way stainless steel cube with two windows for the entry and exit of the laser beam

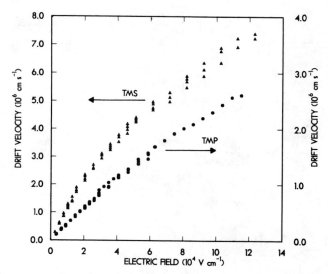

Fig. 7. w versus E for TMS (▲) and TMP (●), Faidas et al. (1989a).
The w(E) values of Faidas et al. (1989a) for TMS are ~ 20
percent higher than those of Döldissen and Schmidt (1979).
See discussion in Faidas et al. (1989a).

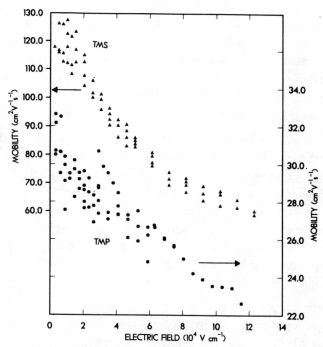

Fig. 8. Mobility of excess electrons versus applied electric field
in TMS (▲) and TMP (●), Faidas et al. (1989a).

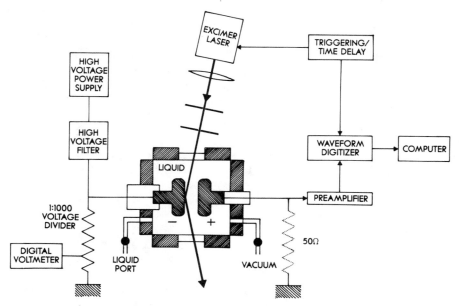

Fig. 9.   Schematic of the experimental set up for the measurement of
w(E) in fast liquids, Faidas et al. (1989a).

and two electrical feedthroughs (one for the high voltage and the other for
the signal) to which the electrodes (two stainless steel parallel-plate
circular disks of 1" in diameter and at a distance of a few mm apart) were
attached.   The beam of an excimer laser (λ = 308 nm, pulse duration ~ 17 ns)
was focused at the center of the negative high voltage electrode in a circular area of ~ 5 mm in diameter.   Voltages in excess of 40 kV could be
applied to the high voltage electrode.

The signal due to the drifting pulse of electrons--generated by the
laser pulse at the cathode--was measured and recorded in two different ways:
(i) In the voltage mode (Fig. 10a) the signal was fed directly to a fast
(response time ~ 3 ns) charge-sensitive preamplifier and was captured,
averaged and stored, by a transient digitizer as a voltage waveform.   (ii)
In the current mode (Fig. 10b) the input of the preamplifier was grounded
through a 50 Ω resistor.   This, in effect, corresponds to electronic dif-
ferentiation of the voltage-mode signal and gives the transient current
waveform.

By numerically differentiating the current-mode signal (Fig. 10c), a
waveform corresponding to the rate of charge injection from the cathode
and collection at the anode is obtained.   The drift time was measured by
determining points A and B on any of the three types of waveforms (voltage, current or charge).

320

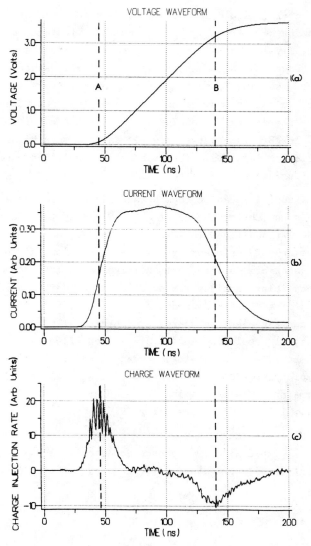

Fig. 10. Voltage (a), current (b), and charge (c) waveforms in TMP.
Drift distance = 2.04 mm, applied voltage = 17,574 V,
drift time = 97.8 ns, see Faidas et al. (1989a).

ULTRAFAST AND ULTRASENSITIVE DIELECTRIC LIQUIDS/MIXTURES

Ultrafast Dielectric Liquids/Mixtures

The fast electron motion in such (cryogenic or room temperature) media finds application in a number of areas such as radiation detectors, see Brassard (1979), Doke (1981), Holroyd and Anderson (1985) and pulsed power switches (Christophorou and Faidas, 1989). In connection with

321

radiation detectors, liquid Ar and liquid Xe are excellent detector media. It can be seen from Fig. 1, for example, that the w in liquid Ar—and especially liquid Xe—is larger than in the corresponding gas over a large range of E/N. This is highly desirable since the magnitude of w determines the time response of radiation detectors and since a large w reduces electron-ion recombination and thus increases the gain of the detector. The saturation of the w versus E/N curve for liquid Xe at comparatively low E/N values (see Fig. 1) and the much higher density of liquid Xe (allowing improved detector spatial resolution) are additional advantages for using Xe as a detector fluid.

The magnitude of w can be considerably increased over a wide E/N range—as in gases (Christophorou et al., 1979)—by the addition of suitable molecular additives to liquid Ar and Xe. Such additives must be nonelectron-attaching, have appropriate cross sections at low energies, and adequate vapor pressure so that additive concentrations of a few percent are possible. Besides enhancing w, such additives would reduce the size and affect the E/N dependence of the electron diffusion coefficient and mean electron energy, which crucially affect the particle detector's spatial resolution and gain and the influence of elec-tronegative impurities. By analogy to gases, small amounts (< few percent) of molecular additives (e.g., CH_4, C_2H_6) were added to liquid Ar and Xe and remarkable increases in w were observed (see Fig. 11 and Yoshino et al., 1976; Shibamura et al., 1975).

Similarly, the large w values for a number of room temperature liquids [Fig. 7; Table 2; Christophorou and Siomos (1984), Christophorou (1988), Schmidt (1984), Freeman (1987), Kunhardt et al. (1988), Holroyd

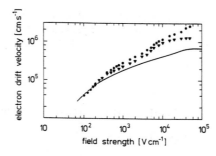

Fig. 11. w versus E in liquid Ar (———) and liquid Ar/CH_4 mixtures; ▼ (N_{CH_4} = 2.6 x 10^{20} molecules cm^{-3}), ●(N_{CH_4} = 6.5 x 10^{20} molecules cm^{-3}). T(Ar) = 87 K, see Yoshino et al. (1976); see this reference for data on other liquid Ar and liquid Xe mixtures).

Table 2. V_o, I_L, and w for Four Fast Room Temperature Liquids

Liquid	$V_o(eV)$	$I_L(eV)$[a]	w (cm s^{-1})[b]
2,2,4,4-Tetramethylpentane	-0.36	8.2	2.4×10^6
Neopentane	-0.40	8.85	$\sim 2.5 \times 10^6$
Tetramethylsilane	-0.57	8.1	6.5×10^6
Tetramethyltin	-0.75	6.9	-

[a]Schmidt (1984), Böttcher and Schmidt (1984) and Buschick and Schmidt (1989).

[b]For $E = 10^5$ V cm^{-1}.

and Anderson (1985)] make such media good candidates for liquid-filled radiation detectors. It is seen from Fig. 7 that the w approaches 10^7 cm s^{-1} at E values of ~10^5 V cm^{-1}. While the w(E) of such room temperature liquids may be further increased by appropriate additives, such w(E) enhancements are not expected to be significant.

Ultrasensitive Dielectric Liquids/Mixtures

In attempting to identify the key physical quantities which determine free electron production (by ionizing particles) and allow the selection and development of "ultrasensitive" liquids, let us refer to the following three simple expressions:

$$G_{fe}^o = P_{esc} G_{te} \tag{1}$$
$$G_{fe}^E = G_{fe}^o + AE \tag{2}$$
$$I_L = I_G + V_o + P^+ \tag{3}$$

In Eq. (1), G_{fe}^o is the yield of free electrons generated in the liquid by the deposition of 100 eV energy by a particular type (e.g., α, β, γ) of ionizing radiation in the absence of an applied electric field (i.e., E = 0); G_{te} is the total electron yield [G_{te} = 100 eV/W, where W is the average energy (in eV) required to produce an electron-ion pair]; and P_{esc} is the escape probability (i.e., the probability that the initial electron-ion pair will separate and not recombine). While $P_{esc} \simeq$ 1 for low-pressure gases (i.e., geminate recombination is unimportant), in liquids normally $P_{esc} \ll 1$ (i.e., most geminate electron-ion pairs recombine with a resultant strong reduction in G_{fe}^o). In Eq. (2), (Freeman, 1987; Onsager, 1938) G_{fe}^E is the free electron yield when an electric field E is applied across the volume in which the electrons are generated; G_{fe}^E exceeds G_{fe}^o by an amount AE where A is a constant that depends on the liquid. In Eq. (3) (see Christophorou and Siomos, 1984;

Christophorou, 1988; Schmidt, 1984; Freeman, 1987; Kunhardt et al., 1988; Faidas et al., 1989b; Faidas and Christophorou, 1988), I_L and I_G are the ionization threshold energies for a species embedded in the liquid and in the low pressure gas, respectively, and P^+ is the polarization energy of the positive ion in the liquid.

The simple expressions (1) to (3) suggest a number of ways of selecting and developing ultrasensitive (large G_{fe}^E) dielectric liquids/mixtures. Clearly, the larger the G_{te} and P_{esc} are, the higher the G_{fe}^E.

Pure Liquids ($V_o < 0$ eV)

The total electron yield G_{te} for pure liquids can be increased if W can be lowered. In view of (3)—and since P^+ and V_o are negative quant-ties—$I_L < I_G$ (see Christophorou and Siomos, 1984; Christophorou, 1988; Schmidt, 1984; Freeman, 1987; Kunhardt et al., 1988; Böttcher and Schmidt, 1984; Buschick and Schmidt, 1989; Faidas et al., 1989b; Faidas and Christophorou, 1988; Tables 2 and 3) and thus, $W_L < W_G$ (Christophorou and Siomos, 1984) and $(G_{te})_L > (G_{te})_G$. This is certainly the case for liquid Ar and Xe (Column 4, Table 3).

While the total electron yield G_{te} can be larger in the liquid than in the corresponding gas, the free electron yield G_{fe}^E is smaller in the liquid—by an amount which depends on the applied electric field—even for the very fast liquids (Column 5, Table 3). This is especially the case for densely ionizing particles (e.g., α-particles; Column 6, Table 3) due to the very low values of P_{esc}. It should be noted that P_{esc} increases with increasing electron drift velocity w; the w—as the electron thermalization length—increases with decreasing electron scattering cross section of the liquid.

For pure liquids ($V_o < 0$ eV), then, a low I_L and large values of w and E are desirable for a large $G_{f}^E e$.

Dielectric Liquids ($V_o < 0$ eV) with Molecular Additives

Clearly the free electron yield can be considerably increased by increasing G_{te} and P_{esc} in liquid rare gas-molecule mixtures. Traces of nonelectron-attaching additives with low I_L in rare gas liquids lower the W_L of the mixture via Penning ionization and other photoionization processes and thus, increase G_{te}. A profound increase in G_{fe}^E can be realized (Fig. 12A, B,; Anderson, 1986; Suzuki et al., 1986) by the use of an additive X to liquid Ar (or Xe) which can absorb efficiently the recombination luminescence which is abundantly produced in rare gas

Table 3. I_L, I_G, and Free Electron Yields for Some Efficient Liquids

Liquid	I_L(eV)	I_G(eV)	Electron Yield (Electrons/100 eV)		
			$G_{te}(e, \gamma)^a$	$G_{fe}^o(e, \gamma)^a$	$G_{fe}^E(\alpha)^b$
Ar	~14.1[c]	15.755[d]	4.4[e] (3.8)$_G$[h]	2.3[f](4.3)[f,g]	0.45[f,g]
Xe	~8.9[c]	12.127[d]	6.5[e] (4.6)$_G$[h]	4.4[i]	–
$C(CH_3)_4$	8.85[j]	10.23[k]	4.30[l]	1.1[f](1.8)[f,g]	0.036[f,g]
$Si(CH_3)_4$	8.1[j]	9.65[k]	–	0.74[f](1.19)[f,g]	0.029[f,g]

[a] For low-ionization density particles (electrons, γ-rays).
[b] For high-ionization density particles (α-particles).
[c] Christophorou and Siomos (1984), p. 307.
[d] Christophorou (1971).
[e] Christophorou and Siomos (1984), p. 304.
[f] Holroyd and Anderson (1985) (T = 296 K).
[g] For E= 10^4 V cm^{-1}, see Holroyd and Anderson (1985).
[h] Gaseous value determined from the W-values in Christophorou (1971).
[i] Huang and Freeman (1977).
[j] From Table 2.
[k] Levin and Lias (1982).
[l] György and Freeman (1987).

liquids, see Doke (1981), and be efficiently ionized. For a liquid Ar, X system these processes can be written as

$$Ar^+ + Ar \longrightarrow Ar_2^+$$
$$Ar_2^+ + e \longrightarrow h\nu_E + Ar + Ar$$
$$h\nu_E + X \xrightarrow{\sigma_{ax}} X^* \xrightarrow{\eta_x} X^+ + e$$

where $h\nu_E$ is the liquid argon excimer (E) recombination luminescence which peaks at ~ 9.55 eV and σ_{ax} and η_x are, respectively, the absorption cross section and ionization efficiency of X^*. To optimize these processes one needs to select an X with an optimum $\sigma_{ax}\eta_x$ product. Such additives (e.g., amines, tetramethylgermanium) (Anderson, 1986; Suzuki et al., 1986) can be added in parts per million levels to binary liquid argon-molecule mixtures already optimized for a maximum electron drift velocity. A large w and a high applied field E will further increase G_{fe}^E by increasing p_{esc} (Fig. 12).

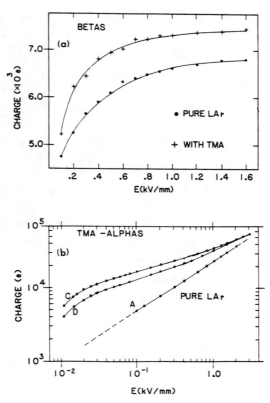

Fig. 12. Charge collected as a function of applied electric field
          for:
          Fig. 12(a): β-particles in pure argon (●) and liquid
                       argon-triethylamine (TEA) mixture.
          Fig. 12(b): α-particles in pure argon (A) and in liquid
                       argon-triethylamine mixtures (Curve C: 4.6 x
                       10^{15} molecules cm^{-3}; Curve D: 1.1 x 10^{15}
                       molecules cm^{-3}). From Anderson (1986); see
                       this reference for details and data on other
                       liquid argon mixtures.

The G_{fe}^{E} of fast room temperature dielectric liquids is limited compared to cryogenic liquid rare gases, especially for densely ionizing
particles (Table 3). For room temperature liquids the electron thermalization length--and hence p_{esc}--are much smaller than for the cryogenic
liquids, internal conversion and dissociation processes rapidly deplete
excited electronic states which might lead to Penning ionization in room
temperature mixtures, and recombination luminescence is too weak or non-
existent to be useful as an additional electron production mechanism in
mixtures. However, it might be possible to improve the efficiency of
fast room temperature dielectric liquids by employing Penning mixtures
where the ionization onset energy, $I_L(X)$, of the impurity X in these fast

liquids is very low [e.g., the I_L of tetrakis(dimethylamino)ethylene (TMAE) in tetramethylsilane has been reported to be 3.54 eV (Nakato et al., 1974) and 3.66 eV (Holroyd et al., 1985) and that of N,N,N', N' tetramethyl-p-phenylenediamine (TMPD) in the same liquid 4.29 eV (Holroyd, 1972) and 4.45 eV (Bullot and Gauthier, 1977)] and by applying very high electric fields.

ACKNOWLEDGMENTS

Research sponsored by the Office of Health and Environmental Research of the U. S. Department of Energy under Contract No. DE-AC05-840R21400 with Martin Marietta Energy Systems, Inc. and by the U. S. Department of Energy and the Office of Naval Research under, respectively, Contracts No. DE-AS05-76ER03956 and N00014-89-J-1990 with the University of Tennessee.

REFERENCES

Allen, A. O., 1976, NSRDS-NBS 58, 1.
Anderson, D. F., 1986, Nucl. Instr. Meth. Phys. Res. A242, 254; ibid, A245, 361.
Bakale, G., and G. Beck, 1986, J. Chem. Phys. 84, 5344.
Bakale, G., U. Sowada, and W. F. Schmidt, 1976, J. Phys. Chem. 80, 2556.
Böttcher, E. -H., and W. F. Schmidt, 1984, J. Chem. Phys. 80, 1353.
Brassard, C., 1979, Nucl. Instr. Meth. 162, 29; J. Engler, and H. Keim, 1984, Nucl. Instr. Meth. Phys. Res. 223, 47.
Bullot, J., and M. Gauthier, 1977, Can. J. Chem. 55, 1821.
Buschick, K., and W. F. Schmidt, 1989, IEEE Trans. Electr. Insul. EI-24, 353.
Christophorou, L. G., 1985, Chem. Phys. Lett. 121, 408.
Christophorou, L. G., 1988, "The Liquid State and Its Electrical Properties," E. E. Kunhardt, L. G. Christophorou, and L. H. Luessen, (Eds.), NATO ASI, Series B, Vol. 193, (Plenum Press), New York, p. 283.
Christophorou, L. G., 1971, "Atomic and Molecular Radiation Physics," Wiley-Interscience, New York.
Christophorou, L. G., and H. Faidas, 1989, Appl. Phys. Lett. 55, 948.
Christophorou, L. G., and K. Siomos, 1984, "Electron-Molecule Interactions and Their Applications," L. G. Christophorou, (Ed.), Academic, New York, Vol. 2, Chapt. 4.
Christophorou, L. G., S. R. Hunter, and J. G. Carter, 1989, Intern. J. Radiat. Phys. Chem. 34, 819.
Christophorou, L. G., D. L. McCorkle, and A. A. Christodoulides, 1984, "Electron-Molecule Interactions and Their Applications," L. G. Christophorou, (Ed.), Academic, New York, Vol. 1., Chapt. 6.
Christophorou, L. G., D. L. McCorkle, D. V. Maxey, and J. G. Carter, 1979, Nucl. Instr. Meth. 163, 141; L. G. Christophorou, D. V. Maxey, D. L. McCorkle, and J. G. Carter, 1980, Nucl. Instr. Meth. 171, 491.
Doke, T., 1981, Portgal. Phys. 12, 9.
Döldissen, W., and W. F. Schmidt, 1979, Chem. Phys. Lett. 68, 527.
Faidas, H., and L. G. Christophorou, 1987, J. Chem. Phys. 86, 2505.
Faidas, H., and L. G. Christophorou, 1988, Rad. Phys. Chem. 32, 433.
Faidas, H., L. G. Christophorou, and D. L. McCorkle, 1989a, Chem. Phys. Lett., 163, 495.
Faidas, H., L. G. Christophorou, P. G. Datskos, and D. L. McCorkle, 1989b, J. Chem. Phys. 90, 6619.
Freeman, G. R., (Ed.), 1987, "Kinetics of Nonhomogeneous Processes," Wiley-Interscience, New York.

Gushchin, E. M., A. A. Kruglov, and I. M. Obodovskii, 1982, Sov. Phys. JETP 55, 650.

György, I., and G. R. Freeman, 1987, J. Chem. Phys. 86, 681.

Holroyd, R. A., 1972, J. Chem. Phys. 57, 3007.

Holroyd, R. A., and D. F. Anderson, 1985, Nucl. Instr. Meth. Phys. Res. A 236, 294.

Holroyd, R. A., S. Ehrenson, and J. M. Preses, 1985, J. Phys. Chem. 89, 4244.

Huang, S. S., and G. R. Freeman, 1977, Can. J. Chem. 55, 1838.

Hunter, S. R., J. G. Carter, and L. G. Christophorou, 1988, Phys. Rev. A 38, 5539.

Kubota, S., T. Takahashi, and J. Ruangen, 1982, J. Phys. Soc. Jpn. 51, 3274.

Kunhardt, E. E., L. G. Christophorou, and L. H. Luessen, (Eds.), 1988, "The Liquid State and Its Electrical Properties," NATO ASI, Series B, Vol. 193, (Plenum Press), New York.

Lekner, J., 1967, Phys. Rev. 158, 130.

Levin, R. D., and S. G. Lias, 1982, "Ionization Potential and Appearance Potential Measurements, 1971-1981," U. S. Department of Commerce, NSRDS-NBS-71, NBS.

Miller, L. S., S. Howe, and W. E. Spear, 1968, Phys. Rev. 166, 871.

Nakamura, S., Y. Sakai, and H. Tagashira, 1986, Chem. Phys. Lett. 130, 551; Y. Sakai, S. Nakamura, and H. Tagashira, 1985, IEEE Trans. Electr. Insul. EI-20, 133.

Nakato, Y., T. Chiyoda, and H. Tsubomura, 1974, Bull. Chem. Soc. Jpn. 47, 3001.

Onsager, L., 1938, Phys. Rev. 54, 554.

Robertson, A. G., 1977, Austr. J. Phys. 30, 39.

Schmidt, W. F., 1984, IEEE Trans. Electr. Insul. EI-19, 389.

Shibamura, E., A. Hitachi, T. Doke, T. Takahashi, S. Kubota, and M. Miyajima, 1975, Nucl. Instr. Meth. 131, 249.

Shibamura, E., T. Takahashi, S. Kubota, and T. Doke, 1979, Phys. Rev. A 20, 2547.

Suzuki, S., T. Doke, A. Hitachi, J. Kikuchi, A. Yunoki, and K. Masuda, 1986, Nucl. Instr. Meth. Phys. Res. A 245, 366.

Yoshino, K., U. Sowada, and W. F. Schmidt, 1976, Phys. Rev. A 14, 438.

DIFFUSION OF ELECTRONS IN A CONSTANT FIELD: STEADY STREAM ANALYSIS

J. H. Ingold

GE Lighting
Cleveland, OH 44112

INTRODUCTION

Diffusion theory of electrons in a drift tube is extended to satisfy
approximate energy balance as the electrons migrate radially and axially in
a steady, uniform electric field. Previously, it was assumed that the characteristic energy is independent of position in the drift region, leading to
the concept of different diffusion coefficients in the radial and axial
directions. However, this assumption is inconsistent with electron energy
balance. In the present paper, it is shown that different transverse and
longitudinal diffusion coefficients are not needed when energy is conserved,
as the electrons migrate radially and axially from a point source. It is
also shown that diffusion theory with energy balance gives results similar
to those obtained from exact solution to the Boltzmann equation by Parker
(1963). In addition, it is shown that the heretofore empirical relation
between characteristic energy and drift tube geometry results naturally from
theoretical analysis.

THEORY

It can be shown that the continuity equation, or diffusion equation,
for electrons of density $n(\rho,z)$ streaming in an electric field $\mathbf{E} = -E\mathbf{k}$ is

$$\nabla \cdot \mathbf{\Gamma} = 0; \quad \mathbf{\Gamma} = -\nabla(nD) - \mu\mathbf{E}n \tag{1}$$

where D and μ are diffusion and mobility coefficients, respectively.
Consistent with Eq. (1), the energy balance equation for the streaming
electrons is

$$\nabla \cdot \mathbf{H} + \mathbf{\Gamma} \cdot \mathbf{E} = -3 \frac{m}{M} \nu_m n\theta; \quad \mathbf{H} = -\nabla \left(\frac{5}{2} n\theta D\right) - \frac{5}{2} \mu\mathbf{E}n\theta \tag{2}$$

where m is electron mass, M is atomic mass, ν_m is momentum transfer collision frequency, and $\frac{3}{2}\Theta(\rho,z)$ is average energy. By assuming $\nabla \cdot \mathbf{H} \ll \Gamma \cdot \mathbf{E}$, and by substituting $p = n\Theta$, $p_q = n\Theta_q$, $r = \frac{W}{D_q}\rho$, $s = \frac{W}{D_q}z$, where $W \equiv \mu_q E$ and the subscript "q" means "equilibrium value." Eqs. (1) and (2) become respectively,

$$\nabla^2 p - \frac{\delta p_q}{\delta s} = 0; \quad -\frac{\delta p}{\delta s} + p_q = p \rightarrow \frac{1}{r}\frac{\delta}{\delta r}\left[r\frac{\delta p}{\delta r}\right] = \frac{\delta p}{\delta s}; \quad p_q = p + \frac{\delta p}{\delta s} .$$

These equations have the solutions

$$p(r,s) = \frac{I_0 \alpha^2}{2\pi e}\frac{\Theta_q}{\mu E}\frac{1}{2s}\exp\left(-\frac{r^2}{4s}\right); \quad p_q(r,s) = p(r,s)\left[1 + \frac{r^2}{4s^2} - \frac{1}{s}\right]$$

for an unbounded gaseous system with a point source of current I_0 at $(0,0)$. The corresponding equations for n and Θ are

$$n = \frac{p}{\Theta_q}\left[1 + \frac{r^2}{4s^2} - \frac{1}{s}\right]; \quad \Theta = \frac{p}{n} = \Theta_q\left[1 + \frac{r^2}{4s^2} - \frac{1}{s}\right]^{-1} \tag{3}$$

The current I_{bh} to a disc of radius b centered on the perpendicular to the z-axis at $z = h$ is

$$I_{bh} = I_0\left[1 - \exp\left(-\frac{Wb^2}{4D_q h}\right)\right]$$

The ratio $R \equiv I_{bh}/I_0$ is the fraction of current collected by the disc:

$$R = 1 - \exp\left(-\frac{\lambda b^2}{2h}\right) \tag{4}$$

where $\lambda \equiv W/2D_q$. Eq. (4) predicts the D_q can be measured by measuring the fraction of current collected by a disc of radius b located a distance h from the point source. Eq. (4) is for a constant collision frequency $\nu(\varepsilon)$ independent of energy ε. When $\nu(\varepsilon) \propto \varepsilon^{\frac{1+1}{2}}$, the corresponding result for the current ratio is

$$R = 1 - \frac{hd_1(d_1 + h) - f^2 b^2/\lambda}{2d_1^3}\exp[-\lambda(d_1 - h)/f] \tag{5}$$

where $d_1 \equiv \sqrt{h^2 + fb^2}$ and $f \equiv \frac{1+1}{1+2}$.

DISCUSSION

Equations 3 predict that Θ, and, therefore, average velocity v_z, are in equilibrium with the field when $s \gg 1$. Comparison of Eqs. (3) with the

330

corresponding equations of Parker shows certain similarities, especially between the equations for Θ. However, Eqs. (3) satisfy Eq. (1) with $D = \mu\Theta$, whereas Parker's equations do not.

An empirical equation similar to Eq. (5) has been used for years to interpret measurements of D_q. See Huxley and Crompton (1974), for a thorough review of theory and comprehensive compilation of data.

CONCLUSION

It is concluded that the heretofore empirical relation between D_q and drift tube geometry results naturally from theoretical analysis when electron energy is conserved.

REFERENCES

Huxley, L. G. H., and R. W. Crompton, 1974, "The Diffusion and Drift of
    Electrons in Gases," (Wiley, New York).
Parker, Jr., J. H., 1963, Phys. Rev. 132, 2096.

DIFFUSION OF ELECTRONS IN A CONSTANT FIELD: TOF ANALYSIS

J. H. Ingold

GE Lighting
Cleveland, OH 44112

INTRODUCTION

Time-of-flight measurement of drift velocity W and diffusion coefficient
D is based on the assumption that when a short pulse of electrons from a point
source is injected into a region of constant electric field, then (a), the
center-of-mass of the pulse travels a distance h in time t_m, where $W = h/t_m$,
and (b), the pulse half-width δt, where $t_m + \delta t$ is the time at which the
intensity of the pulse has fallen to $1/e$ of its maximum value at t_m, is
related to D by an equation such as

$$D = \frac{h^2 \delta t^2}{4 t_m^3} \qquad (1)$$

These relations for W and D are based on the solution

$$I(h,t) \; \alpha \; \frac{1}{\sqrt{4 \pi D t}} \; \exp \left[- \frac{(Wt - h)^2}{4Dt} \right] \qquad (2)$$

of the time-dependent diffusion equation given in Eq. (3) below with constant
D and μ. While measurement of W by this method gives accepted values,
measurement of D by this method generally gives values smaller (Wagner et al.,
1967) than given by the method of lateral diffusion. In other words, values
of D determined from experimental values of e-folding time for decay of a
longitudinal pulse according to Eq. (1) are smaller than those determined from
experimental values of the amount of transverse spreading of a steady stream.
This difference inspired development of the concept of anisotropic diffusion
(Parker and Lowke, 1969; Lowke and Parker, 1969). A review of the theory of
anisotropic diffusion, based on the so-called density gradient expansion
technique, is given in Huxley and Crompton (1974).

Nonequilibrium Effects in Ion and Electron Transport
Edited by J. W. Gallagher *et al.*, Plenum Press, New York, 1990

333

In the present paper, a different approach to TOF analysis is described. This approach is called diffusion theory with energy balance. Diffusion theory of electrons in a constant electric field is based on energy moments of the scalar equation for f_0, the isotropic part of the energy distribution, which is found by the spherical harmonic method of solving the Boltzmann equation. The zeroth-energy moment of the scalar equation gives the particle balance equation, or diffusion equation, and the first energy moment gives the energy balance equation. These two equations are solved numerically in time and one space dimension along the electric field, including non-equilibrium regions near the electrodes, with a pulsed source of electrons at the incoming plane and a perfect sink at the outgoing plane. Calculated pulse widths for constant cross section are significantly less than those obtained from solution of the diffusion equation alone, using equilibrium values of the diffusion coefficient D and drift velocity W. These results suggest that TOF analysis based on Eq. (2) is inadequate, and that the concept of anisotropic diffusion is superfluous when energy balance is satisfied, i.e., when both density and average energy are allowed to vary in time and space.

THEORY

The zeroth moment of the scalar equation for f_0 gives the continuity equation, or the diffusion equation, for electrons of density n streaming with current density Γ in an electric field $\mathbf{E} = -E\mathbf{k}$:

$$\frac{\delta n}{\delta t} = - \nabla \cdot \Gamma; \quad \Gamma = - \nabla(nD) - \mu E n \tag{3}$$

where D and μ are diffusion and mobility coefficients given by

$$D = \frac{2e}{3m} \int_0^\infty \frac{\varepsilon^{3/2} f_0}{\nu(\varepsilon)} d\varepsilon; \quad \mu = - \frac{2e}{3m} \int_0^\infty \frac{\varepsilon^{3/2}}{\nu(\varepsilon)} \frac{\delta f_0}{\delta \varepsilon} d\varepsilon \tag{4}$$

in which m is electronic mass, e is electronic charge, $\nu(\varepsilon)$ is electron-neutral momentum transfer collision frequency, and ε is electron energy.

Likewise, the first energy moment of the scalar equation for f_0 gives the energy balance equation for electrons streaming with heat flow $\mathbf{H}$:

$$\frac{\delta}{\delta t}(n\bar{\varepsilon}) = - \nabla \cdot \mathbf{H} - \Gamma \cdot \mathbf{E} - 2 \frac{m}{M} \nu_\varepsilon n\bar{\varepsilon}; \quad \mathbf{H} = - \nabla(nG) - BEn \tag{5}$$

where M is atomic mass, ν_ε is average energy transfer collision frequency, and it is assumed that average energy $\bar{\varepsilon} \gg \frac{3}{2}kT_g$, where T_g is gas temperature. The quantities G and B are additional transport coefficients given by

$$G = \frac{2e}{3m} \int_0^\infty \frac{\varepsilon^{5/2} f_0}{\nu(\varepsilon)} d\varepsilon; \quad B = - \frac{2e}{3m} \int_0^\infty \frac{\varepsilon^{5/2}}{\nu(\varepsilon)} \frac{\delta f_0}{\delta \varepsilon} d\varepsilon \tag{6}$$

RESULTS

Equations (3) and (5) are solved by finite difference technique in time and one space dimension along the electric field, including non-equilibrium regions near the electrodes, for a pulsed source of electrons at the incoming plane and a perfect sink at the outgoing plane. Numerical solutions for both constant collision frequency and constant cross section are presented. These solutions are compared with numerical solution of Eq. (3) alone for constant average energy and the same equilibrium values of diffusion coefficient D and drift velocity W. The following results are presented:

1. Numerical solutions of Eq. (3) for constant D and W agree closely with the analytic solution given by Eq. (2), giving confidence that the finite difference technique used to solve Eqs. (3) and (5) is accurate.

2. Numerical solutions of coupled Eqs. (3) and (5) for constant collision frequency give pulse widths similar to those obtained from Eq. (3) alone, using equilibrium values of D and μ.

3. Numerical solutions of coupled Eqs. (3) and (5) for constant cross section give pulse widths which are significantly smaller than those obtained from Eq. 3 alone, using equilibrium values of D and μ.

CONCLUSION

These results suggest that TOF analysis based on Eq. (2) is inadequate, and that the concept of anisotropic diffusion is superfluous when energy balance is satisfied, i.e., when both density and average energy are allowed to vary in time and space.

REFERENCES

Huxley, L. G. H., and R. W. Crompton, 1974, "The Diffusion and Drift of Electrons in Gases," (Wiley, New York).
Lowke, J. J., and J. H. Parker, 1969, Phys. Rev. 181, 302.
Parker, J. H., and J. J. Lowke, 1969, Phys. Rev. 181, 290.
Wagner, E. B., F. J. Davis, and G. S. Hurst, 1967, J. Chem. Phys. 47, 3138.

A MONTE CARLO SIMULATION OF ELECTRON DRIFT LIMITED BY COLLISIONS IN GAS
MIXTURES USING THE NULL COLLISION METHOD[*]

David Ramos[a], Edward Patrick[a], Douglas Abner[a],
Merrill Andrews[a] and Alan Garscadden[b]

[a] Wright State University
[b] Air Force Write Research and Development Center

The Monte Carlo method is explored as an alternative to expansion
solutions of the Boltzmann transport equation for determining electron swarm
parameters. This method uses the probability P that the time of flight of
an electron is less or equal to some time T, and the time T. The null
collision method proposed by Skullerud (1968) and developed in detail by
Reid (1979) takes advantage of the simple relationship between the time of
flight T and the probability of having a collision P for a cross section
that is inversely proportional to the velocity. A null collision cross
section is added to real cross sections so as to impose this relation. Once
the time of flight is established, then the position and the energy of the
electron are also defined. A new random number R is generated to determine
which type of collision event takes place. When two gases are present in
the simulation the following conditions determine the type of collision that
takes place. If

$$R > [p_1 \cdot \sigma_1(v)_{total} + p_2 \cdot \sigma_2(v)_{total}] / \sigma(v)_{null+total} = v \cdot \sigma(v)_{total}/k,$$

then a null collision occurs. The p_i's are the partial pressures of gases
involved, $\sigma_i(v)_{total}$'s are their respective total cross sections including
the elastic and inelastic and $\sigma(v)_{null+total} = k/v$ is the null plus total
cross section. In a null collision the energy and the momentum of the
electron remain unaltered, so that its motion is not affected. The null
cross section technique only influences the computation of collision times,
which it considerably simplifies. If the former condition is not satisfied
and we have (do not have)

---

[*] Supported by Air Force Contract F33615-86-C-2720 through SCEEE.

$$R < [p_1 \cdot \sigma_1(v)_{total}] / \sigma(v)_{null+total} = v \cdot p_1 \cdot \sigma_1(v)_{total} / k,$$

the collision is with the first (second) gas. If after generating a new random number we have

$$R < \sigma_i(v)_{elastic} / \sigma(v)_{null+total} = v \cdot \sigma_i(v)_{elastic} / k \quad ; \quad i = 1,2$$

the collision is taken to be elastic; otherwise it is taken to be inelastic. This procedure can be generalized to include additional inelastic cross sections.

The efficiency of this method depends on the amount of null collisions, which can be reduced by dividing the energy range into two or more regions, each with a different total cross section (different k's). This minimizes the increase in null collision as v approaches zero or in a Ramsauer region. When more than one null cross section is used, close attention must be paid to the vicinity of the transition energy. In our simulation we adapted the method used by Maratz (1986) to update the distribution function by the time between collisions. A detailed study reveals that the distribution function is extremely susceptible to developing jump discontinues or spikes unless the updating techniques are carefully refined.

Negative Differential Conductivity (NDC) is usually observed in gas mixtures where one of the gases has a Ramsauer-Townsend minimum and the other has a large inelastic cross section at a slightly larger energy. The application of this modeling technique demonstrates whether or not proposed explanations consistent with the Boltzmann analysis are consistent with the Monte Carlo analysis. The analysis is compared with new experimental data on argon-molecular gas mixtures.

REFERENCES

Maratz, T., 1986, "Simulations of Non-Hydrodynamic Behavior in Gas Discharges," M.S. thesis, University of Pittsburgh.
Reid, I. D., 1979, Aust. J. Phys. 32, 231-54.
Skullerud, H. R., 1968, J. Phys. D 1, 1567-68.

AN ANALYSIS OF TRANSIENT VELOCITY DISTRIBUTION OF ELECTRONS

N. Ikuta, S. Nakajima and M. Fukutoku

Faculty of Engineering
Tokushima University
Tokushima 770, Japan

Despite the remarkable progress of the electron transport analysis in recent years, available methods for the transient analysis seems very few. Kitamori et al. (1980) presented a rigorous method of transient analysis of electron transport "direct estimation of moment" method. However, the procedure is rather complicated and is not easy to utilize it commonly. We used a simple procedure of analysis for the transient transport property of electrons. The procedure is like a simulation of practical behavior of electrons carrying cascading flights under the influence of an electric field and collisions with gas atoms.

We consider here a Boltzmann equation for electrons under conservative conditions assuming the uniform density in real space as the first step.

$$[\partial / \partial t + \mathbf{a} \cdot \partial / \partial \mathbf{v}]f(\mathbf{v},t) = \partial f/\partial t|_{in} - \partial f/\partial t|_{out} \qquad (1)$$

$$= \int \Omega\ (\mathbf{v}' \longrightarrow \mathbf{v})f(\mathbf{v}',t)d\mathbf{v}' - \nu(v)f(\mathbf{v},t)$$

Here, $f(\mathbf{v},t)$ is the density expression of electron velocity distribution function at a small volume element in velocity space $\mathbf{v}d\mathbf{v}$, $\mathbf{a}$ the acceleration rate by the field, $\partial f/\partial t|_{in}$ and $\partial f/\partial t|_{out}$ the rates of electrons incoming and outgoing the volume element $d\mathbf{v}$ and $\Omega\ (\mathbf{v}' \longrightarrow \mathbf{v})$ being an operator of velocity transfer from $\mathbf{v}'d\mathbf{v}'$ to $\mathbf{v}d\mathbf{v}$ by collision, respectively. Electrons at $\mathbf{v}'d\mathbf{v}'$ of $\nu(\mathbf{v}')f(\mathbf{v}',t_n)d\mathbf{v}'$ are lost by collision in dt with azimuthal symmetry and are injected isotropically in lower velocity (energy) element $\mathbf{v}d\mathbf{v}$ considering the kind of collision. These processes of acceleration, ejection and injection of electrons are continued successively by every dt. Thus, the time evolution of the velocity distribution and the transport coefficients defined in velocity space are easily obtained. The starting and colliding rate distributions

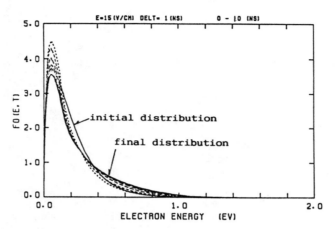

Fig. 1. Relaxation of the velocity distribution from 0 to 10 ns. (in SiH$_4$ at 15 Td).

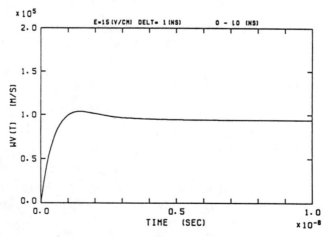

Fig. 2. Transient variation of the drift velocity Wv (0 to 10 ns). (in SiH$_4$ at 15 Td).

$\Psi_s(v_0, t)$ (isotropic) and $\Psi_c(v', t)$ introduced in the "flight time integral" method (Ikuta and Murakami, 1987) are fully obtained in addition to the usual velocity distribution $F(v, t)$ through the passage of time via the transient period. In this calculation, no expansion and no truncation of the velocity distribution are adopted. Polar and cylindrical coordinates in velocity space appropriately divided in v, Θ and in v_x, v_r domains are used. The accuracy of this analysis depends only on the numerical treatments of the velocity shift by the field and of ejection and injection of electrons through collisions. The calculation of the collision rate may be carried more conveniently by using electron energy ε instead of v. We have obtained a good convergence of the velocity distribution and the transport coefficients to those at equilibrium state. Examples of the results are shown in Figs. 1 and 2. In addition, the aim of this research is the analysis of electron behavior under radio frequency electric field at final step.

REFERENCES

Ikuta, N., and Y. Murakami, 1987, J. Phys. Soc. Jpn. 56, 115.
Kitamori, K., H. Tagashira, and Y. Sakai, 1980, J. Phys. D 13, 535.

AN EXACT THEORY FOR THE TRANSIENT BEHAVIOR OF ELECTRON SWARM PARAMETERS[*]

P. J. Drallos and J. M. Wadehra

Department of Physics
Wayne State University
Detroit, MI 48202

We have developed (Drallos and Wadehra, 1988; 1989) a simple, yet exact numerical technique for obtaining time-dependent electron velocity distribution functions (EVDF) from which the transient behavior of various electron swarm parameters can be obtained. Our calculations predict overshoots in the electron drift velocity when the initial average energy of the electron swarm is near or below the final equilibrium average energy of the swarm. Less dramatic overshoots (and undershoots) of the ionization rate and the average energy have also been observed in the calculations when the above conditions were present. Recent experimentally observed (Duffy and Ingold, 1988; Verdeyen et al., 1988) current overshoots in Ar-Hg and Ne-Hg discharges and ionization rate overshoots in N_2 discharges are presumably related to the initial conditions of the electron velocity distribution function. Analysis of the swarm data has shown that, for a given E/N, the time dependence of various swarm parameters can be accurately fitted by a sum of two exponential terms. For a given gas (Ar or Ne, in our case), the decay constants in the exponentials appear to depend only on the value of E/N.

REFERENCES

Drallos, P. J., and J. M. Wadehra, 1988, J. Appl. Phys. 63, 5601.
Drallos, P. J., and J. M. Wadehra, 1989, (to be published).
Duffy, M. E., and J. H. Ingold, 1988, Abstract H-8, 41st Annual Gaseous Electronics Conference, Minneapolis, Minnesota.
Verdeyen, J. T., L. C. Pitchford, Y. M. Li, J. B. Gerardo, and G. N. Hays, 1988, Abstract N-3, 41st Annual Gaseous Electronics Conference, Minneapolis, Minnesota.

---

[*] The support of the Air Force Office of Scientific Research through Grant Number AFOSR-87-03042 is gratefully acknowledged.

ELECTRON TRANSPORT PROPERTY UNDER ELECTRIC AND MAGNETIC FIELDS CALCULATED
BY THE FTI METHOD

N. Ikuta and Y. Sugai

Faculty of Engineering
Tokushima University
Tokushima 770, Japan

Electron transport behavior in a model gas under electric and
magnetic fields is precisely calculated using "flight time integral"
(FTI) method (Ikuta and Murakami, 1987) developed in our laboratory. A
model gas of constant cross section (elastic) with density N, where the
collision probability $Nq(\varepsilon)$ of 10 $[cm^{-1}]$ and the mass ratio between an
electron and gas atom of 0.01 are used. An electric field E of 1 V cm^{-1}
is applied in the -x direction, and the magnetic field B is applied in -x
(parallel) and in -y (perpendicular) directions with values changing up
to 50 Gauss.

In the parallel field, electrons started at the angle Θ_0 from x(-E)
axis fly along screw trajectories of varying pitch with radius r_L =
$mv_0 \sin \Theta_0/eB$ and with Larmor frequency ω_L = eB/m. Since the flight path
length and the displacement in the axial(x) direction does not change by
applying B, the electron velocity (energy) distribution in the form of
Druyvestein is maintained unchanged regardless of the value of B.
Accordingly, transport coefficients, except the transverse diffusion
coefficient D_T, are kept constant. However, due to the fact that the
radial displacement is limited to less than the diameter of Larmor
motion, D_T is decreased in proportion to B^{-2} above a critical value B_C =
$m/e\langle\bar{t}\rangle$, where $\langle\bar{t}\rangle$ is the time average of the flight time (Ikuta et al.,
1985) of electrons.

In crossed fields, electron trajectories projected on the x-z plane
draw trochoidal curves along the z axis. Then, the drift velocity vector
W rotates from x to z(= E X B) direction with increasing B above B_C, the
same as in parallel field, decreasing its magnitude as is seen in
Figure 1. Accordingly, the velocity distribution is suppressed to lower
range by increasing B. However, the variation of W does not agree with

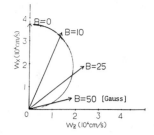

Fig. 1. Rotation of drift velocity vector by increasing B.

that given by the theory hitherto presented. In this situation, it is noted
also that the azimuthal angle symmetry along the direction of W can not be held
and the Legendre expansion loses its validity. Then the velocity distribution
has to be written only as $F(v, \Theta, \phi)$. The diffusion coefficient in each
direction decreases anisotropically with the increase of B above B_C as are seen
in Fig. 2. It is confirmed that the similarity law holds under constant E/N and
B/N conditions, where $F(v, \Theta, \phi)$, W, ND_S, B_C/N are held unchanged.

These results are obtained by using the FTI method. The FTI method can
provide detailed data concerning the flight behavior of electrons in steady
state with the aid of normalized starting and colliding rate distributions and
various transport functions. That is, the velocity distribution obtained by
the FTI procedure has a thickness of time behind it, and the flight behavior
of electrons is precisely described due to this thickness of time even if the
flight path is very complicated such as that in crossed fields (Ikuta and
Sugai, 1989).

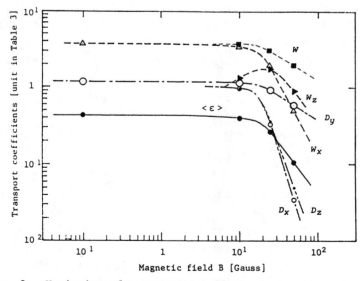

Fig. 2. Variation of transport coefficients for the value B.

REFERENCES

Ikuta, N., H. Itoh, N. Okano, and K. Yamamoto, 1985, J. Phys. Soc. Jpn.
    54, 2485-2493.
Ikuta, N., and Y. Murakami, 1987, J. Phys. Soc. Jpn. 56, 115-127.
Ikuta, N., and Y. Sugai, 1989, J. Phys. Soc. Jpn. 58 (in press).

A MULTIGROUP APPROACH TO ELECTRON KINETICS

S. Clark and E. E. Kunhardt

Weber Research Institute
Polytechnic University
Route 110, Farmingdale, NY 11735

Given initial and/or boundary conditions, the behavior of an assembly of electrons in a gas can be obtained from solution of the kinetic equation for the distribution function, $f(\underline{v},\underline{r},t)$. By expanding the distribution in terms of localized functions in v-space, an equivalent formulation can be obtained in terms of the expansion coefficients. For a set of modulated gaussian functions, the expansion coefficients are proportional to the density of "electron groups" associated with the localized functions. Equations of evolution for these coefficients are explicit derived. With this formulation, the dynamical behavior of the electrons in various regions of velocity space and the influence of the scattering process on these dynamics can be elucidated. This approach is illustrated by numerically solving the initial-value problem for the amplitude equations.

Work supported by the Office of Naval Research.

INTEGRAL EXPANSION OFTEN REDUCING TO THE DENSITY GRADIENT EXPANSION, EXTENDED TO NONMARKOV STOCHASTIC PROCESSES. CONSEQUENT STOCHASTIC EQUATION FOR QUANTUM MECHANICS MORE REFINED THAN SCHRODINGER'S.

G. Cavalleri and G. Mauri

Dipartimento de Matematica dell'Università Cattolica
Brescia, Italy

An integral expansion is obtained which reduces under explicitly given conditions to the density gradient expansion for the number density $\rho(\underline{r},t)$ of stochastic particles. Explicit coefficients in terms of moments are calculated up to and including the fourth order, corresponding to the super-Burnett approximation. The expansion is proved to be valid for both Markov and nonMarkov processes, including those with infinite memory, typical of a stochastic motion with inertia. An application of this expansion is the relation between local average velocities in Markovian stochastic processes and the drift velocity of a probability cloud. In a nonMarkovian stochastic process with inertia the average local velocity $\langle \underline{v} \rangle$ of a spherical cell having a radius equal to the mean free path λ depends on all the preceding history and on the local diffusion velocity $\underline{v}_D$.

In the adjacent spherical cell of radius λ, the diffusion velocity is preserved and another, almost equal, effect doubles the $\underline{v}_D$ value. The average velocity of the two adjacent cells is $\langle \underline{v} \rangle + \underline{v}_D/2$ so that there is a velocity $\pm \underline{v}_D/2$ in the two cells relative to their center of mass. The relevant kinetic energy has to enter the expression of the total energy which is conserved in a nonMarkovian process with inertia. From this condition and the possibility of expressing the local velocity $\langle \underline{v} \rangle$ by a velocity potential ϕ, i.e., $\langle \underline{v} \rangle = \nabla \phi$ (because there is no friction in a motion with inertia), a system of two hydrodynamic equations in ρ and ϕ is derived. They are just equal to Schrodinger's hydrodynamic equation if the diffusion velocity is expressed by the first term only of the gradient expansion, i.e., $\rho \underline{v}_D = -D_2 \nabla \rho$ in an isotropic process. Here $D_2 = h/m$ is the 'inertial spreading coefficient' (which is equal to the usual transversal diffusion coefficient in a Markovian process only). In this case the two hydrodynamic equations are summarized by Schrodinger's single, complex equation in the wave function ψ related to ρ and ϕ by $\psi = \rho^{1/2} \exp (i \phi/2 D_2)$.

Nonequilibrium Effects in Ion and Electron Transport
Edited by J. W. Gallagher *et al.*, Plenum Press, New York, 1990

In an isotropic process the diffusion current relative to the center of mass of two little spheres of radius equal to the mean free path λ of the equivalent random walk turns out to be given by

$$\rho \ \underline{v}_D = -D_2 \ \nabla \ \rho - D_4 \ \nabla \ (\nabla^2 \ \rho).$$

If the second term on the right-hand side of the above equation is retained, a quantum equation more refined than Schrodinger's is obtained.

The correction terms (with respect to the values obtainable from the Schrodinger equation) are of order α^2 (where α is the fine-structure constant) for the stable states.

The correction terms should be greater for scattering problems and in particular when inelastic ro-vibrational cross-sections are calculated. It is known that there are unexplainable differences for these cross-sections between the experimental values obtained by the most reliable and best group led by R.H. Crompton and the theoretical values calculated by M.A. Morrison and co-workers. There is therefore the hope that our correction terms may explain the above differences, and we leave the difficult task of calculating them by our new equation to the very skillful theoretical group of M. A. Morrison.

352

GENERALIZED DIFFUSION COEFFICIENTS AND 1/f POWER SPECTRAL NOISE

G. Cavalleri and G. Mauri

Dipartimento di Matematicia dell'
Università Cattolica
Brescia, Italy

A generic stochastic process can always be represented by a random walk of a particle having velocity v during a free flight between two successive collisions. Let $\nu_0(v)$ be the collision frequency and ν_0^{-1} the average transit time. The result of a random walk with a large number of free flights in a Gaussian process characterized by a power spectral density $D(\omega)$ where $\omega = 2\pi f$ is the angular frequency. It is generally believed that the $1/\omega$ noise cannot be obtained by the usual transport theory down to $\omega \ll \nu_0$. For instance, the $1/\omega$ noise in electronic devices has been measured down to $\omega \simeq 10^{-7}$ while $\nu_0 \simeq 10^{13}$ and this implies that the memory of a fluctuation is remembered for 10^{20} free flights. Notice that in the case the <u>direction</u> of an electron velocity is forgotten after a single flight and the <u>speed</u> (or velocity amplitude) after $F\,\nu_0^{-1}\,m/M < 10^{-7}$ s, where m and M are the masses of an electron and of a scattering center, respectively, and F is a coefficient usually of the order of unity. This is true if in any (however small) v range it is $\nu_0 \propto v^n$ with $n \neq -1$. But $F = F(n)$ and it is shown in the following that $F \to \infty$ when $n \to -1$, so that the memory of a fluctuation becomes infinite and unrelated with ν_0^{-1}. To obtain this result two difficulties have been overcome. The first difficulty is that we need the asymptotic behavior of the correlation function which is rapidly varying for short times, so that the solution of the Boltzmann equation would require a very large number of terms in the Legendre expansion of the velocity distribution function $f(v,t)$. Nobody has ever solved it analytically and with time dependence even with only four terms. This difficulty has been overcome by the mean free path method which allows a separation between short times (where f is rapidly varying) and long times (where f is slowly varying) for which a two term expansion is sufficient (Fokker – Planck approximation). The second difficulty concerns the time-dependent solution of the Boltzmann equation which is very hard to be obtained when $n < 0$ in $\nu_0 \propto v^n$, even with only a two terms expansion. This has been overcome by Stenflo in a rigorous way just with n = -1 in an important paper which apparently passed unnoticed to the people working in the field of $1/\omega$ noise.

The value n = -1 is just at the threshold of the runaway regime, and the physical origin of the 'true' $1/\omega$ noise (i.e., down to $\omega \rightarrow 0$) is here ascribed to the presence of at least a small v range of the equivalent random walk where $\nu_0 = A/v$. Notice that for n = -1 the longitudinal diffusion coefficient $D_x = D\ (\omega{=}0)$ diverges, and this is a necessary condition to have $D_x\ (\omega) \propto \omega^{-1}$ which diverges for $\omega \rightarrow 0$. The origin of a small v range where $\nu_0 \propto v^{-1}$ is ascribed to the Coulomb scattering centers, present in ionized gases and in semiconductors.

FOKKER-PLANCK CALCULATION OF THE ELECTRON SWARM ENERGY DISTRIBUTION FUNCTION

N. J. Carron

Mission Research Corporation
Santa Barbara, CA 93102

An approximate treatment of the Boltzmann collision integral for electrons in a gas, valid for small fractional average energy loss per energy
transfer collision, is presented and studied. It is essentially a Fokker-
Planck expansion in energy space, including mean energy loss (dynamical
friction) and energy straggling (coefficient of diffusion). When applied to
electron swarms in weakly ionized gases, treating angle variables in the
two-term Legendre series, there results a useful, physically meaningful,
differential equation for the time evolution of the energy spectrum in a
time-dependent electric field. Elastic scattering, and inelastic and superelastic energy transfer collisions are included. It is valid for fields
varying slowly compared with the swarm momentum-transfer collision frequency,
but on any time scale relative to the energy-transfer collision frequency.

The time independent solution in a constant field is a simple approximate expression for the steady state energy spectrum of swarm electrons.
Defining the usual mean Loss Function L and a Straggling Function M by

$$L = \sum_{i,k} N_i \left(E_k - E_i\right)\sigma_{ik} \quad , \quad M = \frac{1}{2}\sum_{i,k} N_i \left(E_k - E_i\right)^2\sigma_{ik} \quad ,$$

where N_i is the population density of the ith excited state, E_i its energy,
and σ_{ik} the cross section for the transition from state i to k, the steady
state energy spectrum is

$$F = F_0 \exp\left\{ -\int_0^w \frac{L + \frac{1}{w}\frac{\partial}{\partial w}\left(wM\right)}{\dfrac{\left(eE/N_o\right)^2}{3\sigma_m} + M}\, dw \right\} \quad ,$$

Nonequilibrium Effects in Ion and Electron Transport
Edited by J. W. Gallagher *et al.*, Plenum Press, New York, 1990

where N_o is gas density, w is electron energy, and σ_m is the momentum transfer cross section. The physical meaning of its functional form is made clear by showing its relation to ordinary diffusion-convection theory.

Previous spectra by Pidduck; Druyvesteyn; Davydov; Morse, Allis, and Lamar; Chapman and Cowling; and Wannier are special cases, and the domain of validity of each can be seen from a unified physical perspective. The importance of spread of energy loss about the mean is emphasized, and the physical reasons for the inadequacy of the Continuous Slowing Down Approximation (CSDA) become apparent. It is shown that in the limit of small quantum transition energies the new spectrum is exact. It is further shown that <u>the CSDA violates detailed balance</u>.

The spectrum is used with experimental cross sections to compute swarm transport coefficients in O_2 and N_2, in both of which fractional average energy loss is acceptably small over most energy ranges. Agreement with compiled swarm data is excellent over more than four orders of magnitude in E/N_o for most coefficients, except at certain energies in N_2 which stress the approximation's validity.

The spectrum and time dependent equation should be useful for approximate calculations in both steady state and time dependent cases, for basic theoretical studies, for developing physical insight, for parameter studies, and for scaling into regions not presently covered by experiments.

356

NONLINEAR DIFFUSION

E. E. Kunhardt

Weber Research Institute
Polytechnic University
Route 110, Farmingdale, NY 11735

A continuity equation for the transport of electron density is derived using the concept of a macro-kinetic distribution (MKD) for electrons in a background gas (Kunhardt et al., 1988). In lowest order this distribution is shown to obey an equation that is equivalent to the steady-state Boltzmann equation with an equivalent field that is velocity dependent. An explicit form for the MKD is presented for the case of a quasi-Lorentz gas model. The MKD has been used to evaluate the electron current density and to obtain expressions for the mobility and diffusion coefficients. These coefficients are dependent on the electron density gradient, so that the resulting continuity equation is non-linear. The consequences of these results are illustrated for the case of constant collision frequency.

Work supported by the Office of Naval Research.

REFERENCES

Kunhardt, E. E., J. Wu, and B. Penetrante, 1988, Phys. Rev. A 37, 1654.

SENSITIVE HIGH-TEMPORAL-RESOLUTION TOF ELECTRON DRIFT TUBE; ASYMMETRICAL
CURRENT PULSE OBSERVATION AND DETERMINATION OF V_d, D_L AND D_3*

C. A. Denman and L. A. Schlie

Advanced Laser Technology Division (WL/ARDI)
Air Force Weapons Laboratory
Kirtland AFB, NM 87117-6008

A unique and sensitive time-of-flight (TOF) drift tube has enabled the
first observation of an asymmetric arrival-time-spectra (ATS) of the electron current pulse. From this skewness - a feature predicted by theory but
never experimentally observed - the higher-order diffusion coefficient related to the skewness of the pulse is determined. The drift velocity (V_d),
longitudinal diffusion (D_L) and skewness diffusion (D_S) are reported for the
noble gases as well as for the gases hydrogen azide (HN_3) and hydrogen
sulfide (H_2S). All gases tested exhibit a skewness of the ATS in contrast
to the symmetric Gaussian ATS obtained by a more traditional TOF apparatus.
The asymmetric ATS is well represented for small skewness by extending the
analysis to the second-order diffusion approximation of the density gradient
expansion (DGE) theory. In the case where a large amount of skewness is
observed, such as that seen in the heavier noble gases (Ar, Kr, Xe), the
second-order diffusion approximation analysis of the experimental pulse
diverges, indicating that the next higher-order of the DGE may be necessary.
The highly sensitive current detection and temporal response of this drift
tube has been made possible by combining those components traditionally
found in TOF drift tubes - a pulsed back-illuminated photocathode current
source, guard rings and an anode - with a very sensitive, fast-current
amplifier to resolve the electron current pulse. This virtual ground,
current-to-voltage amplifier has a gain of 4.4×10^7 V/A (23pA/mV) with a
3 dB bandwidth of about 3 MHz. Also, preliminary results for the inference
of HN_3 cross-sections are to be presented.

* Supported by the Air Force Office of Scientific Research, Bolling AFB, D.C.

THE CHARACTERISTIC ENERGY OF ELECTRONS IN HYDROGEN

W. Roznerski, J. Mechlińska-Drewko, K. Leja and
Z. Lj. Petrović[*]

Department of Physics
Technical University of Gdańsk
80-952 Gdańsk, Poland

In the present work some of the suggestions determining the conditions of an accurate measurement of the characteristic energy using Townsend – Huxley method (Huxley and Crompton, 1974) have been applied. Based on the procedure described below, the characteristic energy of electrons in hydrogen at ambient temperature has been determined over the reduced electric field $70 \leq E/N \leq 2500$ Td.

To determine the D/μ coefficient from the measurements of the ratios of currents collected by optional coaxial parts of a divided anode the following form of the expression for the fraction R of the total current falling onto the central part of an anode has been used (Huxley and Crompton, 1974):

$$R = \frac{\left[1 - \left[\frac{b}{r_b'} - \frac{1}{\beta h}\left(1 - \frac{h^2}{r_b'^2}\right)\right]\frac{h}{r_b'}\exp[-\beta(r_b' - h)]\right]\exp[-h(\beta - \lambda_L)]}{\left[1 - \left[\frac{b}{r_c'} - \frac{1}{\beta h}\left(1 - \frac{h^2}{r_c'^2}\right)\right]\frac{h}{r_c'}\exp[-\beta(r_c' - h)]\right]\exp[-h(\beta - \lambda_L)]} . \tag{1}$$

where

$$r_b' = \left[h^2 + \frac{D_L}{D}b^2\right]^{1/2} , \quad r_c' = \left[h^2 + \frac{D_L}{D}c^2\right]^{1/2} ,$$

$$\lambda_L = W/2D_L ,$$

$$\beta = \lambda_L(1 - 2\,\alpha/\lambda_L)^{1/2}$$

The quantities b, c, h, W, α, D and D_L are the radius of the central disc, the external radius of the anode, the length of diffusion space, the drift velocity, the ionisation, transversal and longitudinal coefficients, respectively.

---

[*] Institute of Physics, P.O. Box 57, 11001 Belgrade, Yugoslavia.

Nonequilibrium Effects in Ion and Electron Transport
Edited by J. W. Gallagher *et al.*, Plenum Press, New York, 1990

A suitable combination of the expressions given above allows consideration also of each alternative version of the anode. Having measured the several R values obtained at various experimental parameters, the characteristic energy D/μ and the ratio of longitudinal diffusion coefficient to mobility D_L/μ where both D/μ and D_L/μ are solutions of Eq. (1).

The uncertainties of the D_L/μ coefficient resulting from the procedure applied in this work are approximately ten times larger than those of the characteristic energy.

Table 1 contains the tabulated data of the characteristic energy from 70 up to 2500 Td. The agreement between our results and those by Crompton et al. (1966) over the E/N range from 70 to 200 Td is very good.

Table 1. E/N (Td); D/μ (V)

E/N	D/μ	E/N	D/μ	E/N	D/μ
70	2.18	200	3.87	750	10.1
100	2.71	250	4.40	1000	12.5
120	3.03	300	5.12	1500	16.2
140	3.23	400	6.10	2000	17.7
170	3.46	500	6.97	2500	19.7

The present work was supported in part under project CPEP 01.06.

REFERENCES

Crompton, R. W., B. S. Liley, A. I. McIntosh, and C. A. Hurst, 1965, Proc. 7th ICPIG, Beograd, 1966, Vol. 1, Gradevinsha Knjiga, Publishing House, Beograd, 86.
Huxley, L. G., and R. W. Crompton, 1974, "The Diffusion and Drift of Electrons in Gases," Wiley Interscience, New York.

ELECTRON SWARM PARAMETERS IN KRYPTON AND ITS MOMENTUM TRANSFER CROSS SECTIONS

Y. Nakamura

Department of Electrical Engineering
Faculty of Science and Technology
Keio University
3-14-1 Hiyoshi, Yokohama 223, Japan

INTRODUCTION

The electron drift velocity (W) and the longitudinal diffusion coefficient (ND_L, N: the gas number density) in pure krypton are measured over the range of E/N from 0.3 to 50 Td ($1Td=10^{-17}$ V·cm^2). A momentum transfer cross section (Q_m) for the krypton atom, which is consistent with both of the present swarm parameters, is also derived over the range of electron energy from 1 to 15 eV from a Boltzmann equation analysis.

EXPERIMENTAL

The same double-shutter drift tube that had been used in the previous measurement in argon (Nakamura and Kurachi, 1989) was used in the present study. The gas used here was of the highest purity available (99.99%), and it was introduced into the chamber through a cold trap using dry ice and acetone as coolant. The drift tube was equipped with a nonevaporable getter to minimize impurities in the gas. Performance of the getter was proved satisfactory in the measurements in argon. All measurements were carried out at room temperature.

RESULTS AND DISCUSSIONS

The measured W and ND_L are shown in Fig. 1 by closed circles. Agreement between the present drift velocity and the result of Hunter et al., 1988 (E/N $\leq$3 Td, open circle) was good over the overlapping range of E/N. There were no measurements of ND_L in the gas before. The present swarm parameters were analyzed using a Boltzmann equation analysis. At higher E/N inelastic processes must be included. The ionization cross section reported by Rapp et al., 1965 was included and was fixed throughout the present analysis. The first choice of electronic

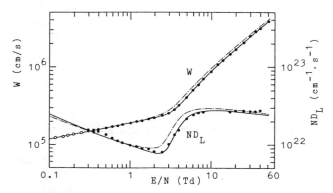

Fig. 1.   The electron drift velocity W and ND_L as a function of E/N
          in krypton.   Curves show the swarm parameters calculated
          with the momentum transfer cross sections shown in Fig. 2.
          Symbols: ●, present; o, Hunter et al., 1988.

excitation cross sections consisted of those for a metastable excitation
with onset energy = 9.91 eV given by Mason et al. (1987), and for an
excitation with onset = 10.03 eV.   The latter was estimated from the
total excitation cross section (Trajmar, et al., 1981) and the above
metastable cross section.   The calculated ionisation coefficient was very
sensitive to the change of the electronic excitation cross sections but
has only a small effect on both W and ND_L (Nakamura and Kurachi, 1989).
Procedure of analysis used in the present study is as follows: W and ND_L
were analyzed and a momentum transfer cross section were at first
modified using the starting inelastic cross sections.   Inelastic cross
sections were then modified until the analyzed ionization coefficient
agreed reasonably with the measurement (Kruithof, 1940).   Finally Q_m was
again modified to give swarm parameters within experimental
uncertainties.

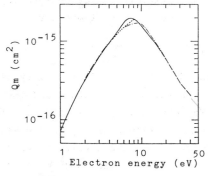

Fig. 2.   Momentum transfer cross sections for the krypton atom.
          Curves: ——, present; - - - - - , Hunter et al., 1988;
          — · — · , Hayashi, (1988).

The final momentum transfer cross sections for atomic krypton are compared with other results (Hunter, et al., 1988; Hayashi, 1988) in Fig. 2. The present result agrees with Hunter's result up to about 3 eV. The calculated swarm parameters given by the present cross section are shown in Fig. 1 by the solid line. Swarm parameters at higher E/N (>1 Td) are sensitive to the change in the momentum transfer cross section around its maximum and the result indicates that ND_L can be a more sensitive test of the Q_m than drift velocity.

CONCLUSION

The drift velocity and the longitudinal diffusion coefficient of electrons in pure krypton were measured over the E/N range from 0.3 to 50 Td. The present swarm data were used to derive a consistent momentum transfer cross section for krypton atom for electron energies between 1 and 15 eV.

REFERENCES

Hayashi, M., 1988, Private Communication
Hunter, S. R., et al., 1988, Phys. Rev. A 38, 5539-5551.
Kruithof, A. A., 1940, Physica 7, 519-540.
Mason, N. J., et al., 1987, J. Phys. B 20, 1357-77.
Nakamura, Y., and M. Kurachi, 1989, J. Phys. D 22, 107-112.
Rapp, D., et al., 1965, J. Chem. Phys. 43, 1464-79.
Trajmar, S., et al., 1981, Phys. Rev. A 23, 2167-77.

THE ELECTRON–MERCURY MOMENTUM TRANSFER CROSS SECTION AT LOW ENERGIES

J. P. England and M. T. Elford

Research School of Physical Sciences
Australian National University
Canberra, Australia

Previous derivations of the momentum transfer cross section (σ_m) for electrons in mercury vapour from drift velocity (v_{dr}) data (e.g. Nakamura and Lucas, 1978; Elford, 1980a) have been restricted to energies above about 0.1 eV due to the relatively high value of the electron mean energy at the lowest value of E/N for which accurate data can be obtained (E is the electric field strength and N, the vapor number density). The energy exchanged in elastic collisions can be increased and thus the mean energy lowered by the addition of helium. Since σ_m for helium is known to high accuracy (Crompton et al., 1970), this is not a significant source of errors in the analysis.

Measurements of v_{dr} have been made for a 46.8% He – 53.2% Hg vapor mixture at 573.1 K by the Bradbury-Nielsen time-of-flight method and the drift tube of Elford (1980b). The pressures, 5.4 to 26.9 kPa, were measured with an uncertainty of < ± 0.2% by a capacitance manometer with the head at the drift tube temperature (Buckman et al., 1984). The values of E/N ranged from 0.08 to 3.0 Td. The effect of dimers, which in pure mercury vapor was found to cause the measured values of v_{dr} to increase linearly with pressure, was about a factor of 5 smaller in the measurements in the He-Hg mixture because in the mixture, the rate of energy loss in collisions with dimers is a very much smaller fraction of the total rate of energy loss. The present data, corrected to remove the effects of diffusion and dimers, are shown in Fig. 1. The estimated uncertainty is generally < ± 1% but rises to ± 3.5% at the lowest value of E/N. The accuracy of the measurement technique was checked by taking measurements in pure He at 293 and 573 K. The results agreed with calculated values using the σ_m of Crompton et al. to within ± 0.1%.

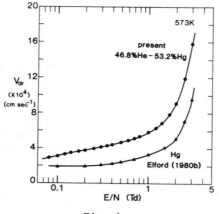

Fig. 1.

The σ_m for mercury (Fig. 2) was obtained over the energy range 0.05 to 4 eV by fitting to the corrected v_{dr} values, using the σ_m for helium of Crompton et al. (1970). All of the 21 data points were fitted to within 0.5%.

The difference between the present cross section and that of Elford (1980a) at low energies is considered to be due to (a) the lack of sensitivity of the experimental v_{dr} values of Elford to σ_m in this energy region and (b) the possibility of large errors at the lowest E/N values when his measured values of v_{dr} were extrapolated to zero pressure to correct for the effect of dimers. The present measured values of v_{dr} in mercury vapor agree with those of Elford (1980b) to within the stated error limits.

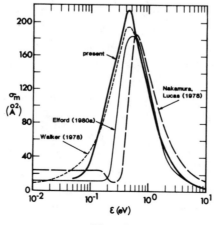

Fig. 2.

The cross section of Walker (1978, personal communication) was obtained using the method of Walker (1975) but with the strength of the polarization contribution to the interaction potential adjusted to give a best fit to the cross section of Elford. The differences between the present cross section and that of Walker (1978) are in general less than 10%.

REFERENCES

Buckman, S. J., J. Gascoigne, and K. R. Roberts, 1984, J. Vac. Sci.
    Technol. A 2, 1599.
Crompton, R. W., M. T. Elford, and A. G. Robertson, 1967, Aust. J. Phys.
    20, 369.
Elford, M. T., 1980a, Aust. J. Phys. 33, 251.
Elford, M. T., 1980b, Aust. J. Phys. 33, 231.
Nakamura, Y., and J. Lucas, 1978, J. Phys. D 11, 325.
Walker, D. W., 1975, J. Phys. B 8, L161.

LONGITUDINAL DIFFUSION TO MOBILITY RATIOS FOR ELECTRONS IN NOBLE GASES

J. L. Pack[*], R. E. Voshall, A. V. Phelps[**] and L. E. Kline

Westinghouse Research & Development Center
1310 Beulah Road, Pittsburgh, PA 15235

The ratio of longitudinal diffusion coefficient to mobility (D_L/μ) for electrons in helium, argon, krypton and xenon have been surveyed experimentally for E/N values between 0.001 and 44 Td at temperatures between 77°K and 373°K. The data are derived from transit times and half widths of current waveforms obtained during earlier measurements of electron drift velocities (Voshall et al., 1965). The scatter in the data is typically ± 20%. All gases were of research grade and the vacuum system was baked out at 250°C for 14 hours giving a rate of rise of less than 10^{-7} Pa/min.

At the lower E/N, the experimental D_L/μ approach the thermal value (kT/e) for all temperatures. At electron energies near those of the Ramsauer minimum and near the onset of excitation, the D_L/μ values for argon, krypton and xenon increase rapidly with E/N. The second region of structure has been observed previously by Kucukarpaci and Lucas (1981) and by Makabe and Shimoyama (1986).

In He, good agreement is obtained with other experiments (Wagner et al., 1967; Kucukarpaci et al., 1981). Our results for D_L/μ for electrons in Ar agree with those of Wagner, Davis and Hurst (1967) for E/N below 3 Td, but are about a factor of two above those of Kucukarpaci and Lucas (1981) for 4 < E/N < 10 Td. For Kr, our results for D_L/μ and electron drift velocity merge with those of Kucukarpaci and Lucas (1981) for E/N near 15 Td. For Xe at 195°K we find a surprisingly large peak in D_L/μ near 0.04 Td, i.e., at average energies near that of the minimum transfer cross section.

---

[*] Present address: 3853 Newton Dr., Murrysville, PA 15668

[**] Present address: Joint Institute for Laboratory Astrophysics
                       University of Colorado, Boulder, CO 80309-0440

Good agreement between measured D_L/μ values and the theory of Lowke and Parker (1969) is found for the common range of E/N (< 3 Td). Our calculated and experimental results for Ar are significantly larger than those of Makabe and Shimoyama (1986). In qualitative agreement with Kucukarpaci and Lucas (1981), our calculations for Ar show a weak maximum in D_L/μ at E/N near 20 Td. However, for Xe we calculate a nearly constant D_L/μ for 7 < E/N < 50 Td and do not obtain the pronounced maximum found experimentally at E/N near 10 Td.

REFERENCES

Kucukarpaci, H. N., and J. Lucas, 1981, J. Phys. D. <u>14</u>, 2001.
Kucukarpaci, K. N., H. T. Salee, and J. Lucas, 1981, J. Phys. D. <u>14</u>, 9.
Lowke, J. J., and J. H. Parker, Jr., 1969, Phys. Rev. <u>181</u>, 302.
Makabe, T., and M. Shimoyama, 1986, J. Phys. D. <u>19</u>, 2301.
Voshall, R. E., J. L. Pack, and A. V. Phelps, 1965, J. Chem. Phys. <u>43</u>, 1990.
Wagner, E. B., F. J. Davis, and C. S. Hurst, 1967, J. Chem. Phys. <u>47</u>, 3138.

RELATIONS BETWEEN ELECTRON KINETICS IN DC ExB AND MICROWAVE DISCHARGES

G. Schaefer and P. Hui

Weber Research Institute
Polytechnic University
Farmingdale, NY 11735

The chief goal in designing discharge devices is to optimize the space and time dependence of the transport parameters and rates for a variety of processes as needed for a specific application. These transport parameters and rates are completely determined by the velocity distribution function subject to the external and boundary conditions. Subsequently, the discharges are totally controlled by the velocity distribution function.

In general, we can shape a distribution function through adjusting external forces and sources, gas mixtures as well as boundary conditions. In the following, we list some means to control external forces.

Some Methods to Control External Forces

Discharge Type	Parameter
DC-Discharge	$(E/N)_{DC}$
Spiking Discharge	$(E/N)_1$, $(E/N)_2$
AC-Discharge:	E/N
(1) Frequency	f
(2) Polarization	Φ
Crossed B-field:	
(1) DC-Discharge	B/N, E/N
(2) AC-Discharge	B/N, E/N, f, Φ

Here we will only study the similarities between DC ExB discharge and discharge generated by circularly polarized microwave with B' perpendicular to E. After studying the trajectory of electron in both field configurations, we found that the distribution functions are the same if the microwave discharge is described in the coordinate system rotating with E and the following relation is satisfied

$$\Omega = \Omega' + \omega .$$

Here

    Ω : electron cyclotron frequency in DC discharge
    Ω': electron cyclotron frequency in microwave discharge
    ω : angular frequency of microwave

To verify the above statement, the energy distribution functions for both DC ExB discharge and circularly polarized microwave discharge were obtained using Monte Carlo simulations. These two energy distribution functions agree very well.

Work supported by the NSF.

COMPUTER SIMULATION OF A DISCHARGE IN CROSSED ELECTRIC AND MAGNETIC FIELDS

G. R. Govinda Raju and M. S. Dincer

Department of Electrical Engineering
University of Windsor
Windsor, Ontario, Canada, N9B 3P4

Application of a steady-state uniform magnetic field ($\underline{B}$) perpendicular to the electrical field ($\underline{E}$) introduces changes in the electron swarm parameters. The Townsend's first ionization coefficient (α/N; N = number of molecules per cm^3), the transverse drift velocity and the mean energy of electrons are of particular interest from the discharge point of view. In addition, the electrons acquire a perpendicular drift velocity depending upon the strength of the magnetic field. The motion of electrons in uniform $\underline{E} \times \underline{B}$ fields is simulated by the Monte-Carlo method, and the swarm parameters are evaluated in nitrogen and sulphur-hexafluoride. The range of parameters for the study are: $540 \leq E/N \leq 900$ Td and $0 \leq B/N \leq 18 \times 10^{-19}$ $T-cm^3$ in SF_6. The corresponding parameters in N_2 are: $240 \leq E/N \leq 600$ Td and $0 \leq B/N \leq 45 \times 10^{-19}$ $T-cm^3$ in N_2. The swarm parameters obtained from the simulation are compared with those obtained using numerical solutions for the Boltzmann equation for the electron energy. Comparisons are also made with the available experimental data. The influence of a magnetic field on the attachment coefficients in SF_6 are also investigated. It is concluded that the Monte-Carlo approach provides an independent means of verifying the validity of the equivalent electric field approach.

Nonequilibrium Effects in Ion and Electron Transport
Edited by J. W. Gallagher *et al.*, Plenum Press, New York, 1990

STRUCTURES OF THE VELOCITY DISTRIBUTIONS AND TRANSPORT COEFFICIENTS OF THE ELECTRON SWARM IN CH_4 IN A DC ELECTRIC FIELD

Naohiko Shimura and Toshiaki Makabe

Department of Electrical Engineering
Faculty of Science and Technology
Keio University
3-14-1 Hiyoshi, Yokohama 223, Japan

INTRODUCTION

There are many theoretical studies of the velocity distribution function of the electron swarm in a DC electric field. The velocity distribution in CH_4 and SiH_4, employed in the plasma processing, has a strong anisotropy because of the energy dependence of the cross section-set. In this work, we have calculated the velocity distribution function, and have shown the structures of the cubic image of the distribution and the swarm parameters in CH_4. The numerical method is based on the finite element method developed in a DC field by Segur et al., (Yousfi et al., 1985), except for the further consideration of the collision term. The velocity distribution function g(r, v, t) of the electron undergoing the binary collision with molecule in a DC electric field is given by the solution of the Boltzmann equation. The direction of the field is taken to be along the-z-axis. The effect of the spatial density gradient of the electrons will be considered on the velocity distribution function as a perturbation on a uniform number density. We can get the hierarchy of the equations for $g^k(v)$, which are tensorial functions of rank k, of the expansion (Kumar et al., 1980). The detail of this scheme is described in Shimura and Makabe (1989).

NUMERICAL RESULTS AND DISCUSSION

We have calculated the velocity distribution function of electrons in CH_4 in a DC field over the wide range 0.10 < E/N < 2820 Td. A set of cross sections is given by Ohmori et al. (1986) and Hayashi (1987). The scattering is assumed to be isotropic in the center of mass system. The gas temperature is taken to be 273 K. In Figs. 1 and 2, we show the

Nonequilibrium Effects in Ion and Electron Transport
Edited by J. W. Gallagher *et al.*, Plenum Press, New York, 1990

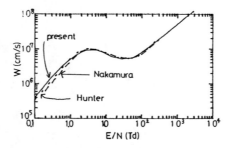

Fig. 1. Drift velocity W.

drift velocity W and longitudinal diffusion coefficient ND_L in CH_4 with the experimental results (Hunter et al., 1986; Nakamura, 1984; Al-Amin et al., 1985). Both results are in good agreement. In Figs. 3-6, we show the three dimensional velocity distributions at several field strengths E/N. The V_T axis in these figures means any axis in the V_X-V_Y plane. The units of the V_Z and V_T axes are taken to be $\sqrt{eV}$. The distribution $g°(v)$ in Fig. 3 is the distribution for the uniform number density of electrons in configuration space. Then, the electrons are transported only by the field. The effect of the field is small in the transverse direction of $g°$. The influence of the field on $g°$ is strengthened according to the decrease in Θ, and the asymmetric profile is realized between the accelerated-region ($\Theta < 90°$) and the decelerated-one ($\Theta > 90°$). The distribution $g_L(v)$ which represents for the longitudinal (field directional) component of the second order coefficient $g^1(v)$ is located on both sides of V_Z-V_T plane in Fig. 4. The high energy electrons contribute to the diffusion. The positive value of g_L is realized in the region grater than the mean energy of the swarm. It means that there is diffusion heating in the V_Z direction of the swarm and the diffusion cooling in $-V_Z$ direction. The picture of $g_T(v)$, which is the transverse component of $g^1(v)$, is more peculiar than those of $g°$ and g_L. The transverse diffusion flux caused by the density gradient of

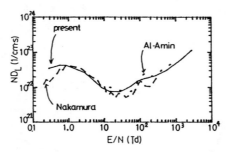

Fig. 2. Longitudinal Diffusion coefficient ND_L.

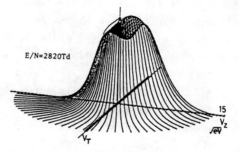

Fig. 3. Velocity distribution $g^o(v)$.

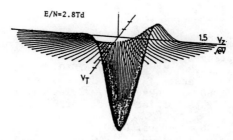

Fig. 4. Velocity distribution $g_L(v)$.

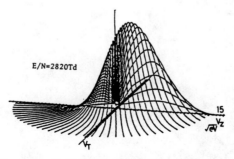

Fig. 5. Velocity distribution $g_T(v)$.

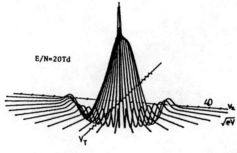

Fig. 6. Velocity distribution $g_{LL}(v)$.

the T direction is affected by the field, and the curvature effect
(Makabe et al., 1986) on the distribution is realized. The curvature
effect is clear at high field strength. The angle Θ, which gives the
maximum value of the distribution, moves toward the V_Z axis with
increasing field strength. The figure of $g^2_{LL}(v)$, which is the longitudinal component of the distribution $g^2(v)$, is more complicated than
other distributions. The third order transport coefficient are mainly
determined g^2_{LL}. The meaning of positive value of the distribution is
'plus' contribution to the coefficient. The detail of the structures in
g^2_{LL} will be discussed at the meeting.

REFERENCES

Al-Amin, S. A. J., et al., 1985, J. Phys. D 18, 1781-1794.
Hayashi, M., 1987, "Swarm Studies and Inelastic Electron-Molecule
    Collisions," Springer, New York.
Hunter, S. R., et al., 1986, J. Appl. Phys. 60, 24-35.
Kumar, K., et al., 1980, Aust. J. Phys. 33, 343-448.
Makabe, T., et al., 1986, J. Phys. D 19, 2301-2308.
Nakamura., H., 1984, "Paper of Tech. Meet. of Elec. Discharges ED-84-28,"
    (in Japanese).
Ohmori, Y., et al., 1986, J. Phys. D 19, 437-455.
Shimura, N., and Makabe, T., 1989, "ICPIG XIX," (Belgrade).
Yousfi, M., et al., 1985, J. Phys. D 18, 359-375.

NEGATIVE ION KINETICS IN BCl$_3$ DISCHARGES

Z. Lj. Petrovic[b], W. C. Wang[a], L. C. Lee[a], J. C. Han[a] and
M. Suto[a]

[a] Molecular Engineering Laboratory
Department of Electrical and Computer Engineering
San Diego State University, San Diego, CA 92182, USA

[b] Institute of Physics, University of Belgrade, P.O. Box 57
11001 Belgrade, Yugoslavia

BCl$_3$ is a gas of great importance for plasma technologies in microelectronics fabrication, especially for etching of aluminum. In this paper
we present some evidence for the three body attachment leading to formation
of BCl$_3^-$ ions at low electron energies and for dissociative attachment at
higher mean electron energies. In the measurements of electron attachment
rate coefficients performed in dilute BCl$_3$ mixtures in nitrogen and argon,
we have observed an increase of electron attachment towards the lowest E/N
(energies) values. This is consistent with the observations of attachment
(Stockdale et al., 1972) at thermal energies and of low photodetachment
thresholds in some RF discharges (Gottscho and Gaebe, 1986).

On the other hand, in the mass analysis of the negative ions present in
discharges containing BCl$_3$, only Cl$^-$ ions were detected. This observation
is in agreement with electron beam studies (Stockdale et al., 1972). The
first set of data was obtained at high pressures (100-400 torr), and the
excited BCl$_3^-$ ions could be stabilized. This might not be the case for the
low pressures which were required to enable the operation of the mass
analyzer. It is worth noting that Stockdale and coworkers (1972) observed a
peak of attachment (two body) at 1.1 eV, while at the same time they
observed a relatively high thermal attachment rate. This observation is
consistent with our data.

Sampling of negative ions from the discharge proved to be less difficult than expected for a relatively narrow range of working conditions
(Lee et al., 1989). In order to establish that negative ions observed by
the mass analyzer originate from the discharge, an excimer laser was used to
detach electrons between the electrodes. This led to a major decrease of

Nonequilibrium Effects in Ion and Electron Transport
Edited by J. W. Gallagher *et al.,* Plenum Press, New York, 1990

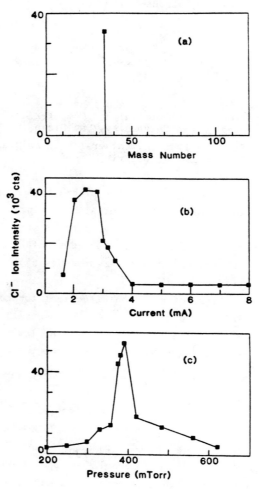

Fig. 1.

the negative count rate, and the time delay between laser pulse and the decrease of the count rate increased as the laser was focused further away from the anode.

In Fig. 1 we present: a) mass spectrum of negative ions sampled from the discharge in 0.5% BCl_3-N_2 mixture; b) dependence of the Cl ion intensity on the discharge current (at pressure of 585 mTorr); and c) dependence of the Cl^- ion intensity on the total gas pressure (for a discharge current of 2mA).

This presentation is based on the work supported by the AFOSR and the SDIO/IST/ONR.

REFERENCES

Gottscho, R. A., and C. E. Gaebe, 1986, IEEE Trans. Plasma Sci. PS-14, 92.
Lee, L. C., J. C. Han, and M. Suto, 1989, Sixth International Swarm Seminar, Glen Cove, N. Y.
Stockdale, J. A., D. R. Nelson, F. J. Davis, and R. N. Compton, 1972, J. Chem. Phys. 56, 3336.

ON THE MECHANISM OF THERMAL ELECTRON ATTACHMENT TO SO_2

H. Shimamori and Y. Nakatani

Fukui Institute of Technology
3-6-1 Gakuen, Fukui 910, Japan

Low-energy electron attachment to SO_2 has been studied by several workers (Rademacher et al., 1975; Bouby et al., 1971). There is, however, a discrepancy in the interpretation of the attachment mechanism, especially at low pressures. The following two different processes have been proposed to explain the experimental results; (i) a radiative attachment ($e^- + SO_2 \longrightarrow SO_2^- + h\nu$) (Rademacher et al., 1975), and (ii) a two-step three-body process involving both the formation of stable negative ions and the collisional electron detachment (Bouby et al., 1971). Since there have been no other reports of radiative attachment to molecules except at extremely low pressures and also the mechanism (ii) seems to violate the principles of detailed balancing as pointed out for electron attachment to NO_2 (Klots, 1970), further studies should be necessary to clarify the true mechanism.

We have examined the attachment mechanism by measuring the pressure and temperature dependences of attachment rates for pure SO_2 and several SO_2-M mixtures (M: Ar, N_2, CO_2 and $n\text{-}C_4H_{10}$) using the pulse radiolysis-microwave cavity method. The results are analyzed in terms of an effective two-body attachment rate constant k_{eff} for a reaction $e^- + SO_2 \longrightarrow SO_2^-$.

The result for pure SO_2 at pressures between 0.4 and 5 Torr indicates that k_{eff} increases with pressure but tends to saturate at higher pressures. This does not agree with the data of Bouby et al. (1971), which showed a constant value of k_{eff} over a pressure range similar to ours. In SO_2-N_2 mixtures k_{eff} similarly increases with buffer-gas pressure and tends to saturate, but in this case k_{eff} still continues to increase gradually at much higher pressures. Similar pressure dependences have been observed for other SO_2-M (M: CO_2, $n\text{-}C_4H_{10}$) systems, and yet the degree of increase in the higher pressure region depends drastically upon the nature of buffer gas. We conclude that the attachment mechanism at low pressures does not involve the radiative attachment and the mechanism (ii) is not possible either. As a consequence, we propose a simple two-step three-body process (1)-(3).

$$e^- + SO_2 \underset{k_2}{\overset{k_1}{\rightleftharpoons}} SO_2^{-*} \quad (1)$$
$$(2)$$

$$SO_2^{-*} + M \xrightarrow{k_3} SO_2^- + M \quad (3)$$

A kinetic analysis has shown, however, that the mechanism (1)-(3) holds only at relatively low pressures, where the value of k_1 is determined to be $(3.6 \pm 0.2) \times 10^{-12}$ $cm^3 molecule^{-1} s^{-1}$. At higher pressures, the pressure dependence of attachment rates evidently indicates the presence of additional attachment processes. Such a behavior might be explained by the appearance of electron attachment to van der Waals molecules at high pressures (Hatano and Shimamori, 1981). In order to check this possibility we measured temperature dependence of the attachment rates because a "negative" temperature dependence should be observed if such a process is contributing. Unfortunately, however, we could not observe any definite negative temperature dependence over the temperature range between 252 and 343 K. As a result, we tentatively suggest for the mechanism at higher pressures that another two-step, three-body process becomes important which involves an SO_2^{-*} ion having a lifetime shorter than that in the mechanism at low pressures; the estimated lifetimes are $\gtrsim 1 \times 10^{-10}$ and $\gtrsim 2 \times 10^{-7}$ s, respectively.

REFERENCES

Bouby, L., F. Fiquet-Fayard, and C. Bodere, 1971, Int. J. Mass Spectrom. Ion Phys. 7, 415.
Hatano, Y., and H. Shimamori, 1981, "Electron and Ion Swarms," 103 (Pergamonn).
Klots, C. E., 1970, J. Chem. Phys. 53, 1616; H. Shimamori, and H. Hotta, 1986, J. Chem. Phys. 84, 3195.
Rademacher, J., L. G. Christophorou, and R. P. Blaunstein, 1975, J.C.S. Faraday Trans. II 71, 1212.

ELECTRON ATTACHMENT TO NF_3

Scott R. Hunter[*]

Health and Safety Research Division
Oak Ridge National Laboratory
Oak Ridge, Tennessee 37831

Nitrogen trifluoride has been used for over a decade as a fluorine donor in rare gas-fluoride excimer lasers, and more recently has found increasing use in plasma etching discharges for the fabrication and semi-conductor materials. Modeling of the NF_3 discharges in these applications requires a knowledge of the electron scattering and electron gain and loss processes for NF_3. Several electron attachment studies using electron beam and electron swarm techniques have been performed in NF_3, but the uncertainty in the rate of electron attachment from these studies is large.

In the present study a high pressure swarm technique (Hunter, 1984) has been used to measure the electron attachment rate constant $k_a(E/N)$ for NF_3 in dilute NF_3/Ar gas mixtures for NF_3 concentrations from 1.6×10^{-7} to 3.6×10^{-6} over the E/N range $6.2 \times 10^{-20} \leq E/N \leq 4.7 \times 10^{-17}$ V cm^2, and in NF_3/N_2 gas mixtures for NF_3 concentrations from 7×10^{-6} to 3×10^{-5} over the E/N range $3.1 \times 10^{-19} \leq E/N \leq 1.24 \times 10^{-16}$ V cm^2. The electron attach-ment cross section $\sigma_a(\varepsilon)$ has been obtained from these measurements using an unfolding procedure (Hunter, 1984) and compared with the previous literature values (Chantry, 1982; Harland and Franklin, 1974). The present cross section (peak value $\sigma_a = 2.06 \times 10^{-16}$ cm^2 at $\varepsilon = 1.6$ eV) is considerably larger than that obtained by Chantry (1982) (peak value $\sigma_a = 1.4 \times 10^{-16}$ cm^2 at $\varepsilon = 1.6$ eV) and Harland and Franklin (1974) (peak value $\sigma_a = 6 \times 10^{-17}$ at 1.7 eV). The present thermal electron attachment rate constant at 300K, $(k_a)_{th} = 5.5 \times 10^{-12}$ cm^3 s^{-1} is between 1 and 3 orders of magnitude smaller than all the previous literature estimates. Impurities in the NF_3 samples used in the previous measurements are thought to be responsible for the higher values.

---

[*] Present address: GTE-Sylvania, 100 Endicott St., Danvers, MA 01923.

The influence of inelastic electron scattering and electron loss by attachment on the electron energy distribution function (and hence on the measured electron transport and rate coefficients) in the previous NF_3/buffer gas measurements (Nygaard et al., 1979; Lakdawala and Moruzzi, 1980) has also been investigated. Incorporating two vibrational excitation cross sections ($\nu_3 = 0.112$ eV and $\nu_7 = 0.128$ eV) in a ratio of 15 to 1 (peak value $_3\sigma_\nu \simeq 6 \times 10^{-15}$ cm^2 at $\varepsilon \simeq 1.8$ eV) and the present attachment cross section gives agreement between the NF_3/He measurements of Nygaard et al., (1979) and the present Boltzmann calculated values to within $\pm$ 20%. In contrast, considerable differences still exist between the experimental measurements of Lakdawala and Moruzzi (1980) in the NF_3/He, NF_3/Ar and NF_3/N$_2$ gas mixtures and the present calculated values, particularly at low E/N values where impurities in the gas sample and difficulties in the experimental technique were significant. The influence of electron loss by attachment on the calculated transport and rate coefficients has been investigated by modifying the Boltzmann transport code to include this process (Pitchford and Phelps, 1983). The present calculations indicate that the greatest effect occurs in the experimental measurements of Nygaard et al. (1979) where a gas mixture composed of 0.5% NF_3 in He was studied. The greatest change in k_a(E/N) occurs near the peak in the rate constant and is always <+10%. The influence of attachment on the lower concentration measurements of Lakdawala and Moruzzi (1980) is well within their experimental uncertainty.

REFERENCES

Chantry, P. J., 1982, "Applied Atomic Collision Physics," E. W. McDaniel and W. L. Nighan (Eds.) (Academic, New York), 35.
Harland, P. W., and J. L. Franklin, 1974, J. Chem. Phys. 61, 1621.
Hunter, S. R., 1984, J. Chem. Phys. 80, 6150.
Lakdawala, V. K., and J. L. Moruzzi, 1980, J. Phys. D. 13, 377.
Nygaard, K. J., H. L. Brooks, and S. R. Hunter, 1979, IEEE J. Quant. Elect., QE-15, 1216.
Pitchford, L. C., and A. V. Phelps, 1983, Bull. Amer. Phys. Soc. 28, 182.

ELECTRON-ENERGY DEPENDENCE OF ELECTRON ATTACHMENT TO MOLECULES AS STUDIED
BY A PULSE RADIOLYSIS MICROWAVE CAVITY TECHNIQUE

H. Shimamori and Y. Nakatani

Fukui Institute of Technology
3-6-1, Gakuen, Fukui 910, Japan

Although the heating of electrons with microwave electric fields
has been applied to study the electron temperature dependence of
electron-ion recombination processes by a conventional microwave cavity
technique (Frommhold et al., 1968), no application has been made of the
pulse radiolysis technique which enables measurements on a short time
scale and with a wide gas-pressure range for a variety of gases. We have
attempted to construct an appropriate experimental system by adding a
microwave-heating unit to the circuit used for studies with thermal
electrons. The geometrical structure of the cavity has been changed
accordingly. A shift of resonant frequency of the cavity (the imaginary
part of the conductivity) can be measured in our circuit. The heating
can be achieved up to 600 mW of microwave power. Preliminary measurements with various gases indicated mostly very low sensitivities for
detection. We finally chose xenon as a buffer gas since it produces a
relatively high density of electrons by X-ray irradiation. In order to
obtain a relationship between the applied microwave power and the
resulting mean electron energy, we utilized the dependence of the
time-profile of the detected signal on the heating power. The microwave
conductivity signal is approximately proportional to a factor of
$1/(\langle \upsilon \rangle^2 + \omega^2)$ (Oskam, 1958), where $\langle \upsilon \rangle$ is the effective electron
collision frequency and ω the detecting microwave radian frequency. In
the case of Xe the signal reflects well the course of electron
thermalization as well as the presence of Ramsauer minimum in the
momentum-transfer cross section, as shown in Fig. 1 for 70 Torr Xe.
Thus, a change in the amplitude of the signal as a function of time is
directly correlated with a change in the magnitude of $\langle \upsilon \rangle$, which is a
function of mean electron energy. When electrons are heated the flat
level shown in Fig. 1, which corresponds to thermalized electrons, goes

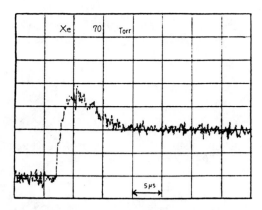

Fig. 1. Output signal for Xe (70 Torr) with no heating power.

up to a level corresponding to that expected from the average electron energy resulting from the microwave heating. Thus we observed signals at different heating powers and found a relation between the applied power and the mean electron energy.

We have applied this technique to electron attachment to chloroform ($CHCl_3$). by adding a trace amount of $CHCl_3$ to Xe we can observe a first-order-decay signal due to the dissociative electron attachment to $CHCl_3$. The attachment rate constant is obtained from the decay signal. By changing the applied heating power we can obtain data on the electron

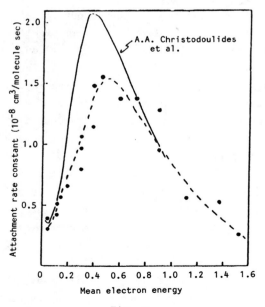

Fig. 2.

energy dependence of the attachment rate constant. The result is shown in Fig. 2. Although the agreement with other data (Christodoulides and Christophorou, 1971) is generally good, there are some discrepancies in the absolute magnitude of the rate constant and in the electron energy giving the maximum rate constant. Further improvement of the scaling of the electron energy may be necessary. Applications to other electron-attaching compounds are presented.

REFERENCES

Christodoulides, A. A., and L. G. Christophorou, 1971, J. Chem. Phys. 54, 4691.
Frommhold, L., M. A. Biondi, and F. J. Mehr, 1968, Phys. Rev. 165, 44.
Oskam, H. J., 1958, Philips Res. Rep. 13, 335.

NEGATIVE ION PULSES INDUCED BY LASER IRRADIATION OF DC DISCHARGE MEDIA

L. C. Lee, J. C. Han and M. Suto

Molecular Engineering Laboratory
San Diego State University
San Diego, California 92182

Electron transport phenomena in gas discharge media are studied using a double-differential pumping mass spectrometer apparatus shown in Fig. 1. Gas mixtures (in the range of 0.5-1 Torr) are discharged by applying a DC voltage to hollow electrodes in a vacuum chamber. Negative ions produced by discharges flow into a second high-vacuum chamber through a pin hole (0.7 mm diameter). The ions are analyzed and detected by a mass spectrometer. An excimer laser is used to irradiate the discharge media.

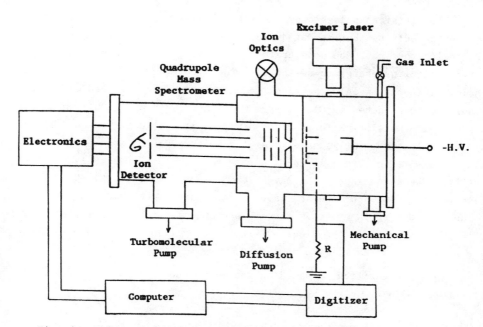

Fig. 1.   Schematic diagram of the apparatus used for the study of electron kinetics in discharges.

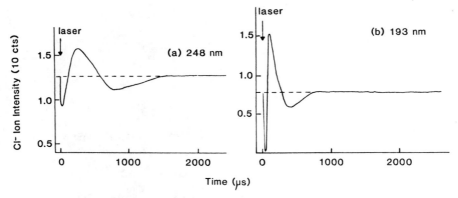

Fig. 2.   Transient negative ion pulses produced by irradiation of
discharge media (Cl_2 in N_2) with (a) KrF laser photons and
(b) ArF laser photons. The laser pulses occurred at t=0.

The Cl^- ions produced from a discharge of trace Cl_2 in N_2 are
studied. When the discharge medium is irradiated by a KrF (248 nm) or
ArF (193 nm) laser pulse, the Cl^- ion intensity decreases right after the
laser pulse and then increases above the DC level as shown in Fig. 2.
The ion intensity decreases again before recovery. Without laser
irradiation, the ion intensity is on the DC level shown as the dashed
lines in Fig. 2. The laser-induced changes (decrease or increase) of the
ion intensity depend linearly on the laser power.

The transient pulses shown in Figs. 2(a) and (b) are likely induced
by photoelectrons that are produced by laser detachment of negative ions.
Photoelectrons could be accelerated by the electric field so that they
have sufficient energy to detach electrons from the Cl^- ions near the pin
hole. This may explain the first ion decrease. The detached electrons
could be re-attached to Cl_2 to form Cl^- ions. This may result in an ion
increase. The slow photoelectrons may also attach to Cl_2 to contribute
to this increase. It is noted that the pulse duration induced by the ArF
laser pulse [Fig. 2(b)] is much shorter than that of the KrF laser [Fig.
2(a)]. The pulse shortening may be due to the fact that the photoelectrons produced by the ArF laser photon have higher kinetic energy
than that of KrF laser, thus the ArF photoelectrons move faster than the
KrF photoelectrons. The electron kinetics involved in the laser-induced
negative ion pulses will be discussed.

This presentation is based on the work supported by the AFOSR and
the SDIO/IST/ONR.

THE ATTACHMENT OF ELECTRONS IN WATER VAPOUR AT LOW VALUES OF E/N

J. C. Gibson and M. T. Elford

Res. School of Physical Science
Australian National University
Canberra, Australia

Moruzzi and Phelps (1966) and Wilson et al (1975) found by sampling ions from drift tubes that no negative ions are observed in water vapour until dissociative attachment becomes significant at values of E/N greater than about 33 Td. Moreover Parr and Moruzzi (1972) found that the attachment coefficient was zero to within their experimental error below 39 Td. On the other hand Bradbury and Tatel (1934) and Kuffel (1959) reported associative attachment at E/N values less than 15 Td which they attributed to the formation of water molecule cluster ions. Gallagher et al (1983) have suggested that a more likely explanation is that their water vapour samples were contaminated by air. The present study was undertaken to determine if and under what conditions attachment occurs in water vapour at values of E/N less than 30 Td.

A new and simple technique, shown schematically in Fig. 1, was used in this work. Electrons emitted from a heated platinum filament F drift in a uniform electric field through a grid G1 to a second grid G2. The analyzer region, where the value of E/N was held at 5 Td, contains a Bradbury-Nielsen grid consisting of parallel and coplanar nichrome wires,

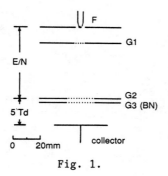

Fig. 1.

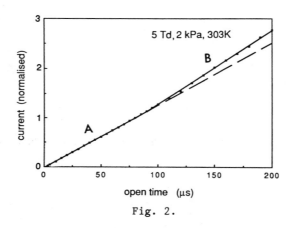

Fig. 2.

with alternate wires connected. Elford (1987) has shown that an incident
current of electrons and negative ions can be analyzed by applying square
wave voltage pulses of variable duration (the open time) between adjacent
wires of the grid. The variation of the transmitted current with open
time (Fig. 2) consists of two linear parts. At small open time only
electrons are transmitted (A) and it is not until the open time exceeds
some critical value that the much slower negative ions are transmitted in
addition to electrons (B). The ratio R of the slope of the negative ion
current part of the plot (obtained by subtracting the extrapolated
electron current) to that for the electron current part is assumed to be
equal to the ration of the ion to electron currents incident on the grid.

At pressures up to 1.33 kPa (295 and 298 K), the measured values of
R decreased as E/N was decreased below 30 Td. A plot of this type is
shown in Fig. 3a. However when the pressure was increased to 2 kPa, R

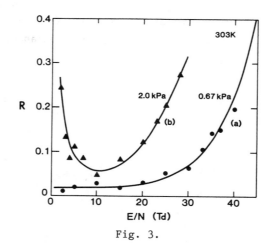

Fig. 3.

increased at low values of E/N (Fig. 3b), in the same manner as observed by Bradbury and Tatel and also Kuffel for the attachment coefficient. It is speculated that the negative ions observed at low E/N values and high vapour pressures are due to associative attachment to water dimers. Since the electron binding energy to $(H_2O)_2$ is estimated to be only 3–6 meV (Wallqvist et al, 1986) electron detachment by ion-molecule collisions is also expected to be a significant process.

REFERENCES

Bradbury, N. E., and H.E. Tatel, 1934, J. Chem. Phys. 2, 835.
Elford, M. T., 1987, Proc. 18th. Int. Conf. on Phen. in Ion. Gases.,
    Hilger, Swansea, p. 130.
Gallagher, J. W., E. C. Beaty, J. Dutton, and L. C. Pitchford, 1983,
    J. Phys. Chem. Ref. Data 12, 109.
Kuffel, E., 1959, Proc. Phys. Soc. 74, 297.
Moruzzi, J. L. and A. V. Phelps, 1966, J. Chem. Phys. 45, 4617.
Parr, J. E., and J. L. Moruzzi, 1972, J. Phys. D, Appl. Phys. 5, 514.
Wallqvist, A., D. Thirumalai, and B. J. Berne, 1986, J. Chem. Phys. 85,
    1583.
Wilson, J. F., F. J. Davis, D. R. Nelson, R. N. Compton, and O. H.
    Crawford, 1975, J. Chem. Phys. 62, 4204.

ISOTOPE STUDIES AND ENERGY DEPENDENCES OF RATE CONSTANTS FOR THE REACTION
O^- + N_2O AT SEVERAL TEMPERATURES

Robert A. Morris[*], A. A. Viggiano and John F. Paulson

Ionospheric Physics Division
Geophysics Laboratory
Hanscom AFB, MA 01731-5000

Rate constants for the reaction O^- + N_2O were measured using a
temperature variable-selected ion flow drift tube (SIFDT) instrument.
The reaction was studied as a function of kinetic energy at four
temperatures: 143, 196, 295, and 515 K. The product branching ratios of
the reactions $^{16}O^-$ + $^{14}N^{15}N^{16}O$ and $^{18}O^-$ + $^{14}N^{15}N^{16}O$ were measured at two
temperatures: 143 and 298 K.

The experiment employs a temperature varible flow tube with which
to control the thermal energy of the reagents as well as an electric
drift field to increase independently the kinetic energy of the ionic
reactant. This permits fixing the overall center of mass collision
energy while varying the relative contributions to that collision energy
from the thermal energy of the reagents (due to temperature) and from the
additional ion kinetic energy due to the electric field of the drift
tube. Thus, reactions of monatomic ions (no internal modes) can be
probed for the effects of the internal energy of the reactant neutral
(Viggiano et al., 1988).

Figure 1 shows the rate constants for the reaction O^- + N_2O plotted
versus center-of-mass collision energy. The energy dependences of the
rate constants measured at different experimental temperatures fall on
the same curve within experimental uncertainty. Therefore, there appears
to be little or no effect of the internal energy of the N_2O reactant on
the rate constant. In other words, the rate constant appears to depend
only upon total energy and not on the particular type of energy. The
pure temperature dependence of the rate constant is shown in the figure
as a solid line and can be represented as $T^{-0.5}$.

---

[*] Air Force Geophysics Scholar

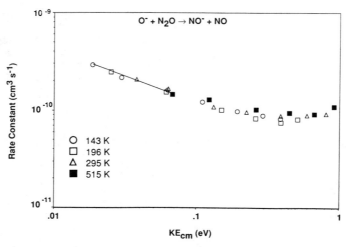

Fig. 1. Rate constants for the reaction $O^- + N_2O$ plotted as a function of center-of-mass kinetic energy.

The dominant product channel at the energies accessible in the experiment is that producing $NO^- + NO$. For the reaction of O^- with $^{14}N^{15}NO$, the ratio of the ionic products, $^{14}NO^-/^{15}NO^-$, approached unity at zero $^{14}N^{15}NO$ flow rate but decreased below unity with increasing $^{14}N^{15}NO$ flow rate. It is suggested that the decrease is due to the secondary reaction $^{14}NO^- + ^{14}N^{15}NO \longrightarrow ^{15}NO^- + ^{14}N^{14}NO$, which we have measured to occur with a rate constant of 1×10^{-11} cm^{3}s^{-1}.

For the reaction $^{18}O^- + ^{14}N^{15}N^{16}O$, the O^- exchange product, $^{16}O^-$, was observable and represents a major product channel along with that producing $NO^- + NO$. All four NO^- isotopic products were observed: $^{14}N^{16}O^-$, $^{15}N^{16}O^-$, $^{14}N^{18}O^-$, and $^{15}N^{18}O^-$. The secondary reactions of both the exchange product $^{16}O^-$ and the various NO^- products with N_2O complicate the interpretation of the product spectra. Computer modeling of the coupled reactions and the resulting product branching ratios measured at the two experimental temperatures will be discussed.

REFERENCES

Viggiano, A. A., R. A. Morris, and J. F. Paulson, 1988, J. Chem. Phys. **89**, 4848; ibid. (in press, 1989).

RECENT FALP STUDIES OF DISSOCIATIVE RECOMBINATION AND ELECTRON ATTACHMENT

Nigel G. Adams, David Smith and Charles R. Herd

School of Physics and Space Research
University of Birmingham
Birmingham B15 2TT England

The Flowing Afterglow/Langmuir Probe (FALP) apparatus was designed to study plasma reaction processes at thermal energies. It has been used to study several such processes including positive ion/negative ion mutual neutralization (Smith et al., 1978; Smith and Adams, 1983), positive ion/electron dissociative recombination (Adams et al., 1984; Adams and Smith, 1988) and electron attachment (Smith et al., 1984; Adams et al., 1986), and the rate coefficients for many such reactions have been determined, including some over appreciable temperature ranges.

Recently, we have used laser induced fluorescence (LIF) and vacuum ultraviolet (VUV) spectroscopic techniques in conjunction with the FALP apparatus to determine the neutral products of the dissociative recombination reactions of some polyatomic ions (Adams et al., 1989). Using this technique, the fractions of the dissociative recombination reactions of the ions HCO_2^+, N_2OH^+, O_2H^+ and H_3O^+ that result in the production of OH radicals both in the ground and vibrationally excited levels of the ground electronic state have been determined. The contribution of OH to the product distribution varies from ~30% for N_2OH^+ to 65% for H_3O^+, the fractions in the ground vibrational level being equal to or greater than the fractions in vibrationally excited levels. This latter observation, although initially unexpected, can be readily explained as being a consequence of a very similar O-H bond length in the ion and in the OH radical. These data have resolved a longstanding discrepancy between two very different theoretical predictions of the products of H_3O^+ recombination (Herbst, 1978; Bates, 1986). The significance of the results to interstellar ion chemistry has also been discussed (Herd et al., 1989). Preliminary results on the production of H and O atoms from the above reactions and some other recombination reactions have also been obtained.

Most dissociative electron attachment reactions do not result in sequential reactions; for example the reaction of CH_3Br with thermal electrons produces Br^- ions and CH_3 radicals. No further reaction then occurs because dissociative attachment to CH_3 radicals is endothermic. Recently, however, we have shown that the attachment reaction of CCl_4 proceeds firstly, as expected, to produce Cl^- and CCl_3 radicals, but then we observed that the product CCl_3 radicals undergo rapid dissociative attachment to produce Cl^- and CCl_2 radicals (Adams et al., 1988). Thus each CCl_4 molecule can remove two electrons from a plasma, and this has implications to the sensitivity of electron capture detectors. Similarly, the CCl_2Br radicals formed in the dissociative electron attachment reactions of CCl_3Br molecules also rapidly dissociatively attach electrons (Adams, et al., 1988). Temperature dependences have been obtained for dissociative electron attachment to the haloethanes CH_3CCl_3, $CH_2ClCHCl_2$, CF_3CCl_3 and $CF_2ClCFCl_2$ and compared with data obtained in non-thermal swarm experiments in which only the electron energy was varied. The differences in the energy dependences obtained using the two techniques are considered to be due to the additional influence of the internal excitation of the reactant neutral in the thermal FALP experiments (Smith et al., 1989).

REFERENCES

Adams, N. G., C. R. Herd, and D. Smith 1989, J. Chem. Phys. 91, 963.
Adams, N. G., and D. Smith, 1988, Chem. Phys. Lett. 144, 11.
Adams, N. G., D. Smith, and E. Alge, 1984, J. Chem. Phys. 81, 1778.
Adams, N. G., D. Smith, and C. R. Herd, 1988, Int. J. Mass Specrom. Ion Proc. 184, 243.
Adams, N. G., D. Smith, A. A. Viggiano, J. F. Paulson, and M. J. Henchman, 1986, J. Chem. Phys. 84, 6728.
Bates, D. R., 1987, Astrophys. J., (Letters) 306, L45; "Modern Applications of Atomic and Molecular Processes," edited by A.E. Kingston (Plenum, London), p 1.
Herbst, E., 1978, Astrophys. J. 222, 508.
Herd, C. R., N. G. Adams, and D. Smith, 1989, Astrophys. J. (in press).
Smith, D. and N. G. Adams, 1983, "Physics of Ion-Ion and Electron-Ion Collisions," edited by F. Brouillard and J.W. McGowan (Plenum, New York) p 501.
Smith, D., N. G. Adams, and E. Alge, 1984, J. Phys. B. 17, 461.
Smith, D., M. J. Church, and T. M. Miller, 1978, J. Chem. Phys. 68, 1224.
Smith, D., C. R. Herd, and N. G. Adams, 1989, Int. J. Mass Spectrom. Ion Proc. (in press).

RECENT SIFT STUDIES OF ION-MOLECULE REACTIONS: THE DEUTERATION OF
INTERSTELLAR MOLECULES

Kevin Giles, David Smith and Nigel G. Adams

School of Physics and Space Research
University of Birmingham
Birmingham B15 2TT England

About eighty molecular species have been detected in interstellar
clouds to date (Irvine et al., 1988). Some of the observed molecules
contain rare (heavy) isotopes, notably deuterium (Millar et al., 1989), in
greater fractional abundances than would be expected from their cosmic or
terrestrial abundance ratios.

Ion-molecule reactions followed by dissociative electron recombination
are considered to be the major production routes to the observed interstellar molecules. Deuterium can be incorporated into interstellar
molecules via reactions such as:

$$XH^+ + HD \underset{\rightarrow}{\leftrightharpoons} XD^+ + H_2 + \Delta E$$

$$XH^+ + D \underset{\rightarrow}{\leftrightharpoons} XD^+ + H + \Delta E$$

The exo- and endo-thermicities (ΔE) of these reactions are largely
determined by the zero-point-energy differences between the reactant and
product species, and thus the reactions are close to thermoneutral. Nevertheless, at the low temperatures in interstellar clouds (down to 10K) the ΔE
are sufficient to significantly inhibit the reverse (endothermic) reactions
and this results in the fractionation of deuterium into the ions.

Reactions of this type can be studied using the SIFT apparatus which
enables the reaction between a swarm of mass selected, thermalized ions and
neutral species to be investigated over the temperature range 80-550K in a
fast flowing helium carrier gas (Smith and Adams, 1987). It has been used
previously to determine the forward and reverse rate coefficients for a
variety of reactions as a function of temperature, and this has indicated
the reaction pathways to several important deuterated species in interstellar clouds, including: HD, H_2D^+, CH_2D^+ and C_2HD^+ (Smith, 1988) into
which D is heavily fractionated.

Nonequilibrium Effects in Ion and Electron Transport
Edited by J. W. Gallagher *et al.,* Plenum Press, New York, 1990

In this study, we have investigated the reactions of CH_3^+/CH_2D^+ and $C_2H_2^+/C_2HD^+$ with various interstellar molecules, including : CH_4, CO, C_2H_2, C_2H_4, H_2O, H_2S, HCN, CH_3OH, CH_3CCH, CH_3CHO, CH_3SH, C_2H_5OH, CH_3CN and NH_3, in order to investigate the partitioning of deuterium within the products. These reactions possess reactive channels and are quite exothermic and therefore do not directly fractionate deuterium. Thus the enhancement of deuterium in the product ions cannot exceed that in the reactant ions.

In the present study, several reactions were observed to incorporate deuterium approximately statistically amongst the product ions (and neutrals), indicating that the H and D atoms are equivalent in the intermediate complex, for example,

$$CH_2D^+ + CH_4 \rightarrow C_2H_5^+ + HD \text{ (30%)[2]}, \quad C_2H_4D^+ + H_2 \text{ (70%)[5]}$$

where the quantities within ( ) and [ ] are respectively the observed product ion percentages and the product distributions expected on statistical grounds.

In some reactions, the deuterium was distributed in a non-statistical fashion; for example,

$$CH_2D^+ + C_2H_5OH \rightarrow C_2H_5^+ + CH_2DOH \text{ (40%)[4]}, \quad C_2H_4D^+ + CH_3OH \text{ (0%)[5]}$$
$$\rightarrow CH_3OH_2^+ + C_2H_3D \text{ (0%)[4]}, \quad CH_2DOH_2^+ + C_2H_4 \text{ (60%)[5]}.$$

From this result, the implied mechanism for the reaction is that in which the CH_3^+/CH_2D^+ attacks the C_2H_5OH at the O atom and no scrambling of the D and H atoms occurs within the intermediate complex, i.e. the H and D atoms are not all equivalent.

In certain reactions incorporation of deuterium was not observed in any of the product ions; for example,

$$C_2HD^+ + CH_3CHO \rightarrow CH_3CO^+ + C_2H_2D \text{ (33%)[1]}, \quad CH_2DCO^+ + C_2H_3 \text{ (0%)[1]}$$
$$\rightarrow CH_3CHO^+ + C_2HD \text{ (67%)[1]}, \quad CH_3CDO^+ + C_2H_2 \text{ (0%)[2]}.$$

Such reactions are expected to occur via weakly bound intermediate complexes where intimate mixing of the atoms cannot occur.

These results are of significance to interstellar chemistry in that they can indicate the distribution of deuterium amongst various interstellar molecules and thus guide observers to the more abundant deuterated species. Additionally, it should be possible to indicate the location of the D atom in the interstellar molecule by the presence or absence of particular product ions.

REFERENCES

Irvine, W. M., L. W. Avery, P. Friberg, H. E. Matthews, and L. M. Ziurys, 1988, Astro. Lett. and Communications 26, 167.

Millar, T. J., A. Bennett, and E. Herbst, 1989, Astrophys. J. <u>30</u>, 906, and references therein.

Smith, D., 1988, Phil. Trans. R. Soc. Lond. A <u>324</u>, 257, and references therein.

Smith, D., and N. G. Adams, 1987, Adv. Atom. Mol. Phys. <u>24</u>, 1.

REACTIONS OF SEVERAL HYDROCARBON IONS WITH ATOMIC HYDROGEN AND ATOMIC NITROGEN

W. Lindinger[a], A. Hansel[a], W. Freysinger[a] and E. E. Ferguson[b]

[a] Institut f. Ionenphysik, Universität Innsbruck
A 6020 Innsbruck, Austria

[b] Physico-Chimie des Rayonnements, Université de Paris Sud,
Orsay, France

Reactions of hydrocarbon ions with atomic hydrogen are of great importance in interstellar molecular synthesis, especially in diffusive clouds (Herbst and Klemperer, 1973). We have therefore investigated the reactions of $C_2H_n^+$, n = 2 to 6, C_3^+, C_3H^+ and $C_3H_2^+$ ions with H, and H_2, and some with D and D_2 in a SIFDT (selected ion flow drift tube) at 300 K in helium buffer gas at low E/N. The rate constants for various processes involved, such as binary reactions, isotope exchange and three-body association have been determined (Hansel et al., 1989). Also investigated were reactions of $C_2H_n^+$ (n = 2 to 4) and of $C_3H_2^+$ with atomic nitrogen as well as several nitrogen and sulfer bearing ions with N (Freysinger et al.). The reactions with H generally lead to a reduction of the hydrogen atoms within the ion in the form

$$C_iH_n^+ + H \rightarrow C_iH_{n-1}^+ + H_2,$$

or perform fast association processes,

$$C_iH_n^+ + H \xrightarrow{He} C_iH_{n-1}^+ + He.$$

Processes involving atomic deuterium usually lead to fast isotopic scrambling but also to association in some cases.

In the case of the $C_3H_2^+$ ion both isomers, cyclic $c\text{-}C_3H_2^+$ and linear $l\text{-}C_3H_2^+$, could be investigated separately. While the energetically higher isomer $l\text{-}C_3H_2^+$ undergoes fast isotopic scrambling with D (1×10^{-9} cm^3 sec^{-1}), the scrambling of the lower $c\text{-}C_3H_2^+$ isomer is slower by an order of magnitude. No scrambling was observed with molecular deuterium, neither for $c\text{-}C_3H_2^+$ nor for $l\text{-}C_3H_2^+$.

That the product of the association reaction between $C_3H_2^+$ and CO remains in the cluster form $C_3H_2^+ \cdot CO$ is indicated by the interesting switching process

$$C_3H_2^+ \cdot CO + H \rightarrow C_3H_3^+ + CO.$$

The investigations of reactions involving atomic nitrogen showed C-N, S-N and N-N bond formation to be generally the dominant processes.

In the case of CO_2^+ reacting with N, the reaction channel into the products $CO^+ + NO$ is strongly favored over the more exoergic channel $NO^+ + CO$, whereas in the case of $N_2O^+ + N$ both channels $NO^+ + N_2$ and $N_2^+ + NO$ are populated at the same rate despite the big differences of their exoergicities.

This work was supported by the Fonds zur Fröderung der Wissenschaftlichen Forschung, Projects P 6696 and P 6933.

REFERENCES

Freysinger, W., A. Hansel, and W. Lindinger, (to be published).
Hansel, A., R. Richter, W. Lindinger, and E. E. Ferguson, 1989, Int. J. Mass Spectrom. Ion Proc. (in press).
Herbst, E., and W. Klemperer, 1973, Ap. J. <u>185</u>, 505.

QUENCHING OF NO^+ (v=1,3) IN LOW ENERGY COLLISIONS WITH CH_4 AND He

A. Hansel, N. Oberhofer and W. Lindinger

Institut für Ionenphysik der Universität Innsbruck
Technikerstr 25 A 6020 Innsbruck, Austria

According to a model developed by Ferguson (1984), the quenching efficiency in low energy collisions between vibrationally excited ions and neutrals depends on the lifetime of the collision complex and thus on the polarizability of the neutral collision partner, which determines the well depth of the interaction potential. This in turn is strongly correlated with the lifetime of the ion-neutral-collision-complex.

In extension of earlier work performed in this laboratory (Federer et al., 1985; Richter and Lindinger, 1988; Lindinger, 1987; Kriegl et al., 1988), we have presently investigated the quenching of NO^+ (v=1) and NO^+ (v $\geq$ 3) in collisions with CH_4 and He, using a selected ion flow drift tube (SIFDT). The quenching rate coefficients obtained presently for NO^+ (v=1) colliding with CH_4 agree well with earlier findings for this process, showing a decline in the regime from thermal to 0.5 eV, followed by an increase at higher collision energies. The first regime corresponds to complex formation, with decreasing lifetime as the relative kinetic energy between the ion and the neutral is increased. The second regime lies energetically at high enough collision energies where complex formation no longer occurs. Therefore, the energy dependence of the quenching of NO^+ (v $\geq$ 1) with CH_4 (k_{q1}) is consistent with the model of Ferguson (1984). The quenching rate constant for NO^+ (v $\geq$ 3) with CH_4 (k_{q1}) lies a factor of three above the one for NO^+ (v $\geq$ 1) at thermal energies and shows a similar relative behavior as a function of collision energies as k_{q1}.

The quenching of NO^+ (v $\geq$ 3) with He shows a similar behavior as the one of N_2^+ (v = 1) with He. Due to the low binding energy of the N_2^+-He system, as well as the NO^+-He system, no complex formation occurs above thermal energies. Thus the observed increases in k_q reflect the ion-neutral interactions at the repulsive part of the potential.

Nonequilibrium Effects in Ion and Electron Transport
Edited by J. W. Gallagher *et al.*, Plenum Press, New York, 1990

REFERENCES

Federer, W., W. Dobler, F. Howorka, W. Lindinger, M. Durup-Ferguson, and
    E. E. Ferguson, 1985, J. Chem. Phys. $\underline{83}$, 3.
Ferguson, E. E., 1984, Vibrational Excitation and De-excitation and
    Charge-Transfer of Molecular Ions in Drift Tubes, 126, "Swarms of
    Ions and Electrons in Gases," Edited by W. Lindinger, T. D. Märk,
    and F. Howorka, (Springer Verlag, Wien, New York).
Kriegl, M., R. Richter, W. Lindinger, L. Barbier, and E. E. Ferguson,
    1988, J. Chem. Phys. $\underline{88}$, 1.
Lindinger, W., 1987, Int. J. Mass Spectrom. Ion Processes, $\underline{80}$, 115-132.
Richter, R., and W. Lindinger, 1988, J. Chem. Phys. $\underline{89}$, 9.

ROTATIONAL ENERGY EFFECTS IN ION-MOLECULE REACTIONS

A. A. Viggiano, Robert A. Morris[*], T. Su[**] and
John F. Paulson

Ionospheric Physics Division
Air Force Geophysics Laboratory
Hanscom AFB, MA 01731-5000

The variable temperature selected ion flow drift tube (VT-SIFDT) has
proven to be a valuable tool for measuring the effect of different types of
energy on the rate constants of ion-molecule reactions. Recently, we have
devised a technique involving a VT-SIFDT to derive the dependence of such
rate constants on the rotational temperature of the reactant neutral
(Viggiano et al., 1988). This technique involves measuring the kinetic
energy dependences of the rate constant at several temperatures. The center
of mass kinetic energy (KE_{cm}) in a VT-SIFDT is given by

$$KE_{cm} = \frac{(m_i + m_b)m_n}{2(m_i + m_n)} v_d + \frac{3}{2} kT$$

where m_i, m_b and m_n are the masses of the ion, buffer and neutral, v_d is the
ion drift velocity, and T is the temperature. A particular center of mass
kinetic energy can be obtained with varying contributions from the thermal
component (second term) and the drift tube component (first term). If the ion
is monatomic and the neutral has no low frequency vibrations, any difference
at a particular kinetic energy between the measured rate constants at various
temperatures can be attributed to a rotational energy effect. Monatomic ions
are required because the internal modes of polyatomic ions are excited in a
drift tube. The temperature dependence at a particular kinetic energy would
then reflect varying amounts of internal excitation of the ion as well as the
rotational temperature dependence of the neutral, a situation that is

---

[*]
    On contract to the Geophysics Laboratory from System Integration
    Engineering.
[**]
Permanent address: Southeastern Massachusetts University, North
Dartmouth, MA 02747.

difficult to interpret. If the neutral has low vibrational frequencies the temperature dependence would reflect the total internal temperature dependenc rather than purely the rotational dependence.

The first series of reactions studied by this technique included the reactions of O^- with CH_4, N_2O, and SO_2. The preliminary results on these reactions reported at this meeting in 1987 indicated that the reaction of O^- with CH_4 had a rotational temperature dependence and the others did not (Viggiano et al., 1988). Since that time we have found that a problem arises with our previous neutral inlet design when the drift field is used, thus invalidating the results (Viggiano et al., 1988). We have remeasured the rate constants for these reactions, and our present results indicate no detectable rotational temperature dependence for any of these reactions.

For polar molecules, the collision rate constant will depend on temperature at a particular center-of-mass kinetic energy. For HCl, theory predicts that at 0.063 eV center-of-mass kinetic energy, the collision rate constant decreases by 18% with increasing temperature over the temperature range 173 K to 488 K (Su and Chesnavich, 1982). This effect reflects the fact that at higher temperatures the locking of the HCl dipole to the ion during a collision is less efficient as the temperature increases at a fixed kinetic energy. We are currently investigating whether this may be seen experimentally for the reactions of several noble gas positive ions with HCl. Preliminary results indicate that a temperature dependence even larger than this is observed for Kr^+ + HCl. These results will be presented.

REFERENCES

Su, T., and W. J. Chesnavich, 1982, J. Chem. Phys. 76, 5183. These are
    the results of trajectory calculations.
Viggiano, A. A., R. A. Morris, and J. F. Paulson, 1988, J. Chem. Phys.
    89, 4848; ibid., (in press).

THE APPLICATION OF A SELECTED ION FLOW DRIFT TUBE TO THE DETERMINATION OF PROTON AFFINITY DIFFERENCES

M. Tichy[a], G. Javahery[a], N. D. Twiddy[a] and E. E. Ferguson[b]

[a] Department of Physics, U. C. W. Wales, Aberystwyth
SY23 3BZ, UK

[b] Laboratoire de Physico-Chimie des Rayonnements
Universite de Paris-Sud, Bat.350, 91405
Orsay, Cedex, France

ABSTRACT

The forward and reverse rate constants for nine proton transfer reactions have been measured as a function of relative kinetic energy using a selected ion flow drift tube (SIFDT). In all but two cases van't Hoff plots of the equilibrium constant against reciprocal centre-of-mass collision energy were linear, and values of the enthalpy and entropy changes were obtained from slope and intercept, respectively. Since ΔH is a measure of the difference between the proton affinities of the two neutral species, the data can be used to provide a proton affinity difference ladder. This ladder agrees extremely well with the established proton affinity scale. The experimental values of entropy change agree well with values calculated from the entropies of the individual ions and neutrals. The agreements of the ΔH's and ΔS's so determined establishes the validity, and utility, of a SIFDT apparatus for proton affinity studies, when linear pseudo-van't Hoff plots are obtained. In the present study two N_2OH^+ measurements gave non-linear Arrhenius and van't Hoff plots and had to be rejected, in agreement with the earlier work. Some speculations of why drift tube measurements lead to reliable thermodynamic data, in spite of the lack of thermodynamic equilibrium between internal and translational modes, are presented.

Values of ΔH and ΔS from $\ln K_{eq} = \Delta S/R - 3\,\Delta H/2E_{cm}$
ΔH is in kcal mol^{-1} and ΔS is in cal K^{-1} mol^{-1}

Reaction	$-\Delta H$	$-\Delta H^*$	ΔS	$\Delta S(calc)$[#]
$H_2Br^+ + CO = HCO^+ + HBr$	2.0	2.6	+1.4	0
$CH_5^+ + HBr = H_2Br^+ + CH_4$	9.1	8.8	-1.7	-1.0
$H_2Cl^+ + HBr = H_2Br^+ + HCl$	6.0	5.8	+0.6	+0.3
$CF_4H^+ + HCl = H_2Cl^+ + CF_4$	7.0	6.5	-0.2	-1.3
$CF_4H^+ + CH_4 = CH_5^+ + CF_4$	4.0	3.5	-1.7	0
$CF_4H^+ + NO = HNO^+ + CF_4$	0.6	0.5	-3.7	-2.5
$CF_4H^+ + CO_2 = HCO_2^+ + CF_4$	1.7	2.0	+0.4	-1.3

[*] Adams, N. G., D. Smith, M. Tichy, G. Javahery, N. D. Twiddy, and E. E. Ferguson, J. Chem. Phys., in press.

[#] Benson, 1976, "Thermochemical Kinetics," 2nd Ed., Wiley, NY.

ENERGY DEPENDENCES OF RATE CONSTANTS FOR THE REACTION $^{22}Ne^+$ + ^{20}Ne AT SEVERAL TEMPERATURES

Robert A. Morris,[*] T. Su,[**] A. A. Viggiano and
John F. Paulson

Ionospheric Physics Division
Geophysics Laboratory
Hanscom AFB, MA 01731-5000

Rate constants for the symmetric charge transfer reaction $^{22}Ne^+$ + ^{20}Ne $\rightarrow$ $^{20}Ne^+$ + ^{22}Ne were measured using a temperature variable-selected ion flow drift tube (SIFDT) instrument. The experiment employs a temperature variable flow tube with which to control the thermal energy of the reagents as well as an electric drift field to increase independently the kinetic energy of the ionic reactant. This permits determining rate constants for ion-molecule reactions as a function of ion kinetic energy at various temperatures in the range 85 - 500 K. In the present work, the reaction $^{22}Ne^+$ + ^{20}Ne was studied as a function of kinetic energy at three temperatures: 193, 297, and 473 K.

The rate constants were found to increase with both increasing temperature and kinetic energy. As shown in Fig. 1, when the rate constants are plotted versus center of mass kinetic energy, the energy dependences measured at the three different experimental temperatures fall on the same curve within experimental uncertainty. This is not surprising since the monatomic ionic and neutral reactants have no internal degrees of freedom and therefore changing either temperature or ion kinetic energy has the same simple effect of changing the overall collision energy. The rate constants vary with collision energy as $E^{0.33}$ and with temperature as $T^{0.28}$.

Rate constants for this reaction were computed via classical trajectory calculations using the probability function of Rapp and Francis (as corrected by Dewangen; Dewangen, 1973) as a function of

---

[*] Air Force Geophysics Scholar.

[**] Chemistry Department, Southeastern Massachusetts University, N. Dartmouth, MA.

Nonequilibrium Effects in Ion and Electron Transport
Edited by J. W. Gallagher *et al.*, Plenum Press, New York, 1990

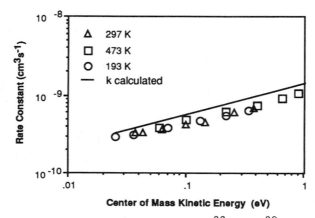

Fig. 1. Rate constants for the reaction $^{22}Ne^+ + {}^{20}Ne$ plotted as a function of center-of-mass kinetic energy.

internuclear separation. The calculations make use of the potential energy function for Ne_2^+ which has been calculated by Michels et al., (Michels et al., 1978). The present method allows the relative velocity to vary in response to the potential energy function and allows the trajectories to depart from rectilinear paths.

Calculated rate constants are also shown in the figure and are in reasonably good agreement with the measured values although the calculated points are systematically higher by 20 to 40%. The magnitude of the calculated energy dependence ($E^{0.42}$) is slightly higher than that determined experimentally, but the values agree within the combined uncertainties.

REFERENCES

Dewangen, D. P., 1973, J. Phys. B 6, L20, and references therein.
Michels, H. H., R. H. Hobbs, and L. A. Wright, 1978, J. Chem. Phys. 69, 5151.

DEEXCITATION OF HE(2^1P) IN COLLISIONS WITH RARE GAS ATOMS[*]

Masatoshi Ukai, Hiroaki Yoshida, Yasumasa Morishima,
Hidenobu Nakazawa, Kyoji Shinsaka and Yoshihiko Hatano

Department of Chemistry
Tokyo Institute of Technology
Meguroku, Tokyo 152, Japan

Deexcitation of lowest excited rare gas atoms is of great importance to elucidate not only gas phase reaction dynamics but also fundamental energy transfer processes in ionized gases. Experimental studies of the optically allowed or the resonance states have been reported in only a few papers (Ukai et al., 1986a; Ukai et al., 1988; Ukai et al., 1986b) because of experimental difficulties. This paper presents deexcitation cross sections of He(2^1P) by Ar, Kr, and Xe obtained using a pulse radiolysis method in a region of mean collisional energy between 13-38 meV. Fairly large cross sections of around or above 100Å^2 and their collisional energy dependence have been obtained and interpreted as Penning ionization cross sections based on a long-range dipole-dipole interaction (Watanabe and Katsuura, 1967; Kohmoto and Watanabe, 1977). Validity of the theoretical formula for Penning ionization cross sections by Watanabe and Katsuura (1967) is discussed. Two kinds of cross sections have also been calculated by means of the impact parameter method with the aid of experimentally simulated classical trajectories; in one procedure, the polarization of the p-state helium has been assumed to rotate in order to keep a collinear or perpendicular configuration with respect to the interatomic axis; in the other procedure, the polarization axis is fixed in a certain direction. The classical motion of the particles have been shown to cause considerable influence on the absolute values and the collisional energy dependence of the cross sections. The influence has increased accordingly to the attractive force of the interatomic potential, i.e., in order of Ar<Kr<Xe. A modified form of dipole-dipole autoionization width partially contributed from the electron exchange interaction is also discussed. It has been suggested that the rotation of the p-state atomic polarization depends

---

[*] A paper on this study is to appear in J. Chem. Phys., 1989.

strongly on the van der Waals interaction with the target atoms. The effect
of the rotation has been shown to be most prominent for Xe but small for Ar.

REFERENCES

Kohmoto, M., and T. Watanabe, 1977, J. Phys. B 10, 1875.
Ukai, M., H. Koizumi, K. Shinsaka, and Y. Hatano, 1986b, J. Chem. Phys.
    84, 3199.
Ukai, M., H. Nakazawa, K. Shinsaka, and Y. Hatano, 1988, J. Chem. Phys.
    88, 3623.
Ukai, M., Y. Tanaka, H. Koizumi, K. Shinsaka, and Y. Hatano, 1986a, J.
    Chem. Phys. 84, 5575.
Watanabe T., and K. Katsuura, 1967, J. Chem. Phys. 47, 800.

MEASUREMENTS OF D_T/K FOR SODIUM IONS DRIFTING IN ARGON

M. J. Hogan and P. P. Ong

Department of Physics
National University of Singapore
Kent Ridge, Singapore 0511

Few measurements have been made to date of transverse diffusion coefficients of ions drifting in a neutral gas and uniform electric field although they are of both experimental and theoretical interest. A new system for making such measurements has been developed, and results for D_T/K of Na^+ ions in Ar at E/N values ranging from 10 to 260 Td are reported here.

The data are obtained by introducing an ion current equal to a few pA into a drift tube through a narrow slit and measuring the current collected by an array of 65 separate rods positioned in a plane perpendicular to the drift tube axis at a fixed distance from the slit. The current to each rod is recorded in turn by a sensitive electrometer under computer control.

It can be shown that the ion current at the rods (in parameterized form) is given by,

$$I(x) = A \{ 1/Q^2 - 3B(x-d)^2/Q^4 \}\exp[-(x-d)^2/Q^2]\exp[B(x-d)^4/Q^4],$$

where x is the perpendicular distance of the rod from the central axis, A and B are parameters, d is an offset of the profile maximum from the center of the drift tube (if any) and $Q^2 = 4zD_T/KE$, where z is the drift distance, E is the electric field, D_T is the transverse diffusion coefficient and K is the mobility. To obtain the value of D_T/K, the data are fitted to the above equation and, since z and E are known, the ratio is determined from Q.

The results obtained the Na^+ in Ar are shown in Fig. 1. A mass spectrum of the ions established conclusively that Na^+ was the only ion

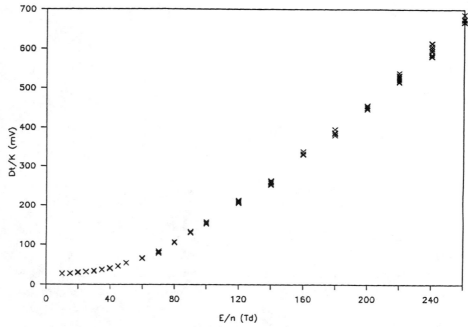

Fig. 1. Ratios of Dt/K for Na$^+$ ions in argon.

present. The standard deviation of the values is less than 3%. Data were obtained at pressure ranging from 20.0 to 33.3 Pa and at temperatures between 294 and 303°C. The drift tube pressure was monitored by a capacitance manometer and controlled by a servovalve to better than 0.5%. All of the values were adjusted to the same temperature (i.e., 297°C) using an approximation given by Stefansson et al., (1988).

There have been no previous direct measurements made with which to compare the present results. However, it is possible to check their validity at low values of E/n. In this range a plot of the values versus $(E/n)^2$ should be linear and extrapolate to the zero field value which equals kT/q. These results when so plotted are fitted well by a straight line and extrapolate to the expected value within the standard error of the linear regression fit.

REFERENCE

Stefansson, T., T. Berge, R. Lausund, and H. R. Skullerud, 1988, J. Phys. D.: Appl. Phys. 21, 1359.

TRANSVERSE DIFFUSION OF NEON IONS IN NEON

Thorarinn Stefánsson

Department of Physics and Mathematics
Norwegian Institute of Technology
N-7034 Trondheim-NTH, Norway

The ratio D_T/μ between the transverse diffusion coefficient and the mobility for $^{20}Ne^+$ ions in neon has been determined from the variance of directly measured transverse current density distributions of mass analyzed ions at 294 K and $20 \leq E/N_o \leq 2000$ Td using a variable length drift tube mass spectrometer. The accuracy in the results, expressed as two standard deviations, is believed to be better than $\pm$ 4% except at $E/N_o \leq$ 150 Td where the accuracy gradually decreases with E/N_o to $\pm$ 8% at 20 Td.

According to diffusion theory, the relation between the variance $\langle x^2 \rangle_J$ of the transverse current density distribution function $J(z;x)$ and the ratio D_T/μ is given by

$$\langle x^2 \rangle_J = 2(D_T/\mu) \, z/E + a$$

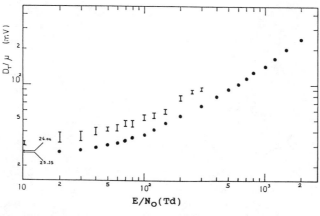

Fig. 1.

where the constant a represents the influence of end effects. By determining $\langle x^2 \rangle_J$ for at least two different drift lengths z, the constant can be eliminated. A detailed description of the apparatus and the method has been given by Thórarinn Stefánsson et al. (1988).

The figure shows the ratio D_T/μ as a function of E/N_o. Full circles: present experiment. Vertical bars: Mårk et al. (1984). The present measurements were performed at 294 K and should be expected to extrapolate to a zero field value of 24.35 mV. The measurements of Mårk et al. were performed at 302 K and should be expected to extrapolate to a zero field value of 26.06 mV. As can be seen from the figure, the present values extrapolate to the expected value. The values of Mårk et al. are higher than the present values by an approximately constant factor and do not extrapolate to the expected zero field value. One is tempted to suspect that there may be a systematic error in the values of Mårk et al.

REFERENCES

Mårk, T. D., G. Sejkora, P. Girstmair, M. Hesche, E. Mårk, H. C. Bryant, and M. T. Elford, 1984, Proc. 4th Symp. of Atomic and Surface Physics, Maria Alm Salzburg, F. Howorka, W. Lindinger, and T. Mark, (Innsbruck: Studia) pp. 144-154.
Stefánsson, Thórarinn, T. Berge, R. Lausund, and H. R. Skullerud, 1988, J. Phys. D: Appl. Phys. 21, 1359-1370.

HELIUM ION CLUSTERS He_n^+ ($n \leq 16$) FORMED IN A LIQUID HELIUM COOLED DRIFT TUBE

Takao Kojima, Nobuo Kobayashi and Yozaburo Kaneko

Department of Physics
Tokyo Metropolitan University
2-1-1 Fukazawa
Setagaya-ku Tokyo 158, Japan

Helium ion clusters $^4He_n^+$ ($n \leq 16$) have been investigated with a selected ion drift tube mass spectrometer.

The drift tube of length 10cm is settled inside a cryogenic system which can be cooled with liquid helium. Reactant ions He^+ are injected into the drift tube which is filled with He gas. The temperature and pressure of He gas are 4.4-4.5 K and 0.02-0.06 Torr, respectively. Colliding with He atoms and forming ion clusters He_n^+, the injected ions drift toward the end of the drift tube. The product ions ejected from the drift tube are focused on the entrance of a quadrupole mass spectrometer and mass analyzed.

Typical mass spectra of product He_n^+ are shown in Fig. 1. These spectra show a little different feature from our previous work (Kobayashi et al., 1988). We have improved the voltage distribution of the focusing lenses set in front of the quadrupole mass spectrometer, so that we can remove the influence of dissociative collisions with background gas molecules in the focusing acceleration region. A maximum appears in the n-distribution at n = 14, 11 and 10 with drift field strength E = 1.0-2.0, 3.0 and 4.0 V/cm respectively.

Drift field dependence of the relative intensity of the mass spectra is shown in Fig. 2. The sum of intensities of He_n^+ is normalized to 1, i.e., $\Sigma I_n = 1$. In Fig. 2, each curve of relative intensities shows a maximum and decreases to disappear in order from larger n when the drift field is increased. The intervals of the decreasing curves between n = 10 and 11, n = 11 and 12, n = 14 and 15 and n = 15 and 16 are much larger than others. Since the average energy of collisions between the ions and He atoms in the drift tube is a function of drift field when pressure

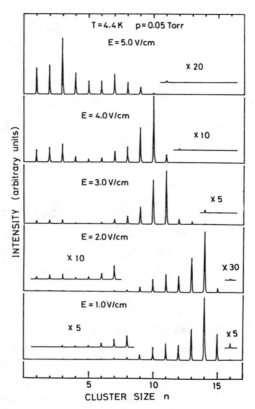

Fig. 1. Mass spectra of He$_n^+$.

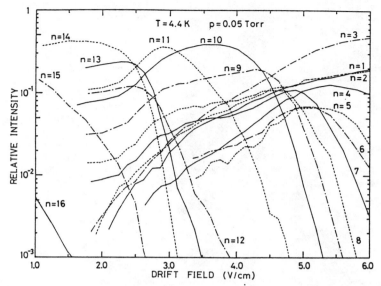

Fig. 2. Drift field dependence of relative intensity of He_n^+.

and temperature of the gas are constants, these features indicate the energetic stability of the helium ion cluster, He_n^+. Thus, in this case, $n = 10$, 11, 14 and 15 are considered as numbers which indicate the size of more stable helium ion clusters. Although $n = 10$ and 14 have also been observed as magic numbers in our previous work (Kobayashi et al., 1988) and in a nozzle beam experiment by King et al. (Stephens and King, 1983), we don't know of any other investigation reporting $n = 11$ and 15 as magic numbers. It is worth noting that $n = 14$ or 15 is just one or two larger than $n = 13$ which is needed to form the first body-centered icosahedral structure. Since molecule-like ions He_2^+ and He_3^+ are stable, we may suppose that the clusters of $n = 14$ and $n = 15$ have He_2^+ centered and He_3^+ centered (distorted) icosahedral structures, respectively.

REFERENCES

Kobayashi, N., T. Kojima, and Y. Kaneko, 1988, J. Phys. Soc. Jpn. <u>57</u>, 1528.
Stephens, P. W., and J. G. King, 1983, Phys. Rev. Lett. <u>50</u>, 1676.

ION AND FAST NEUTRAL MODEL FOR NITROGEN AT VERY HIGH E/N

A. V. Phelps

Joint Institute for Laboratory Astrophysics
University of Colorado and
National Institute of Standards and Technology
Boulder, CO 80309-0440

A model has been developed to explain the measured pressure dependences of emission excited in very low current, uniform electric field discharges in N_2 at very high E/N. Earlier observations by Jelenković and Phelps (1987) of the spatial and E/N dependence of emission from dc discharges for these conditions were interpreted as suggesting that most of the radiation from the triplet states of N_2 results from excitation by fast ions or neutrals, while the emission from N_2^+ resulted from simultaneous excitation and ionization of the N_2 by electrons. More recent observations by Gylys and Phelps (1988) of the time dependent emission from laser-induced, photoelectron-initiated avalanches at voltages below breakdown showed that the emission from the triplet states of N_2 was delayed in time relative to the electron current pulse as expected for heavy particle excitation. However, the models used to analyze the data were not sufficiently detailed to distinguish excitation by the various ions (Gylys et al., 1989) and neutrals likely to be present.

In the experiments (Gylys and Phelps, unpublished) used to obtain data for the present analysis, the technique developed for the transient current growth measurements in N_2 (Gylys et al., 1989) was used to observe the growth of emission with pressure (or voltage) at fixed E/N. The emission signals were integrated over the transient resulting from the release of electrons from the cathode by the pulsed laser. The emission from the 337 nm 2nd Positive band and 670 nm 1st Positive band of N_2 and the 820 nm line of N I were normalized to the emission from the 391.4 nm 1st Negative band.

The model includes the drift of N_2^+, the formation of fast N_2 in charge transfer collisions, the breakup of N_2^+ to form fast N^+, and the charge transfer of N^+ to form fast N. The energy distribution of the drifting N_2^+

determines the initial energy distribution of the fast N_2. Since the mean-free-paths of the N_2, N^+, and the N are of the order of the separation of the electrodes, the transport of these species is treated with simple non-equilibrium models. The available cross section data was summarized earlier (Phelps, 1987). The 820 nm line of N I is assumed to be excited by fast N^+ and N. Using reasonable estimates for the cross sections for excitation of N_2 by the fast ions and neutrals, we obtain satisfactory agreement with the observed intensity ratios for 52 kTd and voltages below 2 kV. At voltages between 3 and 4.7 kV and E/N = 52 kTd, the agreement is significantly improved when ionization of N_2 by fast ions and neutrals is added to the model. Such heavy particle ionization was not in evidence in the analysis of current growth in N_2 (Gylys et al., 1989).

REFERENCES

Gylys, V. T., B. M. Jelenkovic, and A. V. Phelps, 1989, J. Appl. Phys. (in press).
Gylys, V. T., and A. V. Phelps, 1988, Bull. Am. Phys. Soc. 33, 136.
Gylys, V. T., and A. V. Phelps, unpublished.
Jelenkovic, J. M., and A. V. Phelps, 1987, Phys. Rev. A 36, 5310.
Phelps, A. V., 1987, "Electronic and Atomic Collisions," Edited by J. Geddes, et al., (North Holland, Amsterdam) p. 690.

STUDY OF THE ELECTRON TRANSPORT IN AN RF DISCHARGE IN Ar AND CH_4/H_2 BY OPTICAL EMISSION SPECTROSCOPY

F. Tochikubo, T. Kokubo and T. Makabe

Department of Electrical Engineering
Faculty of Science and Technology
Keio University
3-14-1 Hiyoshi, Yokohama 223, Japan

INTRODUCTION

There are many theoretical and experimental studies of an electron swarm in a DC field. The time behavior of electron transport in an RF field is also studied from the Boltzmann equation (Winkler et al., 1985; Makabe and Goto, 1988). But we have few information about the electron transport in an RF discharge plasma, which is mainly used for plasma processing, because of the existence of the sheath and bulk region. Optical emission spectroscopy is one of the methods to know the electron transport with the energy of $\varepsilon > \varepsilon_{th}$ for the emission. In the present work, we have shown the high energy electron transport in the RF (13.5MHz) discharge in Ar and in CH_4 (10%)$/H_2$ by the time- and space-resolved optical emission spectroscopy established by a single photon counting method.

EXPERIMENT

Parallel plate electrodes made of aluminum with a 8.0 cm diameter and with a 2.0 cm spacing are positioned in the center of the chamber. In order to achieve a time-resolved spectroscopy, we use a time-to-pulse height converter (TPHC), which is usually used for the measurement of the radiative lifetime. In this manner, the resolution of ~ 0.2 ns is obtained. The excitation rate to the state (j) by the direct electron impact is expressed as

$$\xi_j(t;z)= \sqrt{\frac{2}{m}}\; n_e(t;z) \cdot N \int_{\varepsilon_{th}}^{\infty} f(\varepsilon,t;z)\sqrt{\varepsilon}\; Q_j(\varepsilon)\; d\varepsilon, \qquad (1)$$

where m and ε are electron mass and energy, N the number density of the molecule, f(ε,t:z) the local instantaneous energy distribution, and $Q_j(\varepsilon)$ the collisional excitation cross section to the state (j) with a threshold energy ε_{th}. The emission intensity $\Phi_{jk}(t;z)$ for transition from the state (j) to (k) is expressed by the convolution integral as

$$\Phi_{jk}(t;z)=k_o \int_{-\infty}^{t} \frac{\xi_j(t';z)+\zeta_j(t';z)}{\tau_{rad}} \exp\left(-\frac{t-t'}{\tau_{eff}}\right)dt', \qquad (2)$$

where $\zeta_j(t;z)$ shows the other production rate (e.g., cascade and reaction processed); k_o is a constant. The effective lifetime τ_{eff} in Eq. (2) is given by

$$1/\tau_{eff} = 1/\tau_{rad} + k_q \cdot N, \qquad (3)$$

where τ_{rad} is a radiative lifetime from the state (j) and kq denotes the self-quenching rate. The excitation rate $\xi_j(t';z)+\zeta_j(t';z)$ is obtained from the emission intensity $\Phi_{jk}(t;z)$ by deconvoluting Eq. (2), and then we will estimate the quenching rate.

In this paper, induced emissions from the RF glow discharge in Ar are mainly discussed. The observed lines are Ar I ($3p_5 \rightarrow 1s_4$: 419.8 nm, τ_{rad} = 98 ns, ε >14.57 eV), Ar II ($4p^4D_{7/2} \rightarrow 4s^4P_{5/2}$:434.8 nm, τ_{rad} = 11.6 ns, ε > 35.05 eV). Therefore, we can observe, in principle, the transport of the electrons with the energy of ε > 14.57 eV and ε > 35.05 eV.

RESULTS AND DISCUSSION

Here, we mainly discuss about the result at 1.0 Torr. Fig. 1 shows the time- and space-resolved emission profile of Ar I ($3p_5 \rightarrow 1s_4$). In this case, the magnitudes of the applied voltage and the current are 40 V and 50 mA, respectively, and the phase of the applied voltage has the delay of 57° from that of the current.

The relative excitation rate is obtained by deconvoluting the emission in Fig. 1, and is shown in Fig. 2. The quenching rate of Ar($3p_5$) is estimated at 6.0×10^{-10} $cm^3 s^{-1}$. Fig. 3 shows (a) the time- and space-resolved emission of Ar II ($4p^4D_{7/2} \rightarrow 4s^4P_{5/2}$) and (b) the relative excitation rate of it near the powered electrode. In this case, the applied peak-to-peak voltage is 245 V with a negative self-bias of 47.5 V and the power is 50 mWcm^{-2}. Ar II ($4p^4D_{7/2}$) is hardly quenched. In the sheath region, the phase of the excitation rate by the electrons with ε > 14.57 eV in Fig. 2 is earlier than that of the applied voltage

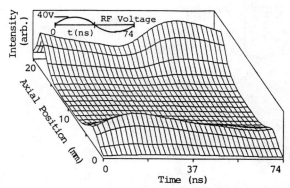

Fig. 1.   Time-resolved emission intensities from Ar I ($3p_5 \rightarrow$
$1s_4 : \lambda$ = 419.8nm, τ_{rad} = 98ns, ε > 14.57eV) in Ar at 13.5
MHz, 1 Torr, 10 sccm and 13 mW/cm^2 as a parameter of
distance from the grounded electrode.

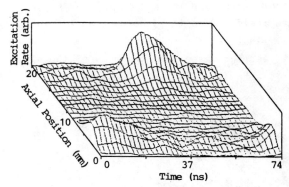

Fig. 2.   Deconvoluted result of the emission from Ar I ($3p_5 \rightarrow 1s_4$)
in Fig. 1, as a parameter of distance from grounded
electrode. τ_{eff} is estimated at 34 ns.

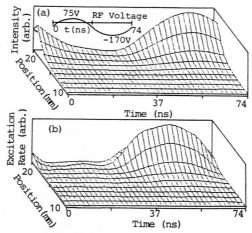

Fig. 3. (a) Time-resolved emission intensity from Ar II ($4p^4D_{7/2} \rightarrow 4s^4P_{5/2}$:434.8nm, τ_{rad} = 11.6ns, ε > 35.05eV) and (b) the deconvoluted result of the emission in Ar at 13.5 MHz, 1 Torr, 10 sccm and 50 mW/cm^2 as a parameter of distance from the grounded electrode.

and almost corresponds to the phase of the current, because the electron density at the phase is higher near the electrode. On the other hand, the phase of the excitation rate by the electrons with ε > 35.05 eV in Fig. 3(b) almost corresponds to that of the applied voltage, because it must be accelerated by the strong sheath field. Then it is noted that the intensity of Ar II is one fiftieth of Ar I. That is, the density of the electrons with the lower energy becomes higher at earlier phase from the peak of the applied voltage.

REFERENCES

Makabe, T., and N. Goto, 1988, J. Phys. D 21, 887-95.
Winkler, R. et al., 1985, Annalen der Physik 42, 537-58.

H_β LINE SHAPES IN RF DISCHARGES IN $CHClF_2$ AND H_2

S. Radovanov, S. Vrhovac, Z. Petrović and B. Jelenković

Institute of Physics
P.O. Box 57, 11001 Belgrade
Yugoslavia

Balmer line shapes in RF glow discharges can be used to study the translational energy distribution of fragments to reveal the excitation mechanisms of molecular excited states and to improve our understanding of processes involved in etching and deposition.

In this work Doppler-broadened H_β line profiles in parallel plate RF (27 MHz) discharges in pure $CHClF_2$ (freon 22) and in pure H_2 were measured. All measurements were done at two different positions: in the vicinity of the powered electrode and in the mid gap. In order to see the effect of the changing pressure we made measurements at 400 and 70 mtorr. After correcting the H_β line profile for the instrumental width, which was found to be 0.03 nm, the mean kinetic energy of the $H^*(n=4)$ fragments was determined from FWHM of the line with the assumption that the fragment energy distribution function is Maxwellian (Tokue et al., 1975).

In $CHClF_2$ discharge we have found single component profiles of more or less the same shape at different positions. Also, at different pressures, profiles were the same even though both peak to peak and bias voltage increased by a factor of 2 (at 70 mtorr). The FWHM was found to be 0.065 nm corresponding to 2.8 eV mean energy of H^* fragments. Experiments with polyatomic atomic molecules containing hydrogen have shown that the energy distribution function is broad and has mean energy of about 2.5 eV (Kouchi et al., 1982), but no results for $CHClF_2$ seem to be available.

In pure hydrogen discharge of 400 mtorr FWHM of 0.06 nm in bulk and 0.07 nm in the sheath of the discharge were obtained. These correspond to energies of 2.5 and 3.6 eV respectively. The first value is in good agreement with the energy of an H(1s) atom after the dissociative excitation of H_2 into $b^3\Sigma\ 2p\sigma$ state. However, profiles obtained in the sheath have a broad component with energies larger than 10 eV (Cappelli et al., 1985).

Nonequilibrium Effects in Ion and Electron Transport
Edited by J. W. Gallagher *et al.*, Plenum Press, New York, 1990

In H_2 discharge at 70 mtorr, operating at very high voltages, FWHM in the sheath increases to 0.08 nm (5 eV). The broad component is present both in the sheath and in the bulk of the discharge.

Our study is aimed at obtaining enough data to determine whether unusually broad Balmer lines originate from direct dissociative excitation, excitation through collisions of fast ions with the surface (Li Ayers and Benesch, 1988) or dissociative excitation from vibrationally excited molecules (Baravian et al., 1987).

REFERENCES

Baravian, G., Y. Chouan, A. Ricard, and G. Sultan, 1987, J. Appl. Phys. 61, 5249.
Cappelli, A. L., R. A. Gottscho, and T. A. Miller, 1985, Plasma Chem. Plasma Process 5, 317.
Kouchi, N., M. Ohno, K. Ito, N. Oda, and Y. Hatano, 1982, Chem. Phys. 67, 287.
Li Ayers, E., and W. Benesch, 1988, Phys. Rev. 37, 194.
Tokue, I., I. Nishiyama, and K. Kuchitsu, 1975, Chem. Phys. Lett. 35, 69.

TIME-RESOLVED INVESTIGATIONS OF H_2 AND $H_2:CH_4$ R.F. PLASMAS

S. C. Haydon, W. Hugrass and H. Itoh

Department of Physics
University of New England
Armidale, NSW 2351 Australia

A pulsed, 10 MHz, variable pulse-length r.f. supply, and image intensi-fier system having gains $\sim 10^7$ and time-resolved photography using 10 ns gating pulses to control the opening time of the first stage shutter, have been used to access information at any phase of a single r.f. cycle and at various stages of the r.f. pulse. This paper reports measurements made in H_2 and $H_2:CH_4$ mixtures.

Previous investigations of this kind (Sato and Haydon, 1984; Kemp et al., 1987) emphasized the r.f. behavior of N_2 and $N_2:O_2$ mixtures using plane-plane and point-plane electrode geometry. Increasing interest in the practical applications of these r.f. techniques has directed our attention to the production of amorphous and diamond-quality carbon deposits, and we report now the results of some preliminary investigations of the many parameters controlling the r.f. processes.

One of the most important of these parameters is atomic hydrogen. Consequently we have looked for evidence of atomic hydrogen concentrations by monitoring the $H_\alpha(\lambda = 656.2$ nm) emissions from the r.f. discharges in both point-plane and plane-plane geometry at various pressures in the range 0-40 torr. Some of the more significant observations that emerged were:

i) with point-plane electrode geometry the H_α-intensities were highest at high pressure in regions close to the sharp point.

ii) with plane-plane electrode geometry the H_α-intensities are highest at large p close to each plane electrode and greater near the copper rather than the stainless steel electrode.

iii) As the pressure is reduced a redistribution of H_α-intensities occurs to the body of the plasma.

iv) H_α-intensities are some one hundred times greater with point-plane electrode geometry at the same gas pressures.

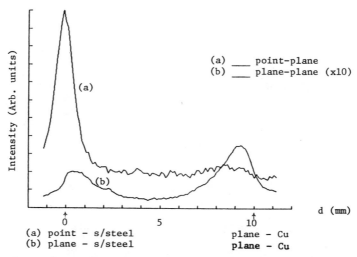

Fig. 1. Comparison of H$_\alpha$ (λ = 656.2 nm) intensities of point-plane and
plane-plane geometries of P = 10 torr.
(a) ——— point-plane, (b) ——— plane-plane (x10).

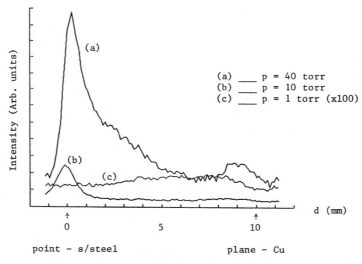

Fig. 2. Comparison of H$_\alpha$ (λ = 656.3 nm) intensities for point-plane
geometry at various pressures. ■ 40 torr, □ 10 torr, ◇ 1 torr
(x 100).

Figures 1 and 2 illustrate some of these salient features.

Interpretation of these preliminary observations has highlighted the importance of a variety of related parameters including the relative sizes of the powered and unpowered electrodes, the effect of de-tuning the r.f. signal, the nature of the coupling of the r.f. power to the interaction chamber, etc. This has prompted modifications to the existing ionization chamber to provide more reliable access to the influence of such parameters and further investigations of both H_2 and $H_2:CH_4$ mixtures are presently in progress. Further details of these studies will be presented at the seminar.

This work has been supported by ARC, AINSE and ERB.

REFERENCES

Kemp, W., N. Sato, and S. C. Haydon, 1987, Proc. XVIIIth Int. Conf. on Phenomena in Ionized Gases, Swansea, U.K., 822-23, July.
Sato, N., and S. C. Haydon, 1984, J. Phys. D: Appl. Phys., 17, 2009-2021 and 2023-2036.

SCATTERING OF ELECTRONS OF HIGH-MOLECULAR RYDBERGS IN DENSE ATOMIC AND MOLECULAR GASES

U. Asaf[*], K. Rupnik, W. S. Felps and S. P. McGlynn

Department of Chemistry
Louisiana State University
Baton Rouge, LA 70803

The effect of rare gases, rare gas mixtures and hydrogen on the high-n Rydberg states of methyl iodide and benzene has been discussed recently (Köhler et al., 1987; Asaf et al., 1989a; Asaf et al., 1989b; Reininger et al.,). According to the experimental evidence, the pressure induced energy shift of these molecular Rydbergs varied linearly with density up to relatively high perturber densities ($\sim 1 \times 10^{21} cm^{-3}$). These linearities can be reproduced accurately by the Fermi model modified by the Alekseev-Sobel'man polarization term (Alekseev and Sobel'man, 1966).

Electron scattering cross sections of rare gases obtained from these experimental results (Table 1) agree well with existing data for "classical" pressure shift studies using atomic Rydbergs. Electron scattering cross sections of rare gases as measured by pressure shift agree with cross sections near or at zero electron energies obtained from swarm data (Gilardini, 1972); however, large differences exist relative to H_2 swarm (Gilardini, 1972) and TOF (Ferch et al., 1980) data which are higher by 50 - 100%.

A recent empirical model (Rupnik et al.) provides a linear correlation of the electron scattering lengths of He, Ne, Ar, Kr and Xe gases and atomic polarizabilities. This model suggests the use of the non-spherical polarizability for hydrogen and predicts a scattering length $a = 0.83a_o$ ($\sigma = 2.43 \times 10^{-16} cm^2$) in agreement with the result of the pressure shift study.

---

[*] The Racah Institute of Physics, The Hebrew University, Jerusalem, Israel 91904.

Nonequilibrium Effects in Ion and Electron Transport
Edited by J. W. Gallagher *et al.*, Plenum Press, New York, 1990

Table 1. Electron scattering cross sections (in units of $\overset{\circ}{A}{}^2$) of rare gas atoms and hydrogen measured by pressure shift studies using atomic and molecular Rydberg transitions

Type of Rydberg	He	Ne	Ar	Kr	Xe	H_2
Atomic[*] (Alkali Metals)	4.81[a]	0.14[a]	7.91[a]	35.5[a]	119.5[a]	5.25[b]
						4.73[c]
						3.73[d]
Molecular (CH_3I and C_6H_6)	4.10[e]	0.09[f]	9.34[e]	32.2[f]	**	3.04[h]
	4.03[f]	0.10[g]	10.40[f]	37.3[g]		3.45[h]
	3.87[g]		8.66[h]			
			10.28[g]			

[a]Tan and Ch'en, 1970.

[b]Ny and Ch'en, 1938.

[c]Amaldi and Segrè, 1935.

[d]Thompson et al., 1987.

[e]Asaf et al., 1989a.

[f]Köhler et al., 1987.

[g]Reininger et al., 1989.

[*] Values for rare gases are selected from an abundance of experimental data. Scattering lengths were recalculated using the Fermi model modified by the Alekseev-Sobel'man polarization term.

** In the spectral range of these high molecular Rydbergs the Xe gas is highly absorbing.

REFERENCES

Amaldi, E., and E. Segrè, 1935, Nuovo Cimento 11, 145.
Asaf, U., W. S. Felps, and S. P. McGlynn, 1989a, Phys. Rev. A. 40, 5458.
Asaf, U., W. S. Felps, K. Rupnik, S. P. McGlynn, and G. Ascarelli, 1989b, J. Chem. Phys. 91, 5170.
Ferch, J., W. Raith, and K. Schröder, 1980, J. Phys. B 13, 1481.
Gilardini, A., 1972, "Low Energy Electron Collisions in Gases," (John Wiley & Sons New York).
Köhler, A. M., R. Reininger, V. Saile, and G. L. Findley, 1987, Phys. Rev. A. 35, 79.
Ny, T. Z., and S. Y. Ch'en, 1938, Phys. Rev. 54, 1045.
Reininger, R., E. Morikawa, and V. Saile, 1989, Chem. Phys. Lett. 159, 276.
Rupnik, K., U. Asaf, and S. P. McGlynn, 1990, J. Chem. Phys. (in press).
Tan, D. K. L., and S. Y. Ch'en, 1970, Phys. Rev. A 2, 1124.
Thompson, D. C., E. Kammermeyer, B. P. Stoicheff, and E. Weinberger, 1987, Phys. Rev. A 36, 2134.
Vlekseev V. A., and I. I. Sobel'man, 1966, JETP 22, 882.

A CONSTANT RATIO APPROXIMATION THEORY OF THE CYLINDRICAL POSITIVE COLUMN OF A GLOW DISCHARGE

T. Dote[a] and M. Shimada[b]

[a] Department of Electrical Engineering
Saitama University, Urawa, Saitama 338, Japan
[b] Department of Physics
Osaka Kyoiku University, Tennohji, Osaka 543, Japan

A novel theory of the positive column of a glow discharge without the assumption of the quasi-neutrality has been proposed, in the place of the conventional quasi-neutral theory. As for this, recently in order to facilitate understanding, the theory with the simple Schottky model has been reported (Dote and Shimada, 1987). Here, the theory with the swarm analysis using the transport equations has been presented.

The electron flow equation in weakly ionized gases is given by

$$n_e v_r = - D_s \frac{\partial n_e}{\partial r} , \tag{1}$$

$$D_s \equiv \frac{\frac{2}{3}(U + U_+)}{\frac{1}{C} M (\nu_{m+} + \frac{1}{C}\nu_i) + m\{\nu_m + \nu_{ex} + (2 - \frac{1}{C})\nu_i\} + (\frac{M}{C^2} + m)\frac{\partial v_r}{\partial r}} , \tag{2}$$

$$C \equiv n_+/n_e$$

where n_e and v_r are the electron density and the radial velocity; n_+ is the positive ion density; m and M are the electron and positive ion masses, respectively; ν_m and ν_{m+} are the momentum transfer collision frequencies of electrons and positive ions, respectively; ν_{ex} and ν_i are the excitation and ionization collision frequencies of electrons, respectively; U and U_+ are the electron and the positive ion random (or average) energies, respectively; D_s denotes the electron effective diffusion coefficient. These equations are derived under the constant ratio approximation, $\nabla n_e/n_e = \nabla n_+/n_+$, according to Allis (1956).

By substituting Eq. (1) into the electron continuity equation, we obtain the equation of the electron velocity as follows,

$$\frac{\partial \phi}{\partial \xi} = \frac{\xi(\phi^2 + \alpha) - \alpha^{1/2}\phi}{\alpha^{1/2}\xi(1 - \phi^2)} .$$

(3)

where $\phi \equiv v_r/v_b$, $\quad \xi \equiv r/\Lambda$, $\quad n \equiv n_e/n_{e0}$ $(n_{e0}=[n_e]_{r=0})$,

$$v_b^2 \equiv 2(U + U_+)/3(M/C^2 + m), \quad \Lambda \equiv (D_{sw}/\nu_i)^{1/2}, \quad \alpha \equiv D_{sw}\nu_i/v_b^2 ,$$

$$D_{sw} \equiv \frac{\frac{2}{3}(U + U_+)}{\frac{1}{C} M (\nu_{m+} + \frac{1}{C}\nu_i) + m\{\nu_m + \nu_{ex} + (2 - \frac{1}{C})\nu_i\}} .$$

When ξ_b denotes the value of ξ at $\phi=1$, which is obtained by solving Eq. (3),

$$(D_{sw}/\nu_i)^{1/2} = R/\xi_b ,$$

(4)

where R is the tube radius. Eq. (4) corresponds to the expression from which the electron temperature is derived. Moreover C is determined from the following equation,

$$\frac{4\pi\sigma_{e0}}{\nu_i} = \frac{\frac{1}{C}\{\frac{M}{m}(\nu_{m+} + \frac{1}{C}\frac{3}{2}\nu_i) - \nu_i\} - (\nu_m + \nu_{ex} + \frac{5}{2}\nu_i)\frac{U_+}{U}}{(C - 1) (\nu_m + \nu_{ex} + 2\nu_i)(1 + U_+/U)} ,$$

(5)

where σ_{e0} is the electron conductivity at the axis. Figure 1 shows one example of the characteristics of the electron temperature T_e versus p_0R (p_0: the pressure reduced to 0°C). The dotted line stands for the case with the quasi-neutral assumption.

It is significant to note that the electron temperature characteristics vary with the electron density at the tube axis.

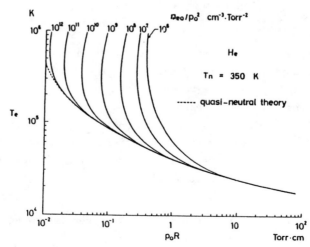

Fig. 1. Characteristics of T_e versus p_0R for values of n_{e0}/p_0^2 at $T_n=350$ K for He.

REFERENCES

Allis, W. P., 1956, Handbuch der Phys., ed S. Flügge, (Springer-Verlag), 21, 397.

Dote, T., and M. Shimada, 1987, Meeting on Atomic and Molecular Processes in Space-Swarm Phenomena and Atomic Collision Processes, 61.

LASER-INDUCED OPTO-GALVANIC STUDIES OF PRE-BREAKDOWN SWARM PHENOMENA

A. Ernest, M. Fewell and S. C. Haydon

Department of Physics
University of New England
Armidale NSW 2351, Australia

Spatial and temporal investigations of ionization growth have been
used extensively to study the swarm properties of electrons in gases.
However, interpretation of the complex behavior, even under well-
controlled pre-breakdown conditions, has often been seriously compromised
by the unsuspected influence of energetic neutral metastable particles
(Haydon and Williams, 1967).

More recently (Brunker and Haydon, 1983) both pulsed and cw tunable
dye laser techniques have been developed in attempts to modify the
populations of selected metastable energy states and so identify their
role in pre-breakdown phenomena. By monitoring the effects of these
population changes on both the spatial and temporal characteristics of
the ionization growth in neon it has been possible to obtain information
about the contributions made to secondary ionization by resonant and
non-resonant photons, monomer and dimer ions as well as the various
metastable particles themselves. The ability to perturb the populations
continuously with c.w. ring dye laser techniques has provided new
opportunities to test the predictions of rate-equation analyses of the
complex phenomena and some preliminary results have been reported
elsewhere (Wang et al., 1987). The dominant influence of impurity
effects were clearly demonstrated, but quantitative agreement between
observations and theory was not entirely satisfactory. The main
limitation centered around the need to operate the ring dye laser under
multimode conditions in order to achieve sufficiently large opto-galvanic
signals.

In this paper we report significant refinements to the procedures
previously used, including the use of an upgraded tuned laser facility

capable of producing adequate single-mode output for continuous perturbation of selected metastable energy states. A 10 W argon-ion pump
laser is used to produce up to 1 W of single mode ring dye-laser emission
using intra-cavity Fabry-Perot etalon techniques. The dye-laser beam is
spatially filtered and beam-expanded to yield a substantially uniform
intensity distribution across that portion of its profile used in the
interaction chamber. A hollow cathode discharge lamp, operated under
pre-breakdown conditions, is used to tune to the desired transition. The
laser beam, operating at constant output power, is attenuated by
appropriate combinations of neutral-density filters in order to investigate the spatial and temporal growth of ionization as a function of laser
power density.

Figure 1 shows typical I vs d and I vs t records for various laser
power density conditions. The temporal traces have been obtained by
signal averaging using a Le Croy 9400 digital oscilloscope. From such
data it is possible to abstract values for the fundamental time constants, $1/\tau$, which are related to the diffusion coefficient, D_m, for the
neon metastable particles and to the volume quenching coefficient G. The
total quenching coefficient G_τ is made up of that due to trace impurities
present in the particular gas sample and that due to the action of the
tuned laser radiation. The major thrust of the investigation now in

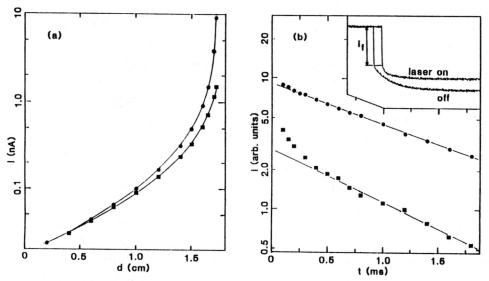

Fig. 1. The effect of laser irradiation on the (a) spatial and (b)
temporal ionization growth in Ne [E/N = 141Td and N = 7.1 x
10^{16} cm^{-3}]. ●- without laser; ■-with laser intensity of
33.5 W m^{-2}. In (b), d = 1.4 cm and the inset shows the
signal-averaged data from which the log I(t) plot was
obtained.

progress is to refine the rate-equation analysis to yield satisfactory agreement between observation and theory. Although the results shown in Fig. 1 are confined to the effects of changes in the $1S_5$ metastable populations caused by laser-induced perturbations to the $2P_4$ levels by radiation at $\lambda = 594.5$ nm, other transitions are under investigation and will be described in detail at the seminar.

Support of the Australian Research Committee, AINSE and the ERB is acknowledged.

REFERENCES

Brunker, S., and S. C. Haydon, 1983, Journal de Physique; Colloque 7, Suppl. 11, 44, 55-63.
Haydon, S. C., and O. M. Williams, 1967, J. Phys. D. 9, 523-36.
Wang, Y., A. Ernest, and S. C. Haydon, July 1987, Proc. XVIIIth Int. Conf. on Phenomena in Ionized Gases, 214-15, Swansea, U.K.

INFLUENCE OF NEGATIVE ION AND METASTABLE TRANSPORT ON THE STOCHASTIC
BEHAVIOR OF NEGATIVE CORONA (TRICHEL) PULSES

S. V. Kulkarni and R. J. Van Brunt

National Institute of Standards and Technology
Gaithersburg, MD 20899, USA

It has been shown (Van Brunt and Kulkarni, 1989) that negative pulsating
point-plane corona (Trichel pulse discharge, Trichel, 1938) is inherently a
stochastic process in the sense that the initiation and development of a
discharge pulse can be significantly influenced by residual ion space charge
and metastables from previous discharge pulses. The stochastic behavior of
this discharge phenomenon has been revealed from a direct measurement of
various conditional discharge pulse-height and time-interval distributions.
The size of any discharge pulse is shown to be strongly correlated with its
time separation from the previous discharge pulse such that the mean pulse
amplitude increases with increasing time separation from the previous pulse.
This behavior for corona pulses in N_2/O_2 and Ne/O_2 gas mixtures is shown to be
due to the influence of moving negative-ion space charge from previous discharge pulses in reducing the electric field strength in the vicinity of the
point electrode at the onset of the next discharge pulse. The correlation
between pulse amplitude and pulse-time separation is found to cease for time
separations that are sufficiently long to allow completion of ion transport
across the gap. For sufficiently short-time separation between pulses, it is
found that for O_2 and N_2/O_2 mixtures, large pulses are preferentially followed
by large pulses and small pulses, by small pulses, whereas the opposite correlation in amplitude is observed in Ne/O_2 mixtures for more than 20% Ne
content. This effect in the former case (N_2/O_2) can be attributed to the
predominant influence of metastables from the previous pulse on enhancing
growth of the subsequent pulse, and in the later case (Ne/O_2) on the predominant influence of negative-ion space charge in inhibiting subsequent
discharge pulse growth. In the case of N_2/O_2, it is found that the larger a
pulse, the shorter the mean time separation to the next pulse, whereas the
opposite effect is seen for Ne/O_2. Again, the behavior in the former case can
be attributed to the effect of metastables in enhancing the probability for

initiation of the next pulse through secondary electron release processes at the cathode. For Ne/O_2, metastables are of much less importance and the predominant effect is that of negative-ion space charge in inhibiting release of electrons from the cathode via field-assisted photoelectric effect. Consistent with the correlations between pulse amplitudes and time separations, it is found that for N_2/O_2 long time separations are preferentially followed by subsequent short-mean-time separations, whereas for Ne/O_2, long time separations are preferentially followed by subsequent long mean-time separations.

REFERENCES

Trichel, G. W., 1938, Phys. Rev. 54, 1078.
Van Brunt, R. J., and S. V. Kulkarni, 1989, Rev. Sci. Instrum. (in press).

IMPROVEMENT IN THE BREAKDOWN STRENGTH OF SF_6/N_2 MIXTURES-A PHYSICAL APPROACH

D. Raghavender and M. S. Naidu

Department of High Voltage Engineering
Indian Institute of Science
Bangaloro 560 012, India

ABSTRACT

It is well known that the use of SF_6/N_2 mixtures in SF_6 gas insulated equipment could solve the liquification problem, reduce the cost of the gas and to some extent lessen the sensitivity of the dielectric strength to local field enhancement.

Most of the earlier studies in SF_6/N_2 mixtures have been concentrated essentially to determine the breakdown strength by illustrating the breakdown characteristics using a limited number of concentrations of SF_6 gas in N_2, at one or two gap spacings. However, very little information is available on the basic physical processes that are responsible for improving the breakdown strength of N_2 with the addition of various concentrations of SF_6 gas in SF_6/N_2 mixtures.

In view of this, investigations were undertaken in SF_6/N_2 mixtures in which the SF_6 content was varied from 0.1% to 20% to determine the lightning impulse (1.2/50μs) breakdown strength (V_{50}) using a rod–plane electrode geometry (rod diameter of 5 mm and a plane electrode of diameter of 230 mm). The gas mixture pressures used were in the range of 0.1 to 0.5 MPa (1 to 5 bar), while the gap separations were varied from 5 to 100 mm. The results thus obtained at all pressures and gap separations could always be reproduced to within ±2%.

The results obtained indicate that at a gap spacing of 40 mm and corresponding to a pressure of 0.1 MPa, the positive V_{50} of SF_6/N_2 mixtures (0.1% to 20% SF_6 content) are higher than the pure N_2 values by about 90% to 120%. However, at the highest pressure studied, i.e., 0.5 MPa, the corresponding variations are from 1.6% to 23.0%. On the other hand, under negative polarity the V_{50} are generally higher than the corresponding positive values by 20% to 30%.

The above observations in which the breakdown strength of N_2 with the addition of small quantities of SF_6 show substantial improvements, may be explained as follows.

The N_2 molecules, although not as complex as SF_6 molecules, have metastable and active states which would affect the secondary ionization (namely, photoionization) process (Hartman and Gallimberti, 1975). It has been shown that the radiation produced by the electrical discharges in N_2, SF_6 and SF_6/N_2 mixtures have wave lengths in the region of 900 to 1800Å, and are capable of causing photoionization in the gas medium (Blaire et al., 1976).

In 0.1% SF_6/99.9% N_2 mixtures, during the initial ionization process by electron collision, in the region of the highest field strength where $\alpha > \eta$, the presence of photons with energies of about 10 eV was observed in the neighborhood of the ionizing zone. These photons will be absorbed within a distance of $\lambda = 1/\mu$, where μ is the photoabsorption coefficient. The photoabsorption coefficient for pure N_2, for 10%SF_6/90%N_2 and for 20%SF_6/80%N_2 mixtures lies between $0 - 1.3$ μm^{-1}, $0.6 - 4.5$ μm^{-1}, and $2 - 15$ μm^{-1} respectively. As stated, the photons produced during the initial ionization will be absorbed in the neighborhood of the ionizing zone where $\alpha > \eta$, and after absorption these photons will cause the excitation of neutral gas molecules or create new electrons by ionizing the already excited gas molecules. The electrons thus generated will cause further ionization by collision in the region $\alpha > \eta$, and thereby generate new photons away from the rod electrode tip (Yializis et al., 1979) (also see Bhalla and Craggs, 1962 for similar observations in SF_6 gas).

The electrons thus generated, because of the increased photoionization, will undergo capture in the presence of SF_6 molecules and create negative ions through an associative or dissociative attachment process. This explains the observed increase in the breakdown strength (V_{50}) of SF_6/N_2 mixtures when small quantities of SF_6 were added to N_2 under both the positive and negative polarity impulse voltages.

REFERENCES

Bhalla, M. S., and J. D. Craggs, 1972, Proc. Phys. Soc. 80, 151.
Blaire, D. T. A. et al., 1976, Proc. 4th Int. Conf. on Gas Discharges, London, p. 401.
Hartman, G., and I. Gallimberti, 1975, J. Phys. D 8, 670.
Yializis, A., et al., 1979, IEEE Trans., PAS-98, 1932.

GEOMETRY-DEPENDENT DISPLACEMENT CURRENT IN THE PULSED-TOWNSEND DRIFT TUBE[*]

Edward Patrick[a], Douglas Abner[a], David Ramos[a],
Merrill Andrews[a] and Alan Garscadden[b]

[a] Wright State University
[b] Air Force Wright Research and Development Center

The method of images was used to determine the displacement current as a function of distance for the case of a point charge approaching a guarded anode of finite size. The analysis was extended to include the effect of an adjacent cathode as a boundary condition. This has produced a simplified model for the current response of a pulsed-Townsend drift tube as a function of the drift-tube geometry. Sources of error are presented that can arise purely from the drift tube geometry even when guard rings are used to establish a uniform electric field. However, these effects occur regardless of the uniformity of the field supplied by the drift-space bias voltage. The results of the model agree well with experimental mobility data obtained by the authors. The analysis also explains the origin of peculiarities in the current transients recorded in some previous pulsed-Townsend experiments.

The principles used are from the original image charge analysis of displacement current by Maxwell (Maxwell, 1891). The treatment of displacement current on an anode of finite size is common to the theory and design of particle detectors (Rossi and Staub, 1949). In the study of electron mobility in gases the exclusion of displacement current analysis can cause error and confusion (Dickey, 1952) in the experiment due to unsatisfactory drift tube geometry.

It is demonstrated that displacement current induced on a disk-shaped, guarded anode due to an electron moving along the axis of the anode (Crowley, 1986) is approximated by

$$I = (qw_D/d) \sum_n (d/a)\{[1 + (nd - z)^2/a^2]^{-3/2} + [1 + (nd + z)^2/a^2]^{-3/2}\}$$

---

[*] Supported by Air Force Contract F33615-86-C-2720 through SCEEE.

where q is the elementary charge, a is the anode radius, w_D is the electron drift velocity, d is the cathode-anode separation, and z is the distance from the cathode along the z-axis.

For a pulsed-Townsend drift tube of appropriate geometry, the above expression reduces to the approximation $I = qw_D/d$. The validity of this "steady state" approximation is given in terms of the dependence of the current response on the drift tube geometry. The "aspect ratio" is defined as the ratio of the drift length to the anode radius. Recommendations are given for the drift tube aspect ratio required for a constant-current response as collected on the anode when a shielded anode collector such as a Faraday cup is not utilized. The effects of using a measured current that includes the displacement current on the interpretation of the current pulse shape, diffusion of the swarm, and position of the swarm centroid are also demonstrated. Supporting experimental data by the authors and other experimenters will be presented (Hornbeck, 1951).

REFERENCES

Crowley, J. W., 1986, "Fundamentals of Applied Electrostatics," John
    Wiley & Sons, N.Y., N.Y.
Dickey, F. R., 1952, J. Appl. Phys. 23, 1336.
Hornbeck, J. A., 1951, Phys. Rev. 83, 374.
Maxwell, J. C., 1891, "A Treatise on Electricity and Magnetism," 3rd ed.,
    Vol. 1, Clarendon Press, London, U.K.
Rossi, B. B., and H. H. Staub, 1949, "Ionization Chambers and Counters,"
    1st ed., McGraw-Hill, N.Y., N.Y.

LIST OF ATTENDEES

SWARM SEMINAR
August 2-5, 1989

Harold Anderson
University of New Mexico
Chem./Nuclear Eng. Dept.
203 Farris Eng. Center
Albuquerque, NM 87131
(505) 277-5661

Merrill Andrews
1031 Wenrick Drive
Beavercreek, OH 45385
(513) 873-2954

Gianni Ascarelli
Purdue University
Physics Department
W. Lafayette, IN 47906
(317) 494-3014

Steven Bajic
University of Birmingham
Physics & Space Research
P.O. Box 363
Birmingham B15 2TT England

Michael Barnes
IBM Corporation
East Fishkill Facility
Route 52 - B/300, 2/48A
Hopewell Junction, NY 12533
(914) 894-4405

H. A. Blevin
Flinders University
School of Physical Science
Bedford Park
S. Australia 5042 Australia
(08) 388-2176

Michael Brennan
Australian National University
Res. Sch. of Phys. Sci.
GPO Box 4
Canberra 2601 Australia

Voitek W. Byszewski
GTE Laboratories, Inc.
40 Sylvan Road
Waltham, MA 02254
(617) 466-2470

Neil Carron
Mission Research Corp.
P.O. Drawer 719
Santa Barbara, CA 93102
(805) 963-8761

Giancarlo Cavalleri
Dept. di Matematica
University Cattolica
via Trieste, 17
25121 Brescia Italy
030-57-286

Max Chung
Polytechnic University
Weber Research Institute
Route 110
Farmingdale, NY 11757
(516) 755-4344

Steve Clark
Polytechnic University
Weber Research Institute
Route 110
Farmingdale, NY 11735
(516) 755-4219

Robert W. Crompton
Australian National University
Res. Sch. of Phys. Sci.
GPO Box 4
Canberra A.C.T.2601 Australia

D. Kenneth Davies
Westinghouse S&T Center
1310 Beulah Road
Pittsburgh, PA 15235
(412) 256-1009

J. De Urquijo
Instituto de Fisica,
UNAM
62190 Cuernavaca, Mor.
P.O. Box 139-B Mexico

Craig Denman
Air Force Weapons Lab.
ARDI/Bldg. 243
Kirtland AFB,
NM 87117-6008
(505) 844-0883

Paul J. Drallos
Dept. of Physics
Wayne State University
Detroit, MI 48202
(313) 545-1974

Malcolm T. Elford
Australian Nat. Lab.
Res. Sch. of Phys. Sci.
GPO Box 4
Canberra A.C.T.2601 Australia
(062) 492402

Homer Faidas
University of Tennessee
Physics Department
Knoxville, TN 37996
(615) 574-6203

John Forster
IBM Corporation
East Fishkill Facility
Route 52 - B/300, 2/48A
Hopewell Junction, NY 12533
(914) 894-4405

Jean W. Gallagher
National Institute of
Standards & Technology
A323 Physics Building
Gaithersburg, MD 20899
(301) 975-2204

Andrew M. Garvie
Polytechnic University
Weber Research Institute
Route 110
Farmingdale, NY 11735
(516) 755-4330

Kevin Giles
University of Birmingham
Physics & Space Research
P.O. Box 363
Birmingham B15 2TT England

Claudine Gorse
Department of Chemistry
Univ. 4 Trav.
Re David, 200
70126 Bari Italy

David B. Graves
University of California-Berkeley
Department of Chem. Eng.
Berkeley, CA 94720

Eric Grimsrud
Montana State University
Department of Chemistry
Bozeman, MT 59717
(406) 994-5418

Yoshihiko Hatano
Tokyo Inst. Technology
Department of Chemistry
Meguro-ku, Tokyo 152 Japan

S. C. Haydon
University of New England
Department of Physics
Armidale NSW 2351 Australia

M. A. Hayes
SERC Daresbury Lab.
Warrington
WA 4 4AD U.K.

Mark J. Hogan
Nat. University of Singapore
Dept. of Physics
Kent Ridge
Singapore 0511
Republic of Singapore

David F. Hudson
NSWC Code F43
10901 NH Avenue
Silver Spring, MD 20903-5000
(202) 394-1248

Ping Hui
Polytechnic University
Weber Research Institute
Route 110
Farmingdale, NY 11735
(516) 755-4371

Scott R. Hunter
GTE Sylvania
100 Endicott Street
Danvers, MA 01923
(508) 750-2524

N. Ikuta
Tokushima University
Department of Elec. Eng.
Minami-josanjima,
Tokushima 770 Japan
0886-74-0006

John Ingold
GE Lighting
Cleveland, OH 44112
(216) 266-2121

456

B. M. Jelenkovic
Institute of Physics
Belgrade Yugoslavia
011 107864

Rainer Johnsen
University of Pittsburgh
Department Physics/Astronomy
Pittsburgh, PA 15260
(412) 624-9285

Roger Jones
Polytechnic University
Weber Research Institute
Route 110
Farmingdale, NY 11735
(516) 755-4246

John Keller
IBM Corporation
East Fishkill Facility
Route 52 - B/300, 2/48A
Hopewell Junction, NY 12533
(914) 894-4405

Hulya Kirkici
Polytechnic University
Weber Research Institute
Route 110
Farmingdale, NY 11735
(516) 755-4368

Larry Kline
Westinghouse R&D
1310 Beulah Road
Pittsburgh, PA 15235
(412) 256-2689

Takao Kojima
Tokyo Metropolitan University
Department of Physics
2-1-1 Fukazawa,
Setagaya-ku Tokyo 158 Japan

Sanjay V. Kulkarni
NIST
Electricity Division
Bldg. 220, Room 344
Gaithersburg, MD 20899

Erich Kunhardt
Polytechnic University
Weber Research Institute
Route 110
Farmingdale, NY 11735
(516) 755-4250

Mark J. Kushner
University of Illinois
607 East Healey Street
Champaign, IL 61820
(217) 244-5137

Yan Ming Li
GTE Laboratories
40 Sylvan Road
Waltham, MA 02254
(617) 890-8460

Nathan Marcuvitz
Polytechnic University
Weber Research Institute
Route 110
Farmingdale, NY 11735
(516) 755-4327

Jane Messerschmitt
Polytechnic University
Weber Research Institute
Route 110
Farmingdale, NY 11735
(516) 755-4238

Robert A. Morris
AFGL/LID
Hanscom AFB,
MA 01731-5000

Yoshiharu Nakamura
Dept. of Elec. Eng.
Keio University
3-14-1 Hiyoshi
Yokohama 223 Japan
044-63-1141

Kevin F. Ness
Parks College of
Saint Louis University
Cahokia, IL 62206
(314) 241-0280

James Olthoff
National Institute of
Standards & Technology
B344/220
Gaithersburg, MD 20899
(301) 975-2431

P.P. Ong
National University of Singapore
Physics Department
Singapore 0511
Republic of Singapore
065-7722810

John L. Pack
3853 Newton Drive
Murrysville, PA 15668
(412) 327-3741

Edward Patrick
1534 S. Smithville Road
Dayton, OH 45410
(513) 253-2001

457

Bernie Penetrante
Lawrence Livermore Lab.
L-417
P.O. Box 808
Livermore, CA 94550
(415) 423-9745

Zoran Petrovic
JILA
University of Colorado
Campus Box 440
Boulder, Colorado 80309-0440
(303) 492-8442

Arthur V. Phelps
JILA
University of Colorado
Campus Box 440
Boulder, CO 80309-0440
(303) 492-7850

Leanne Pitchford
CPAT
118 Rte de Narbonne
31062 Toulouse
Cedex France

Darvke Raghavender
Department of High Voltage
Engineering
Indian Inst. of Science
Bangalore India 560012

G. R. Govinda Raju
University of Windsor
Department of Elec. Eng.
Windsor, Ont., Canada N9B 3P4
(519) 253-4232

Orlando Ramos
Apt. 101
4399 Riverside Drive
Dayton, OH 45405
(513) 275-4935

Kresimir Rupnik
Louisiana State University
Dept. of Chemistry
Baton Rouge, LA 70803-1804
(504) 388-2310

L. A. Schlie
WL/ARDI
Air Force Weapons Lab.
Kirtland AFB,
NM 87117-6008
(505) 844-0883

Daniel Schweickart
Wright Res.& Dev. Center
WRDC/POOC-4/Bldg. 450
Wright Patterson AFB,
OH 45433
(513) 255-3835

Hiroshi Shimamori
Fukui Inst. of Tech.
3-6-1 Gakuen,
Fukui 910 Japan
0776-22-8111

Naohiko Shimura
c/o T. Makabe
Dept. of Elec. Eng.
Keio University
3-14-1 Hiyoshi
Yokohama 223 Japan

David Smith
University of Birmingham
Physics and Space Res.
P.O. Box 363
Birmingham B15 2TT   England
021-414-3344

Thorarinn Stefansson
Norwegian Institute of Technology
Division of Physics
7049 Trondheim-NTH,
Norway
07593645

Fumiyoshi Tochikubo
c/o T. Makabe
Department of EE
Keio University
3-14-1 Hiyoshi Yokohama 223 Japan
044-63-1141

N. D Twiddy
University College of Wales
Department of Physics
Aberystwyth, SY 23 3BZ
U.K.
(0970) 622803

Richard Van Brunt
National Institute of
Standards & Technology
Bldg. 220, B344
Gaithersburg, MD 20899
(301) 975-2425

458

Larry A. Viehland
Parks College of
Saint Louis University
Cahokia, IL 62206
(618) 337-7500

Albert Viggiano
Geophysics Lab./LID
Hanscom AFB,
MA 01731-5000
(617) 377-4028

Ming Cheng Wang
Polytechnic University
Weber Research Institute
Route 110
Farmingdale, NY 11735
(516) 755-4247

Malcolm Wright
Rockwell Power Systems
Kirtland AFB,
NM 87185-6008
(505) 844-0883

INDEX

Accuracy
  cross section, 245
Association
  ion-atom, 261
  three-body, 261
Attachment
  in H_2O, 395
  in SO_2, 385
Attachment coefficient
  in CCl_4, 177
  in CO_2, 177
  in Dry air, 177
  in HCl, 177
  in Humid air, 177
Attachment rate
  in $CHCl_3$, 389
  in liquids, 313
  in NF_3/Ar mixtures, 387
  in NF_3/N_2 mixtures, 387

Beam-in-a-box (BIB), 143
Boltzmann equation, 49, 83
      99, 121, 337, 377
  time-dependent, 1, 37
Breakdown
  in SF_6/N_2 mixtures, 451
  low pressure, 121

Characteristic energy
  in H_2, 361
  in liquids, 313
  in Ar, 313
Charge transfer, 261
  negative ions in SF_6, 229
Cleanliness effects, 143
Cluster formation, 423
Collision frequency, 83
Continuous slowing down
      approximation, 355
Conversion
  negative ions in SF_6, 229
Corona impulses
  in Ne/O_2 mixtures, 449
  in N_2/O_2 mixtures, 449
  in O_2, 449

Cross section, electron impact
  attachment in NF_3, 387
  elastic scattering in Hg, 11
  excitation
    He^+, 245
    ions, 245
  momentum transfer
    in Ne, 11, 49
    in He, 49
    in H_2-Kr mixture, 11
    in Kr, 11, 363
    in Ar, 11, 49
    in Hg, 367
  rotational excite in N_2, 11
  total scattering
    in C_6H_6, 439
    in CH_3I, 439
    in Ne, 11
    in Ar, 11
    in RG, 439
  vibrational excite in H_2, 11
Cross section, photon
  photoionization, 245

Data
  electron transport, 121
  for plasma processing, 121
  ion transport, 121
Dense media, 275
Density gradient expansion, 49
Density fluxuations
  in liquids, 291
  static, 291
Deposition, plasma, 121
Dielectric mixtures, 313
Dielectric liquids, 313
Diffusion theory with
  energy balance, 337
Diffusion, electron
  anisotropic, 49, 337
  in Kr, 363
  in constant field, 337
  isotropic, 329
  longitudinal, 49, 329, 363, 377

461

longitudinal
  in He, 371
  in CH_4, 377
  in Kr, 371
  in Xe, 371
  in Ar, 371
  skewness coefficient, 359
  time dependent, 337
  transverse, 329
Diffusion, ion
  longitudinal coefficient, 197
    in SF_6, 211
  transverse, 419
  transverse Ne^+ in Ne, 421
Displacement current, 453
Dissociation
  Cl_2 in N_2, 393
  electron impact cross
    section, 121
Dissociation recombination
  in H_3O^+, 401
  in O_2H^+, 401
  in HCO_2^+, 401
  in N_2OH^+, 401
Distribution, spatial
  emitted radiation, 67
Distribution, temporal
  emitted radiation, 67
Distribution function, electron,
    99, 49, 355
  dc, 121
  ensemble-averaged, 83
  in SiO_4, 339
  in CH_4
  macrokinetic, 83
  non equilibrium, 1, 67, 83, 121,
    143, 157
  non Maxwellian, 1
  time dependent, 343
Distribution function, ion, 229
Drift velocity, electron
  calculated, 337, 343
  general remarks, 1
  in Ar, 49, 313
  in CH_4, 377
  in He, 49
  in Hg-He mixtures, 367
  in H_2-Kr mixture, 11
  in Kr, 11, 363
  in liquids, 313
  in Ne, 49
  in N_2, 11
  in $SF6$, 197
  in SiO_4, 339

Effective mass
  density dependence, 291

Electrode sheath characteristics,
  121
Electron transport,
  see Transport, electron
Electron detachment
  negative ions in SF_6, 229
Electronic excitation
Emission
  in $CHClF_2$, 433
  in H_2, 433
  in H_2, 435
  in H_2/CH_4, 435
Energy balance equation, 99
Energy loss function, 99
Etching plasma, 121
Experimental technique
  conductivity cell, 275
  drift tube, 229, 419, 359
    double shutter, 197, 211
    mass spectrometer,
      211, 421, 423
    pulsed Townsend,
      177, 211, 453
    pulsed time-of-flight, 67
    shielded collector, 177
    uniform field, 395
  electron beam, 11
  FALP, 401
  gas discharge, 375, 381,
    427, 445
  high pressure swarm, 387
  laser irradiation, 445
  mass spectrometry, 381, 393
  microwave discharge, 373
  optical emission spectroscopy,
    429, 433, 435
  photon emission, 67
  pulse radiolysis, 385, 389, 417
  RF discharge, 429, 433, 435
  SIFDT, 399, 403, 407, 409,
    413, 415
  steady-state current growth, 177
  steady-state Townsend, 67
  swarm, 11
  time-of-flight spectra, 197
  time-of-flight, 359
  Townsend-Huxley method, 361

Flight time integral (FTI)
  method, 345

Gas discharge
  DC, 37
  low pressure, 157
  magnetic multicusp, 37
  RF, 37
Gryzinski method, 37

Hall mobilities, 291
High field regime
    electron transport, 1, 99
Hydrodynamic regime, 83

Ion transport, see Transport, ion
Ionization growth dynamics, 99

Kinetics
    electron nonequilibrium, 37
    electron theory, 349
    heavy component electronic, 37
    heavy component vibrational, 37
    macro, 83
    negative ion in BCl_3
    negative ion in gas
        discharges, 381
Kurtosis coefficient, 49

Liquids
    dielectric, 313
    electron mobility in, 291
    recombination in, 275
    transport in Ar, 291
Local field model, 157
Local field equilibrium, 1, 157

Macrokinetic distribution, 357
Macrokinetics, 83
Macroscopic dynamical variables, 83
Master equation, 37
Mean electron energy
    in liquids, 313
    in Ar, 313
    in Xe, 313
Mobility, electron
    density dependence, 275
    drift, 291
    Hall, 291
    in liquids, 291
    phonon-limited, 291
    temperature dependence, 275
    time-of-flight, 291
    zero-field, 291
Modeling conditions
    anisotropic scattering, 157
Mobility, negative ion, 197
    in SF_6, 211, 229
Mobility, positive ion
    in SF_6, 211
Modeling technique
    beam-in-box, 143
Modeling conditions
    current density, 143
    full forward scattering, 157
    isotropic scattering, 157
    local field equilibrium 157

Modeling
    gas discharges, 143, 157
    high-pressure discharge, 427
    nonequilibrium, 83
    particle-in-cell, 157
    plasma processing, 121
    rf gas discharges, 121
Monte Carlo calculation, 337
Monte Carlo simulation, 1, 49, 67,
    157
Multibeam model, 157

Non-equilibrium effects, 143
Non-hydrodynamic behavior, 49
Non-Markovian stochastic
    processes, 351
Nonlinear continuity
    coefficient, 49
Null collision method, 337
Numerical method
    finite element, 377

Particle-in-cell (PIC)
    technique, 157
Phonons
    in liquids, 291
Photoionization
    atoms, 245
Plasma processing
    deposition, 121
    etching, 121
Plasmas
    high pressure, 261
    DC, 157
    rf, 157
    weakly ionized, 157
Proton affinities, 413

Quasi-Lorentz gas model, 83
Quenching
    He^* + RG, 417
    $NO^+(v = 1.3)$, 409

Radiation distribution
    spatial, 67
    temporal, 67
Rate coefficient
    general, 1, 245
    in SF_6, 229
    ionization, 99
    ionization in Ne, 445
Reaction rate
    general remarks, 121
    atomic nitrogen, 407
    atomic hydrogen, 407
    deuteration, 403
    electron, 143

Ne$^+$ + Ne, 415
negative ion-neutral, 197
O$^-$ in N$_2$O, 399
pressure dependence, 261
Reaction rate
   recombination
      in dense media, 275
      three-body, 261
Reactions
   O$^-$, 411
Recombination
   electron-ion, 261
      in liquids, 275
      in dense media, 275
   electron-ion temp.
      dependence, 275
   ion-ion, 261
Rydberg states, 439

Scaling laws
   use in modeling, 143
Secondary electron production, 99
Simulation, 375
Single-beam model, 157
Skewness coefficient, 49
Spectral noise
   theory, 353
Steady stream analysis, 329
Stochastic processes, 353, 449
   non-Markovian, 351
Surface reaction rates, 121

Theoretical method
   Boltzmann equation, 49, 121,
      337, 339, 355, 377
   close-coupling calculation, 245
   continuous-slowing-down
   approximation, 355
   density-gradient expansion, 49,
      83, 351
   energy balance equation, 99

flight-time-integral (FTI)
   method, 345
Gryzinski method, 37
hard-sphere estimate, 49
macrokinetic distribution, 357
Master equation, 37
mean-free-path method, 353
modified effective range
   (MERT), 11
Monte Carlo, 337
multibeam model, 99
null collision method, 337
Random walk, 353
Schottky model, 441
steady stream analysis, 329
time dependent Boltzmann
   equation, 37
Theory
   electron diffusion, 337
   electron
      transporthydrodynamic, 1
   of electron in drift tube, 329
TOF analysis
   electron, 337
Transport, electron, 67
   calculated coefficients, 49,
      83, 121
   dense media, 275
   high-field, 1, 99
   hydrodynamic, 1
   non equilibrium, 1, 99
   rate coeffcient (modeling), 143
   theory, 339, 345
Transport, ion, 197, 211
   in SF$_6$, 211

Ultrafast dielectric
   liquids/mixtures, 313

Weak localization phenomena, 291